D1074342

PHYSICS, PHILOSOPHY, AND PSYCHOANALYSIS

BOSTON STUDIES IN THE PHILOSOPHY OF SCIENCE

EDITED BY ROBERT S. COHEN AND MARX W. WARTOFSKY

VOLUME 76

PHYSICS, PHILOSOPHY AND PSYCHOANALYSIS

Essays in Honor of Adolf Grünbaum

Edited by

R. S. COHEN
Boston University

and

L. LAUDAN
Virginia Polytechnic Institute

D. REIDEL PUBLISHING COMPANY

A MEMBER OF THE KLUWER ACADEMIC PUBLISHERS GROUP

DORDRECHT / BOSTON / LANCASTER

Library of Congress Cataloging in Publication Data
Main entry under title:

Physics, philosophy, and psychoanalysis.

(Boston studies in the philosophy of science ; v. 76)
Bibliography: p.
Includes index.
1. Physics–Philosophy–Addresses, essays, lectures. 2. Philosophy–Addresses, essays, lectures. 3. Psychoanalysis–Addresses, essays, lectures. 4. Grünbaum, Adolf. I. Grünbaum, Adolf. II. Cohen, Robert Sonné. III. Series.
Q174.B67 vol. 76 [QC6.2] 501s [530'.01] 83-4576
ISBN 90-277-1533-5

Published by D. Reidel Publishing Company,
P.O. Box 17, 3300 AA Dordrecht, Holland.

Sold and distributed in the U.S.A. and Canada
by Kluwer Boston Inc.,
190 Old Derby Street, Hingham, MA 02043, U.S.A.

In all other countries, sold and distributed
by Kluwer Academic Publishers Group,
P.O. Box 322, 3300 AH Dordrecht, Holland.

Printed in The Netherlands.

EDITORIAL PREFACE

To celebrate Adolf Grünbaum's sixtieth birthday by offering him this bouquet of essays written for this purpose was the happy task of an autonomous Editorial Committee: Wesley C. Salmon, Nicholas Rescher, Larry Laudan, Carl G. Hempel, and Robert S. Cohen. To present the book within the *Boston Studies in the Philosophy of Science* was altogether fitting and natural, for Grünbaum has been friend and supporter of philosophy of science at Boston University for twenty-five years, and unofficial godfather to the Boston Colloquium. To regret that we could not include contributions from all his well-wishers, critical admirers and admiring critics, is only to regret that we did not have an encyclopedic space at the committee's disposal. But we, and all involved in this book, speak for all the others in the philosophical, scientific, and personal worlds of Adolf Grünbaum in greeting him on May 15, 1983, with our wishes for his health, his scholarship, his happiness.

Our gratitude is due to Carolyn Fawcett for her care and accuracy in editing this book, and for the preparation of the Index; and to Elizabeth McMunn for her help again and again, especially in preparation of the Bibliography of the Published Writings of Adolf Grünbaum; and to Thelma Grünbaum for encouraging, planning, and cheering.

Boston University R.S.C.
Center for the Philosophy and History of Science M.W.W.

TABLE OF CONTENTS

ADOLF GRÜNBAUM
(Photo by permission of Nathaniel Braverman)

ADOLF GRÜNBAUM: A MEMOIR

Due to the instinctive genius of the New York City school system, I (from Manhattan) met Adolf Grünbaum (from the farthest reaches of Brooklyn) at the DeWitt Clinton High School (in the northernmost Bronx) about forty-five years ago. Coming with parents, sister and brother, he was a very recent refugee from Cologne. We had two-hour subway rides to school, more than ten thousand fellow students (all male), a splendid four years of Latin, and world politics going down toward disaster; but to my delight I found a friend who had read Bertrand Russell and who already understood that a life of the mind and a life of action were possible together, and even more, that love of physics and love of philosophy could be joined. I went on to Wesleyan University in 1939, and Adolf joined me there a year later; we went to Yale Graduate School briefly for master's degrees, and then again after the Second World War, to complete our doctorates, to live in a cooperative house named Steamview, to study with such great teachers and scholars as Brand Blanshard, Peter Hempel, Henry Margenau, Paul Weiss, Leigh Page, Gregory Breit, F. S. C. Northrop, the young Fred Fitch, Charles Hendel. His parents, Benjamin and Hannah, became my dear friends; later, our wives and children too formed a beloved circle. Our philosophical comradeship was from the beginning a humane and enduring affectionate friendship. Shall I prepare an objective, impersonal memoir? Certainly not.

Adolf Grünbaum was born May 15, 1923. He came to the United States in 1938, and was naturalized in 1944, while serving in the United States Army. He married Thelma Braverman on June 26, 1949, and they have a daughter, Barbara. Adolf had absorbed the shock of American life quickly, while his parents met their mid-life challenge of this new world with Stoic courage. And courage was needed for the hard labor of the immigrant worker who had been a comfortable middle-class German Jewish housewife, for the fortitude of the interminably ill father, and the poverty for all of them, the two younger children, and the other relatives who had escaped from Germany. Grünbaum looked ahead to a life of learning, but the family could not help; with luck, the American mixture of self-help and charitable scholarship stipends worked. He came to Wesleyan, to the small Connecticut city of Middletown, a mixture of Yankees and Sicilians who in their different ways were so very far from either German or New York cosmopolitan life: there he studied mathematics, physics, and philosophy at the small, quiet, private undergraduate college, itself still in the glow of its Methodist origins, still predominantly Protestant, white, all-male, a genteel middle-class cousin to the wealthier élite Ivy League universities. But the isolated college world was being shaken by the outside world, by the financial rigors of the great depression that ended only in the swirl of war already underway in Europe, by the obvious intrusion among all the students of the coming American role in the war, by the sweet and intelligent sole Jewish faculty member, the political sociologist Sigmund Neumann, a refugee from Berlin.

Wesleyan had a tradition of scholarly research, and Grünbaum plunged into his courses, as if each of them were somehow to be transmuted into a seminar on the logical and philosophical foundations of its subject, whether the calculus or classical mechanics or an innocent elementary laboratory experiment. Adolf's reputation for endless philosophizing came early. His main philosophical teacher at Wesleyan, the wise and gentle Quaker, Cornelius Krusé, once met us on the street, and when, on inquiry, he learned that we were so deep in argument about a calculus problem that we barely noticed him, he spread the word that we were debating about Berkeley, or perhaps Cantor, or at any rate about infinity and Zeno's paradoxes. We were in fact going to the movies.

Grünbaum was dogged as well as deep. In high school, scarcely in America, studying Latin, barely fluent in English, he decided that a civilized human being must also know French, and so he set himself that task too. He may be the only American, German refugee or not, who learned French by patiently, doggedly, reading *Les Misérables.*

His war service, and his graduate work at Yale, came quickly since the undergraduate curriculum was greatly compressed, accelerated to meet manpower needs. Grünbaum worked for a time in a radar-related military research group of the Columbia University Division of War Research, located to our pleasure (I was there too) in the upper reaches of the Empire State Building. But soon he was in the United States Army, and like so many young German-speaking refugees, he was assigned to combat intelligence, and then occupation service. How strange it was for him to interrogate German academic prisoners, the young American army philosopher confronting the members of the established German professoriat. I recall some trophies he sent me from Hitler's *Reichskanzlei* in Berlin, especially a set of the great Muret–Sanders encyclopedic English-German dictionary, and more vividly his letters concerning the interrogations of such Nazi scientists as the mathematician Bieberbach and the physicist Lenard. But also the splendid surgeon Sauerbruch. He also was very happy to find a beautiful set of the papers of Helmholtz for me, and a fascinating unpublished manuscript on 'Scheinprobleme der Wissenschaft', written during the last years of war by the aged and noble Max Planck.

Back at Yale, Grünbaum completed his M.S. in physics, and then undertook studies and dissertation research for the doctorate in philosophy. That dissertation, on 'The Philosophy of Continuity' [5 - see Bibliography] was inspired (as I believe Grünbaum would happily agree) by Hans Reichenbach's great treatise, the *Philosophie der Raum-Zeit-Lehre* of 1928, and written with Reichenbach's student at Berlin, C. G. Hempel, as Adolf's faculty advisor. It went directly to conceptual criticism of previous work on Zeno's paradoxes; an impressive result was Grünbaum's own analysis, a model of mathematical and philosophical argument, of which a major part was published as 'A consistent conception of the extended linear continuum as an aggregate of unextended elements' [9].

Grünbaum was married by then to the extraordinarily intelligent and lovely Thelma whom he had first met during his high school years; he received his first faculty appointment at Lehigh University, in Bethlehem, Pennsylvania, in 1950. And then he set to work.

Looking over the thirty-three years since 1950, we find that Adolf Grünbaum has written five books, and more than a hundred articles. His research may be described under three headings:

(1) space and time,
(2) scientific rationality, and the falsifiability criterion,
(3) the cognitive status of psychoanalytic theory.

But his thought and influence reach further. Ethical and religious issues have always demanded his concern and his analysis. His early paper, 'Causality and the science of human behavior' [8], also published in 1952, has been reprinted in psychological, psychoanalytic and philosophical journals and anthologies, revised and developed over the years. I think also of his courses on 'science and religion', his clarifying research on Freud's methods and theories, his life-long interest in the rational understanding of social conflicts, his vigorous recruitment of superb moral theorists among his colleagues. Especially, I enjoy his caustic delight and incredulity at exposure of a particular piece of institutional obfuscation or individual wishful thinking. His style is not Bertrand Russell's limpid grace, and his philosophical method is closer to technical physics than Russell's (and less that of the logician) but Grünbaum's spirit soars like that of Russell; they laugh and weep for human joys and sufferings; alike, they seek rational understanding, and alike they reject claims for an absolute, for a super-human moral source, for a Mystery which might console. The contingent about us is mystery enough, and for Grünbaum as for Einstein, its rational understanding is inspiration enough.

Twenty years ago, the first edition of Grünbaum's magisterial treatise on the *Philosophical Problems of Space and Time* [67], was published. Ten years ago the second edition appeared, as Volume 12 of *Boston Studies*, so greatly enlarged that some of us called it Adolf's 'cubic book' [130]. Reviews have shown the respect for a classic of the epistemology of theoretical physics, and comparison with Reichenbach's book was quickly and repeatedly made. The Soviet edition, to which Grünbaum contributed enough new materials for me to call it the 1½-edition [99], was warmly received by the demanding philosophical and scientific readership there; and French, German, Polish, Italian, Australian, and British reviews too have hailed the book. Within a few years, and with further books and papers on the philosophy of space and time, Grünbaum had been recognized as '. . . more or less the dean of American philosophers interested in space and time', his book described as 'the most comprehensive treatment of the philosophical problems of geometry in this century' and his research accomplishments as 'the single most valuable body of work in the contemporary literature on space-time'. By the time Adolf was called to become the Andrew Mellon Professor of Philosophy at the University of Pittsburgh in 1960, he was informally tagged as the 'Mr. Space and Time' of American philosophy.

Two noteworthy responses to his work on space and time took place during the late 60s and early 70s. The journal *Philosophy of Science* devoted an extraordinary symposium in volume 36, the issue of December 1969, to a 'Panel Discussion of Grünbaum's Philosophy of Science'. Characteristically responsive, respectful of critics, enthusiastic, and forthcoming, Grünbaum was moved by the careful work of the panelists and by other commentaries on his work to prepare a wide-ranging and detailed essay which he called a 'critical exposition and reply'; the first installment [117] took 120 pages of the next volume of the journal. Then in 1974, the Institute of Relativity Studies and the Center for Philosophy and History of Science at Boston University held a conference on Absolute and Relational Theories of Space and Space-Time to mark the publication of the second edition of *Philosophical Problems of Space and Time*, and Adolf presented the title paper and took part in a similar conference that same year at the Minnesota Center for the Philosophy of Science [148].

Modern Science and Zeno's Paradoxes [85] caught the attention of philosophers, mathematicians and physicists when it was published in 1967, for in it Grünbaum had, as one specialist wrote 'done more than anyone else since Bertrand Russell to enhance our understanding of the paradoxes and various suggested solutions'. Adolf's two monographic studies, one of 'Carnap's views on the foundations of geometry' [68], written in the 50s, and the other of 'Geometry, chronometry and empiricism' [56], prepared in 1959 during one of his happy and productive periods with Herbert Feigl when Adolf was Visiting Research Professor at the Minnesota Center for the Philosophy of Science, were but two of dozens of his research papers on the foundations of physics. Especially noted among physicists was his paper on 'The anisotropy of time' [88] at Thomas Gold's conference on the nature of time held at the Cornell University Center for Radiophysics and Space Research in 1967. There were other delightful philosophical conundrums of physics to be seriously treated. Who has not wondered about the possible action of a future event on the past or present? and whether modern or even classical physics might provide the basis, in view of mathematically acceptable time-reversed solutions to our physically descriptive equations? Adolf provoked interesting responses with his insightful analysis of 1976: 'Is pre-acceleration of particles in Dirac's electrodynamics a case of backward causation? The myth of retrocausation in classical electrodynamics' [146]. Earlier, in 1957, he had responded to speculative flights about Bohr's principle of complementarity with sober cautions in his 'Complementarity in quantum physics and its philosophical generalization' [39]. Again and again he revised and elaborated his physics-based conception of the mind-dependent status of what has so easily been described as 'temporal becoming'.

Grünbaum, as epistemologist, has clarified some of the most troubling of classical concerns about scientific rationality, and indeed about the quality of reasoning in human affairs generally. In 1954, he published his incisive paper on 'Science and ideology' [21], in 1962 his 'Science and man' [62], in 1969 the influential 'Free will and the laws of human behavior' [107], in 1978 an interesting lecture on 'The role of psychological explanations of the rejection or acceptance of scientific theories' [157].

Running through his philosophical work has been a theme central to serious philosophy of the sciences in every age, namely to distinguish what is contingent or conventional from what is cognitively of greater weight.

Grünbaum has labored over the conventionalist thesis since the mid-fifties, going beyond the technically fascinating problem of his 'Conventionalism in geometry' [44] in 1959, to his now classic discussion of Pierre Duhem in his paper on 'The Duhemian argument' [50] of 1960, which itself has been a focus of debate among his colleagues, revised as late as 1976 as a centerpiece [141] in the collection of Sandra Harding, *Can Theories be Refuted?* He continued his critical research on this controversial thesis of Duhem and Quine, the denial of the feasibility of crucial experiments in science (which Adolf calls the 'D-thesis') for some years. Perhaps the most significant development is his assessment of Popper's theory of science, first presented in his (as yet unpublished) lecture to the Heisenberg *Fest* at Pennsylvania State University in 1971, a symposium on 'The role of crucial experiments in science'. His lecture is now widely known from his paper in our volume of *Boston Studies* published in memory of Imre Lakatos, 'Is falsifiability the touchstone of scientific rationality? Karl Popper vs. inductivism' [140]. Adolf's attack was then elaborated in a lively and provocative trio of papers in the *British Journal for the Philosophy of Science* that same year: (1) 'Can a theory answer more questions than one of its rivals?' [143], (2) 'Is the method of bold conjectures and attempted refutations *justifiably* the method of science?' [144], and (3) '*Ad hoc* auxiliary hypotheses and falsificationism' [145]. The issue was practical as well as of intrinsic interest, for it led to Grünbaum's examination of Karl Popper's popular critique of the scientific standing, indeed the meaningful status, of Freud's psychoanalytic theory, and beyond that to the warrant to be ascribed to clinical evidence for that theory.

When it comes to Freud, we find Adolf Grünbaum at his most respectful and most disturbing. In psychology, the deepest of human science, the theoretically-conceived explanatory entities and processes of Sigmund Freud's psychoanalytic psychology have posed once again the epistemological problems of all sciences that claim to go beyond their observational data. For Grünbaum it was of greatest importance to examine what was functionally analogous in Freud's work to experimental inquiry in the other empirical sciences; this meant not the several experimental tests of psychoanalytic theories but what Freud himself considered central, clinical evidence from the analyst's personal interaction with the patient. Going beyond his critique of Popper's Freud-rejection (wherein Adolf established the scientific testability of Freud's theory), he wrote a series of lectures and papers on the logic of psychoanalysis. From 1982, one that stimulated both philosophers and psychologists has the saucy title 'Can psychoanalytic theory be cogently tested "on the couch"?' [178]. But it was preceded by Adolf's pioneering paper of 1979 on the 'Epistemological liabilities of the clinical appraisal of psychoanalytic theory' [167] and by his dissection of 'The placebo concept' [176] in 1981, a paper propadeutic to further work throughout the methodological aspects of medical inquiry. And now Grünbaum is presenting his major critical examination of Freud's *corpus* in his treatise on *The Foundations of Psychoanalysis: A Philosophical Critique* [181].

Adolf left Lehigh University in 1960. He had been appointed to a chair at Lehigh, as William Wilson Selfridge Professor of Philosophy, in 1956 at age 33. He went to the University of Pittburgh as the new Andrew Mellon Professor of Philosophy, and he undertook the task of establishing there the new Center for Philosophy of Science and a reconstituted Department of Philosophy. Indeed Adolf was an astounding organizer, the master recruiter of the most extraordinary faculty of philosophy since Whitehead collaborated with the young Russell. Monographs, treatises, book series, new journals, hundreds of articles, research conferences, dozens of the stalwart creative new philosophers of the past quarter century, have come from the Pittsburgh philosophers. And they have received research students and colleagues from every part of the philosophical and scientific world. The assessment of that Center, of the Department of Philosophy, and its younger sibling Department of History and Philosophy of Science, will someday receive the attention of historians of philosophy, and sociologists of institutional development; for with all their differences of scale and goals, the story of Grünbaum's philosophical Pittsburgh – a little jewel of imaginative organization – belongs with the stories of such ambitious places as Flexner's Princeton Institute, and Hutchins's Chicago.

Adolf Grünbaum was appointed Research Professor of Psychiatry at the University of Pittsburgh in 1979. He is a Fellow of the American Association for the Advancement of Science, and a Fellow of the American Academy of Arts and Sciences. He has been elected Gifford Lecturer at the University of St. Andrews for 1984. Among his other academic honors, Grünbaum was the Arnold Isenberg Memorial Lecturer at Michigan State University in 1965, Louis Clark Vanuxem Lecturer at Princeton University in 1967, Mahlon Powell Lecturer at Indiana University in 1968, Thalheimer Lecturer at the Johns Hopkins University in 1969, Everett W. Hall Lecturer at the University of Iowa in 1973, Visiting Research Professor at the University of Minnesota in 1956 and again in 1959, recipient of the L. Walker Tomb

Prize from Princeton University in 1958 for his work on the philosophy of time, cited for scholarschip by Wesleyan University as distinguished alumnus of 1959, the inaugural lecturer giving the first series of the newly established 'Konstanzer Dialoge' at the University of Konstanz in 1983, twice elected President of the Philosophy of Science Association (1965–70), vice president of the American Association for the Advancement of Science for Section L (History, Philosophy and Sociology of Science), President of the American Philosophical Association, Eastern Division (1982), Chairman of the Section on the Philosophy of the Physical Sciences for the 1964, 1971, and 1983 International Congresses of the Logic, Methodology and Philosophy of Science. With Larry Laudan, Grünbaum edits the *Pittsburgh Series in the Philosophy of Science* (University of California Press). At last count, he has been a member of the editorial boards of twelve philosophical or scientific journals in five countries. He was a pre-doctoral Fellow of the American Council of Learned Societies in 1948–50, a Faculty Fellow of the Ford Foundation in 1954–55, and is Honorary Fellow of the Yale Graduate School. Far from least of these honors, in his own view, I am certain, he was chosen for recognition as 'Great Professor' for 1967 by *THE OWL*, the undergraduate student yearbook of the University of Pittsburgh, for "ability as a teacher in the classroom, willingness to involve yourself with your students, and interest you have generated in your classes and students".

May a personal memoir of a friend say something of his style? Not that Buffon was altogether correct to say that style is the man, but in Adolf's case the man's character seems almost visibly before me in his manner and tone, in his earnestness of purpose, and in his striking endeavor to communicate all the way into the minds and feelings of his readers and listeners. He *will* make a difference to you. If the reader backs away at first, puzzled by all those *italics* and **boldface**, CAPITALS and underlining, parenthetical phrases, sub-tones crescendos and obbligatos, I must say that Adolf is *simply talking* to us; and he uses every which way of setting forth the nuances and complexities as well as the main structures of his arguments. What is the problem, what are the sorts of evidence, the alternative hypotheses, the previous assumptions made and explanations offered? what are the stages of argumentation, the hidden presuppositions, the blunders, the evasive tactics, the misunderstandings of those he confronts? What is this list, you may say, is Grünbaum a trial lawyer, debating each case as if life and treasure depend on the outcome? or is he a philosopher? And of course he is the debating, searching, dialectical philosopher, good-humored and serious at once, for what in life does not depend upon Adolf's interests: our understanding of rationality and experience, of the cosmic reality of space and time, and of the human reality of thought and emotion?

Grünbaum dearly wants to persuade, but he makes little compromise with standards of argument. To his admiring students and colleagues, there is no greater respect than Adolf's imposing upon them the rigorous demands for technical competence he has set upon himself. Critique of science must come *fully* informed by work within the citadel of scientific theory itself. For Grünbaum as with Reichenbach "there is no separate entrance to truth for philosophers . . . the path of the philosopher is indicated by that of the scientist."

And so Adolf troubles and irritates people, deliberately I suppose, by his fusion of tolerance and severity, his sweetness of humane disposition combined with a tough rejection of the ill-informed, or the superficial, of concealed ignorance, prejudice and cliché, of wishful thinking, sloppy argument, incomplete quotation, and above all, of high-phrased obfuscation. Of course, we might wonder who would defend all or any of these illogicalities which seem to come from the index of that fine old textbook of 'straight and crooked thinking', but the wonder, for Adolf, has been how often they occur.

In Grünbaum's exchanges, his struggle for clarity, indeed for a fundamental understanding of just what is what, is sometimes muscular and passionate. Who will forget the exchange between two splendid philosophers, Grünbaum and Hilary Putnam, over the nature of modern geometry. To Putnam's detailed 50-page 'Examination of Grünbaum's Philosophy of Geometry' in 1963, we ourselves published Adolf's 150-page essay of reply, rebuttal and elaboration of 1969 in our *Boston Studies 5* [102]. Muscular, yes. But Adolf wrote there: "I am most grateful to Hilary Putnam . . . his work has been a valuable stimulus to me to clarify my views both to others and to myself". And he goes on to say what I find essential to Grünbaum's sweetness and honesty: "Critical severity is linked here with friendly respect". I think he would say the same were he to converse with Kant and Leibniz.

Grünbaum's personal sweetness, so immediately clear face-to-face, may not always be directly recognizable in his writings; shall we demand that sweet passion be evident in the search for truth? But beyond that, Adolf is so eager to reach out, to *show*, to *teach*, to have the reader join in the fun, that he may go at times beyond what the reader, who has just cautiously dipped into the fray, expected. And there is affectionate amusement in Adolf's admiring circle when he overdoes it. Daniel Dennett, an admirer, and a wise observer of the philosophical profession, caught the ever-youthful energetic sincerity of Adolf Grünbaum, in the entry for Adolf in his dictionary of current philosophers:

Grünbaum, noun (in German folklore): a tree which, when one of its fruits is bruised, produces another of the same shape, taste, and texture, but five times as large.

Grünbaum's friends are a mighty legion. Some must be named in this memoir: besides those mentioned already, his teachers were the mathematician Burton Camp and the physicist Walter Cady; his long-time philosophical colleagues have included Nicholas Rescher, Wesley Salmon, Wilfrid Sellars, Allen Janis, Larry Laudan, Richard Gale, Kurt and Annette Baier, Alan Anderson, Ted McGuire, and his beloved C. G. Hempel; Russ Hanson, Imre Lakatos, Paul Feyerabend; and among his students, Bas van Fraassen, Brian Skyrms, Alberto Coffa, Philip Quinn. Adolf's correspondence is legendary, world-wide, compulsively responsible; and his friends affectionately take care that when Adolf travels, he should know the pick-up schedule at local post-boxes for miles around. Who else among professors joins the university postal workers' Christmas party each year? His consistency is noted too. They say Paris felt secure, knowing from her blue light shining that Colette was at work; and I well recall the light from Adolf's dormitory room on winter college evenings, reflecting over the snow, reassuring his fellow students that one of us was faithful to the life of the mind. But progress too: he may not have been the Saturday afternoon enthusiast for college football, but now he roots for the Pittsburgh Steelers *malgré lui.*

Boston University ROBERT S. COHEN

ALBERTO COFFA

GEOMETRY AND SEMANTICS: AN EXAMINATION OF PUTNAM'S PHILOSOPHY OF GEOMETRY

There are many ways to shed light on how and why our conception of geometry changed during the last two centuries. One fruitful strategy is to relate those changes to contemporary developments in other fields, from physics to epistemology. Within this general framework, there is one project that I find both enlightening and not sufficiently pursued: that of examining the link between philosophies of geometry and the semantics they presuppose. This standpoint is particularly appropriate for an evaluation of Putnam's views since few philosophers of science have been more keenly aware of the central character of semantic issues in philosophy, or more productive in bold, insightful ideas in that field.

Geometry is no small topic in Putnam's estimation of things. He once observed that the overthrow of Euclidean geometry is "the most important event in the history of science for the epistemologist" (1975b, **I**, p. x); he also thinks that the received account of that overthrow is a scandal, and that a correct account is bound to give us an entirely new perspective on the character of scientific knowledge. Our strategy in this paper is quite simple: in Section I we shall examine Putnam's enemy, the conventionalist; more precisely, those semantic points in conventionalism that Putnam means to challenge. In Section II we turn to Putnam's own proposals on geometry and semantics.

I. SEMANTICS AND GEOMETRY IN THE NINETEENTH CENTURY

As is well known, concepts (not essences) are what meanings were before they were wedded to the word. If we want to know what our ancestors thought about meanings, we should not read their books on essences but their logic books, where they talked about concepts and judgments, meanings and propositions.

Praiseworthy as it was in so many respects, modern philosophy (rationalism *and* empiricism) contributed little more than confusion and superficiality to the field of semantics. Their doctrine of 'ideas' or 'representations' and of the judgments which they constituted, was grounded on the artful confusion of knowledge and its objects, of psychological epistemic states and their content

R. S. Cohen and L. Laudan (eds.), Physics, Philosophy and Psychoanalysis, 1–30.

and, in the end, of "what really is the case and what one judges to be the case," the key distinction of what Putnam now calls metaphysical realism (1981, p. 71). Sloppy semantics thus provided the premise that naturally led to the idealistic debauch.

One of the offspring of this style of semantics was Kant's conception of the synthetic *a priori* and, in particular, his interpretation of geometric knowledge. Let me review the central facts.

The root of all semantic evil, we have often been told, is the analytic-synthetic distinction, which was invented by Kant. Surprisingly enough, this turns out to be true. The problem began with the often unnoticed fact that Kant introduced two different notions of 'analytic' (and three of 'synthetic'). There is, to begin with, the official or 'nominal' sense of analytic according to which an assertoric judgment of the form "All *A*'s are *B*'s" is analytic when the predicate-concept *B* is thought, however implicitly, in the subject-concept *A*. Modern readers get nervous when they hear this sort of definition and refuse to take it at face value because of the appeal to implicit thought and because of its limited scope. It is, however, essential to take it quite seriously if we want to know what exactly lies behind this remarkably influential idea.

Like many of his predecessors Kant endorsed what we might call a 'chemical' picture of the concept – indeed, of all representations – according to which concepts consist of constituents that could be identified by means of a process called 'analysis'. Inspired perhaps by Leibniz's "*petites perceptions*" Kant thought that when we look at the Milky Way with the naked eye we actually see all of its stars, and that the telescope merely helps us see distinctly what was already obscurely present in the original representation.[1] This holds not only for intuitive representations, as the one just considered, but for concepts as well. Thus, except for the limiting *simple* cases, concepts always have constituents of which we may not be aware but which are 'tacitly thought' in the concepts we have, and which may be uncovered by means of the intellectual telescope of analysis. It is within this picture of the concept that it makes sense to define a judgment as analytic when it is a statement reporting the outcome of a process of conceptual analysis. An assertoric judgment will be synthetic when it is not analytic.[2]

Next to this characterization of 'analytic' there is an entirely different one in Kant's writings, namely, the distinction between judgments of definition or clarification (*Erläuterungsurteile*) and ampliative judgments (*Erweiterungsurteile*).[3] Kant, his followers and many commentators regard this distinction as obviously identical with the preceding one at least as far as their extension is concerned. Analytic judgments are thereby assumed to be precisely those that clarify the nature of the concepts involved in the judgments under consideration and synthetic judgments are similarly presumed to be those which extend our knowledge beyond the information contained in their constituent concepts. In other words, Kant is assuming that all claims grounded only on our understanding of the concepts involved must be analytic in the first, nominal, sense. As he puts it, "it is evident that from mere concepts only analytic knowledge . . . can be derived" (A 47/ B 64–5). In particular, if *A* is a simple concept, "All *A*'s are *B*'s" could never be grounded on our understanding of the concepts *A* and *B*.

This assumption leads to one of the central principles of Kant's philosophy: the so-called "principle of synthetic judgments". What this principle says is that in all synthetic judgments the ground of the synthesis between the subject- and the predicate-concept must lie in an intuition. (Implicit is a third notion of 'synthetic' as a judgment whose ground must lie in an intuition.) The reasoning for this principle is easy to reconstruct. The ground of the synthesis must lie in a representation other than those contained in the judgment. Now, representations are for Kant either intuitions or concepts. If the grounding representation were a concept, we would then be dealing with three concepts, but "from mere concepts only analytic knowledge can be derived." So the representation which provides the required link must be an intuition.

We are now a very small step away from the centerpiece of Kant's picture of geometry, the idea of pure intuition. Since there obviously is synthetic necessary knowledge, there must be a form of intuition capable of providing a necessary ground for such judgments; clearly empirical intuition can't do the job, so there has got to be some other, more powerful sort of intuition; let us call it 'pure intuition'.

The existence of this form of intuition allows Kant to draw the famous distinction between necessities of thought (*Denknotwendigkeiten*) and necessities of intuition (*Anschauungsnotwendigkeiten*). Synthetic propositions are never true purely in virtue of concepts and it is therefore always possible to *think* their negations; but some synthetic propositions, those which are *a priori* true, are such that their negations, although thinkable, are not intuitable. According to Kantian doctrine, I can think Lobachevskian geometry without any difficulty, but I cannot associate objects with its axioms in an intuition (hence, anywhere). I can think, Kant explains, the concept of a diangle (two straight lines which enclose a space), but since the concept in question cannot be constructed in intuition, I cannot give myself an object for it and therefore the concept cannot be involved in any knowledge claim.

Progress in foundations-research during the nineteenth century was largely the process of extirpating pure intuition from every field of scientific knowledge. Early in the century Bolzano pushed space, time and Kant out of the calculus when he started that great movement now known as the rigorization of analysis (Cauchy, Weierstrass, Dedekind, Cantor, etc.). Decades later Frege and Russell would push the pure intuition out of arithmetic. Only geometry remained a Kantian stronghold throughout the century. The only alternative to Kant appeared to be an implausible empiricism of the J. S. Mill variety, which, as Putnam has argued, was hard to take very seriously owing to its inability to explain the sense in which geometry is necessary. But, more on this later.

Let me side-track from history for two paragraphs in order to consider briefly which of Kant's assumptions was being challenged in this process of de-Kantization of science. Note that his conclusion that there must be a pure intuition grounding the judgments of mathematics and geometry follows from two premises: (i) that those disciplines contain synthetic *a priori* judgments and (ii) the principle of synthetic judgments. Different philosophers challenged different principles, of course, but the interesting point is that most of the members of the analytic-positivist tradition, including the conventionalists, accepted (i); and that the most fruitful challenge to Kant derived from rejecting *not* (i), but (ii).

It is well known that Frege and Russell were ardent supporters of the synthetic *a priori* (in Russell's case, even in the field of logic) but it is less well known that the dominant figures in the Vienna Circle also accepted Kant's thesis. Many things have prevented us from seeing this – sloppy or hostile reading, mostly; but also Schlick's and other people's concentration on the *third* sense of 'synthetic' in Kant, and the entirely justified positivistic wrath against the Kantian answer to the question of how *a priori* knowledge is possible. Whatever the reasons, it is undeniable that it was part and parcel of the positivist philosophy that there are judgments that cannot be derived by analysis or definition and logic (hence they are 'nominally' synthetic even in Frege's sense) but that are nonetheless, in a sense to be specified, necessary. Far from refusing to cross the doorstep of Kant's system, positivists such as Carnap, Wittgenstein and Schlick accepted the Kantian premise that there are synthetic *a priori* judgments and set out to explain their possibility. There is, of course, another empiricist line extending from Mill to Quine, that refuses to take the idea of necessity seriously and thereby does not even grant the basic premise of Kant's challenge. But virtually all of the analytic philosophy that is worth reading and that was written before Quine belongs to the opposite tradition.[4] Let us now return to our fast-forward history.

As Kant and his followers were reducing rationalist semantics to the absurdity known as idealism, making the object of knowledge an element of our mental world, a small but powerful movement developed in Europe inspired by the idea that at the root of the current philosophical cataclysm was semantic confusion. The leading figures of this group were Bolzano, Frege, and Russell. While idealists and neo-Kantians dealt with 'big' problems, the semanticists chose to concentrate on what appeared to greater minds as trifling matters: concepts and propositions. The best-known outcome of these researches was the birth of logic. But this was only a corollary of a much broader semantic program intended to provide the intellectual tools with which one could eventually build a sane account of the link between knowledge and the world.

One thing this account made possible was to recognize the fallacy implicit in Kant's principle of synthetic judgments. Bolzano may have been the first one to realize that there are synthetic and necessary propositions grounded on concepts alone and hence that the principle is quite false. His penetrating analysis of the nature of logic prepared the ground for the recognition that analytic knowledge, far from being identical with conceptual knowledge, is an almost insignificant subclass of it. Analytic knowledge, in Kant's or Frege's sense, is knowledge grounded on a handful of concepts, the so-called logical ones. But if the concept of conjunction, for example, can provide a sufficient ground for "*A* & *B* implies *A*," surely the color concepts can provide a sufficient ground for "if this is red it isn't blue." The idea that non-logical concepts are not well equipped for this sort of thing, popular as it is these days, cannot be taken seriously. Bolzano didn't; and he concluded that synthetic *a priori* judgments can be grounded on concepts alone.[5] Going beyond philosophical pronouncement, he formulated with stunning insight a technically worked out program for the development of all mathematics (including geometry) on a purely conceptual, intuition-free basis. The link between this project and conventionalism will soon become apparent.

The philosophical populace is fond of making fun of the Platonic element in the theory of the concept which emerged from this tradition. But under the guidance of this Platonic picture semantics became of age. The positivists eventually noticed that one need not be a Platonist to develop a semantics that stands at arm's length from psychology and also from syntax and pragmatics. But the idea of semantics as an independent domain provided the ground for both a non-idealist theory of knowledge and for an account of the *a priori* and the necessary that avoids idealism. Whether Quine's destruction of this field of semantics still leaves room for those cherished goals is a question we shall have reason to raise at a later stage in this paper.

The progress elicited by this new way with meanings was enormous, but there were problems too. Leaving the matter of Platonism aside, the semantic picture which we encounter in Frege's and Russell's writings proved incapable of explaining certain crucial features of mathematical and empirical knowledge. The first field where its inadequacies became transparent and, indeed, obstructive to the progress of science was, coincidentally, that of geometry. Conventionalism emerged in geometry late in the nineteenth century partly as a program to correct these shortcomings in the realists' semantics which, as it turned out, were due to an excessive reliance on the Kantian doctrines of analysis and definition.

I know no better way to bring forth the semantic root of the conflict between these two viewpoints than to examine the debates which took place near the end of the century between the two leaders of the realist approach to semantics, Russell and Frege, and the two leaders of the emerging conventionalist philosophy, Poincaré and Hilbert. The situation is far too complex to be analyzed here in detail.[6] It will suffice if we indicate the central issue of disagreement and its sources.

The subject of both debates was definition. Frege and Russell had heard their respective duel-partners say that geometric axioms are definitions (in disguise) and they could hardly control their rage. "It is high time that we began to come to an understanding of what a definition is," Frege explained to Hilbert; "since mathematicians almost invariably ignore the role of definitions," complained Russell to Poincaré, "I will allow myself a few remarks on this topic"; and he concluded his patronizing lecture saying

> All these truths are so obvious that I would blush to recall them, were it not that mathematicians insist on ignoring them.[7]

The subject was, indeed, definition, but in not quite the sense in which Frege and Russell thought of it.

What Frege and Russell meant by definition was, in effect, a method to construct new meanings from old. Like Kant, they viewed concepts as complexes and their definitions as analysis. To define a term was to exhibit its construction from simple(r) concepts. Frege had given a precise logical expression to this traditional picture, and Russell would soon do likewise. Having produced a wonderful theory of definition, they assumed that there was no other legitimate sense of that word, and concluded that their opponents had to be talking about it. Thus, when Poincaré challenged Russell to define 'distance' without using that word, Russell replied that this request was like that of a schoolteacher asking his student to spell the letter A without using it. But Poincaré was not asking Russell to analyze the unanalyzable, only to explain how he knew what the word means.

Although commentators have been as successful as Frege and Russell in ignoring this fact, there can be little doubt that Poincaré and Hilbert were using the word 'definition' in its dictionary sense. This sense happens to be the one that Mill had called "the simplest and most correct notion of a definition," to wit, "a proposition declaratory of the meaning of a word" (*A System of Logic*, VIII, Section 1). To define an expression is to say what it means. In this sense, although not in Frege's and Russell's, it makes perfect sense to ask for the definition of the primitives or, if we confuse use and mention, of the indefinables.

One of the central questions which concerned the young Russell was what he called "the problem of indefinables". This problem arises in a standard (formal or informal) axiomatic construal of knowledge, and it concerns the primitive terms and their associated indefinable notions. Definition in the logician's sense tells us how to produce new meanings from old, but it tells us nothing about how to get our original stock of meanings. The problem of indefinables is the problem of how we ever come to be in possession of that stock, or more generally, of what exactly is the link between the primitive terms of our theories and what they mean and stand for. The question does not call for a baptism – like calling the relation 'designation', 'reference' or 'connection function'; it calls rather for an account that explains how human beings can establish the link in question.

In idealistic philosophies there is no problem of indefinables since discourse, its topic and its meaning – to the extent that these distinctions are drawn at all – remain entirely within the domain of mind. In the context of a philosophy such as Frege's and Russell's, on the other hand, the problem becomes acute since their strategy to avoid psychologism is to offer a semantics based on a domain of meanings which stand aloof from the range of things mental and, indeed, human. The problem of indefinables is here the problem of the semantic pineal gland: how do Fregean or Russellian simple concepts ever manage to penetrate the frontiers of the mind? How do we come to know them and use them for the purposes of constructing knowledge claims?

Most philosophers didn't have much to say on how to solve this problem;[8]

but virtually all of them agreed on how *not* to solve it. Frege, for example, explained to Hilbert that

> it can never be the purpose of axioms and theorems to establish the meaning of a sign or word occurring in them; rather, the meaning must *already* be established (1899, pp. 62–63);

and Russell explained to Poincaré that, concerning indefinables such as *distance* and *straight line*

> any proposition, whatever it may be, in which these notions occur, is either an axiom or a theorem ... When I say the straight line is determined by two points, I assume that [these] are terms *already* known and understood, and I make a judgment concerning their relation (1899, pp. 701–702).

Geach has traced back to Buridan a version of the principle underlying this attitude:

> the reference of an expression *E* must be specifiable in some way that does not involve first determining whether the proposition in which *E* occurs is true;

and he calls this Buridan's law.[9]

Is Buridan's law true? It depends on how you read it. The version of it that was widely endorsed before conventionalism was this: that there is no proper use for a sentence unless its constituent expressions have been given a meaning and perhaps a referent *before* they join their partners in the sentence. We might express this as the claim that 'semantic force' (meaning and reference) is theory-independent. The violation of this principle by conventionalists may account for most of the wrath which they elicited among philosophers. And yet, their challenge had good reasons.

Helmholtz may have been the first one to observe that there seems to be something wrong with Buridan's law in the field of geometry, for geometric axioms display a feature reminiscent of Kant's idea of how experience is constituted. In order for an object to qualify as a possible confirmer or falsifier of geometric claims it must, of course, be among the things the claims are about. But, Helmholtz noted, the only effective way to tell whether an object is in the range of reference of geometric axioms is to see whether the axioms are true of it.[10] Indeed, he detected a similar constitutive role in other propositions such as the principles of measurement and of causality; and he tried to express their peculiar character by saying that these propositions are such that "until we have them, we cannot test them."[11]

Some neo-Kantians (for example, Riehl (1904)) saw in this doctrine an oblique vindication of Kant. But the move can be so interpreted only if one ignores the essential ingredient of this geometric doctrine that different ('conflicting') sets of regulative maxims are equally legitimate. Geometric axioms are like Kantian rules in that they are logically prior to and constitutive of their objects, but unlike them in that we have a choice.

What Poincaré and Hilbert add to Helmholtz's point is this other one: that the root of the constitutive character of geometric axioms is not epistemic but semantic. By calling them "definitions" they were expressing, however imprecisely, their intention to place them in the territory of purely conceptual truths from which they had been excluded by Kant's confusion between non-analytic and ampliative. And to Helmholtz's remark "we cannot test them until we have them", they will merely add "and once we have them, it is too late to test them."

Once the realists' semantics removed the old idealist confusion between concept and object, it became possible to understand the role of regulative maxims as constituting not the objects of knowledge – whatever that might mean – but rather the *concepts* in terms of which we think of them. The endorsement of these rules is a crucial part of what is involved in the mastery of a set of concepts. Once those concepts are at hand one may, once again, look at the expressions of the rules from the standpoint the new concepts provide. Then the rules will appear to make trivially true claims – claims true in virtue of the concepts in question, as Bolzano hoped, or true in virtue of meanings, in the jargon soon to be favored by positivists. In this way conventionalists came to think of geometry as a purely conceptual discipline, and to account for both its necessary and its conventional aspects. They also solved Russell's problem by noting, in effect, that the axioms of geometry define its indefinables.

Before we leave these historical matters, I should like to add two remarks that should help draw more sharply the doctrines we are examining.

First of all, it is instructive to examine the conventionalist stand vis-à-vis Buridan's law. If we mean by *claim* a statement that conveys information about concrete or abstract matter of fact, then the conventionalist agrees that a claim must be expressed by means of sentences whose words have acquired semantic force *before* they join their partners in the sentence. The conventionalist grants, in fact, insists that if a sentence were used with the purpose of fixing the meaning or reference of some of its words, then the sentence in question couldn't possibly convey any information. But he does not thereby conclude that endorsing a sentence of that sort would make no sense. His point is that sentences may be endorsed not only for the purposes of displaying what Russell called propositional attitudes but also for that of fixing semantic content, thus displaying our decision to use certain words in certain ways.[12]

The second point is this. In the course of examining the recent history of attempts to deal with the nature of scientific testing Glymour noticed that in its earliest stages the logical positivist movement tended to endorse a position that – like his own – does not draw a distinction between conventional and factual statements but attributes instead factuality to all scientific statements, even if each statement in a class may have to be assumed in the process of testing the remaining ones. And then he observed that at one point – Reichenbach in the twenties and Carnap shortly thereafter – positivists turned to the task of detaching the conventional or analytic from the factual or synthetic. Glymour sees this as a step in the wrong direction, at least on the subject of testing, and wonders why it was taken (1980, p. 53).

My guess, as you may gather from the preceding, is that the step was taken as an attempt to deal with a problem that our positivistic forefathers regarded as prior to that of testing: the problem of meaning and reference, the problem of how a statement gets to say something about the world, and of what exactly it is that it says. Since positivists were much less committed to verificationism than Quine and his followers, they believed that the problem of determining what we are saying is prior to that of how we test whether what we are saying is true. The network of bootstrap strategies that Glymour has so artfully uncovered presupposes that all statements being tested convey information. The conventionalists' problem was to determine when that assumption is justified.

One of their central premises is stated and endorsed in this statement of Putnam's:

That empirical laws cannot be confirmed or disconfirmed, and, indeed, cannot be empirical laws at all unless the words occurring in them have a meaning is trivial (1963, p. 121);

but, as already noted, it seemed equally trivial to them that those words must have semantic force before they gather together into sentences *if* those sentences are to express claims. This doctrine, sometimes called *semantic atomism*, was the common ground between conventionalists and their realist opponents. There can be no doubt that semantic atomism is a tacit, essential premise in the inference from the failure of alternative proposals to solve the problem of indefinables, to the conventionalist solution. Whether any sense can be made of the denial of semantic atomism is one of the underlying themes of the second part of this paper.

The developments reviewed above are unified by their link with the semantic issues which concern us in these pages. As is well known, in our century conventionalism developed along non-semantic dimensions that were also implicit in the writings of the founding fathers: first Reichenbach extended the doctrine in an epistemological direction, then Grünbaum proceeded along an ontological course. Yet, at the basis of these developments there was always a commitment to the central semantic insights that conventionalism had reached by the end of the century.

And then came Quine, his slings and arrows aimed at the semantic heart of conventionalism. For decades the best philosophers this side of the Atlantic and of the Rio Grande refused to take seriously talk about meaning and conventions. When Putnam entered the scene it was widely believed that Quine had refuted the linguistic version of logical conventionalism put forth by Carnap and Wittgenstein;[13] but conventionalism still thrived in the philosophy of geometry. Putnam decided to lead the Quinean forces in a final charge against that rebellious philosophical colony. Whether he conducted a successful operation or the philosophical equivalent of the charge of the Light Brigade, is what we now turn to see.

II. PUTNAM'S PHILOSOPHY OF GEOMETRY

There are two topics that dominate Putnam's thought on geometry and, indeed, on most everything else: apriority and semantics. The core of his conception of geometry can perhaps be stated in two claims: that Euclidean geometry was virtually *a priori*, and that it has been shown to be false. Both points call for new theories of the *a priori* and of semantics.

II.a. *The* a Priori

In the debate between necessitarians and their opponents presented on p. 4, Putnam's heart is clearly with the former, even if his mind may end up elsewhere. Again and again he tells us that the feeling of necessity or apriority associated with some theoretical claims must not be dismissed as a merely psychological phenomenon but must be recognized as a symptom of the fact that the targets of those feelings play a rather extraordinary role in the associated knowledge situations. These statements, which he often calls "framework principles", must never be confused with the more vulgar category of the "merely empirical" claims. In fact, what makes the overthrow of Euclidean geometry so important for epistemology is that it "represents

. . . an unprecedented revision of framework principles" (1963, p. 109). In that episode Science offered a subtle clue to one of its deepest secrets, a clue we are bound to miss if we ignore the virtual apriority of Euclidean axioms at that time. It is therefore, a "philosophical scandal" that the received account of the nature of geometry has been blind to the role of the *a priori* in empirical knowledge:

It is the task of the methodologist to explain this special status [of framework principles], not to explain it away (1965b, p. 92).

If it was just an illusion that the statements of Euclidean geometry were necessary truths . . . then philosophers of science owe us a plausible explanation of this illusion (1975e, p. ix).

Apparently Putnam thinks that Russell's if-thenism (Putnam's nice name), or the version of it implicit in Reichenbach's "normative function of intuition" (Reichenbach, 1928, Section 9), is the only serious attempt ever made by analytic philosophers to explain the feeling of necessity associated with geometric and other principles. This explanation, as he points out, is "simply absurd" (1975e, p. ix), for it can only account for the feeling of necessity of the inference and not for the equally obvious necessity associated with its premises. Thus blinded to the element of apriority in geometry, the conventionalists (actually, their opponents) are said to have thought that the truth or falsehood of geometry is a "merely empirical" matter (e.g., 1963, Section 3), thereby turning a Wagnerian opera into a nursery rhyme. Geometry is, of course, empirical (for Quine will put his foot down in the end); but not *merely* empirical.

Since the difference between the empirical and the merely empirical is equal to Putnam's conception of the *a priori*, one naturally wonders what that conception is. But before we start our fateful search for that doctrine in Putnam's writings, we should first notice how unfair is his assessment of his predecessors' standpoint.

As we have seen, conventionalists had a pretty honorable explanation of the necessity of geometric axioms. This explanation is hard to recognize if you think that convention is the opposite of necessity; but convention is only the opposite side of necessity: what looks conventional from the outside looks necessary from the inside, as it were. Are the axioms of Euclidean geometry conventional? Before we have adopted them, of course they are – says the conventionalist – since we can take them or leave them as we please. Are they necessary? After we have adopted them, of course they are; since by adopting them we have displayed our decision to employ certain concepts which involve those axioms essentially. It isn't really that the endorsed axioms are true in any ordinary sense of the word. Axioms so construed are really not claims but linguistic devices whose syntactic appearance (a "disguise") is exactly like that of claims, but whose semantic and pragmatic role is that of helping implement our decision to use certain concepts.

This explanation of necessity is, no doubt, vague and imprecise. Lovers of clarity, like myself, would much rather do without it. So one turns hopefully to Putnam's writings in search of a clearer and more plausible theory.

Even though Putnam has invested a great deal of energy in trying to convince his Quinean audience that we need a theory of framework principles, what is offered as such is hardly a theory. Apart from a few side-remarks, a few examples and repeated claims that the denial of framework principles is "literally inconceivable",[14] all we have to go by is Putnam's characterization of those principles vis-à-vis alternative frameworks.

Putnam's idea is this: that there are some claims – precisely these framework principles – which are such that no matter how much empirical evidence against them may be gathered, no one does or should drop them until an alternative theory is developed (e.g., 1962a, p. 46). Observations in the neighborhood of this one are familiar from the writings of several philosophers of science, but Putnam deserves great credit for having been, together with Kuhn, the first one to have stressed the role of alternative theories in theory-change. Whether this insight can be made to play a role in the theory of the *a priori* is an entirely different matter.

Putnam clearly hopes it can. This becomes apparent on those rare occasions when he decides to say something concerning the character of frameworks, for example, in the field of geometry. What are geometric framework principles?

> An answer would be difficult to give in detail, but I believe that the general outlines of an answer are not hard to see. Spatial locations play an obviously fundamental role in all of our scientific knowledge and in many of the operations of daily life. The use of spatial locations requires, however, the acceptance of some systematic body of geometrical theory. To abandon Euclidean geometry before non-Euclidean geometry was invented would be to 'let our concepts crumble' (1962b, p. 243; see also 1962b, p. 248–249).

And that is all.

Notice first the closing point: that when we give up certain principles, certain concepts will crumble. The conventionalist can make some precise sense of this metaphor, as we know; Putnam cannot. Metaphors aside, this explanation of the feeling of necessity and inconceivability has its facts upside down. For what we are asked to suppose is that the modal feelings emerge after the theory is introduced and as a consequence of the fact that there are no alternatives to it. But the process clearly followed the reverse course: people decided first which statements appeared to be geometrically necessary and constructed their geometric theory in conformity with such feelings. The absence of alternative frameworks, far from being the cause of our modal inclinations, is usually its effect.

The problem is that Putnam has tried to extract from a sound methodological premise a modal or semantic conclusion that is simply not there. One might readily concede that no matter how many experiments of the Michelson–Morley variety had been performed after Maxwell, and no matter how many planets with square orbits had been discovered after Kepler, the corresponding theories would not have been abandoned by the scientific community. It may even be true that in situations of this sort no theory should be abandoned until someone puts forth an alternative theory that does no worse than the preceding one. But to think that the scientific community would not – let alone *should* not, not to mention *could* not – think that the theory in jeopardy is false until alternatives develop, is something that could be seriously entertained only by someone who, like Kuhn, does not think that talk of truth makes much sense in the case of scientific theories. Those of us who can still distinguish between abandoning a theory and believing that it is false have no difficulty in recognizing the relevance of Putnam's point to the sociology of knowledge as well as its irrelevance to the semantic and epistemological issues he has raised. After all, one does not have to be Bellarmine to notice the difference between sticking to a theory that works better than any other around, and believing that what it says is true.

The bottom line is that Putnam has no theory of the geometric *a priori* – not yet anyhow. His *second* point, as you recall, was that what was virtually *a priori* has now been shown to be false. The centerpiece of his case for this conclusion is a new semantics intended to ground the claim that Euclid, Hume, Riemann and Einstein referred, convention-independently, to the very same things when they used the expression 'straight line' (e.g., 1962a, p. 50).

II.b. *Semantics*

One can find in Putnam's writings a variety of principles or maxims intended as parts of a doctrine of semantic force and, particularly, of reference. At different times he has told us that reference is determined (i) by coherence,

(ii) implicitly, by our theories, (iii) by causality, via introducing events, (iv) by Nature, via a relation $same_{NK}$ (same-natural-kind) and (v) by the division of linguistic labor. It appears that Putnam regards all of these as aspects or elements of a unified semantic doctrine. For the purposes of our discussion I will divide these theories into two groups, depending on where they appear to stand on the issue of semantic atomism. We start with those that appear to challenge it.

II.b.1. *Implicit Semantics*. Let us return to the passage which we started to quote on p. 10.

That empirical laws cannot be confirmed or disconfirmed, and, indeed, cannot be empirical laws at all unless the words occurring in them have a meaning is trivial. That the words must *first* be given a meaning by the laying down of *definitions* is not trivial, and indeed, in the opinion of most philosophers of science today, is not true. Yet it is just this that Reichenbach is concerned to assert and in no uncertain terms. He asserts *both* that *before* we can discuss the truth or falsity of any physical law all the relevant theoretical terms must have been *defined* by means of 'coordinating definitions' ... (1963, p. 121; Putnam's italics).

The italics suggest that Putnam has two main problems with Reichenbach's position on this point: that semantic force must be available *before* the appropriate expressions can enter into a claim, and that the way to generate it is via *definition*. The latter objection, as should be clear by now, is merely verbal and motivated only by Putnam's "negative essentialism" (cf. 1975d) concerning definitions. Like Poincaré and Hilbert, Reichenbach meant by 'definition' a process through which we assign semantic force to an expression; and he surely was entitled to mean that. Thus, to the extent that there is a disagreement with conventionalism, it must be on the first issue. Putnam must think that words acquire semantic force through the claims in which they occur; and he does, indeed, say, that reference is fixed implicitly by theory, and also by coherence. Thus, his alternative to Reichenbach's proposal is to say that

the extensions of theoretical terms are in practice determined only by the correspondence rules together with the theoretical postulates together with the requirement that further postulates and singular statements be accepted only to the extent that they are compatible with the requirements of inductive *and* descriptive simplicity (1963, p. 123; Putnam's italics).

More generally, he explains, the referents of physical theories

are not literally operationally defined; they are in a very complicated sense 'defined' by systems of laws (1975b, I, p. x).

In the case of geometry, for example,

> the metric is implicitly specified by the whole system of physical and geometrical laws and 'correspondence rules'. No very small subset by itself fully determines the metric (1963, pp. 93–94; see also 1965b, p. 83).

and the reference of 'distance'

> need not be fixed by a convention. It can be fixed by *coherence* (1975d, p. 165; Putnam's italics.).

Putnam adds that Reichenbach, not having read Quine's 'Two Dogmas',

> could not have foreseen that it would be possible to provide an answer – what we have called a 'coherence account' – to the question of reference of scientific terms which not only did not presuppose the analytic-synthetic distinction, but which was in spirit fundamentally hostile to that distinction (1975d, pp. 175–176).

In spite of their coordinated appearances, I frankly don't know whether the remarks on fixing reference by coherence and those on fixing it implicitly through laws are intended as parts of the same doctrine. I will discuss them separately, starting with the coherence account.

Even though Reichenbach may, perhaps, be excused, those of us who have held Quine's hand and failed to feel the shock of recognition are left in a somewhat uneasy position. As we turn hopefully to Putnam for further explanations of the role of coherence in semantics, all we find are applications of that supposedly clear idea, so that we are left to infer from them what the idea is. Putnam's main application of the coherence theory is against conventionalism, particularly against the introduction of "universal forces" as conventions. His reasoning is as follows.

Reichenbach and Grünbaum have argued that the introduction of universal forces does not alter the content of a physical theory since such an operation, as Grünbaum has put it, amounts to no more than a change in metric standards. Putnam, on the other hand, thinks that the coherence of a theory deteriorates significantly when we introduce hypotheses of universal forces, and that this loss of coherence allows us to conclude that de-universalized forces are what 'force' really refers to in standard scientific theories. As far as I can gather, the underlying premise is that given two statements of the form "force is such and thus", the one that joins most coherently with the body of background knowledge is the true one (or, better yet, the more coherent, the 'truer'). Putnam does not say this explicitly, but readers are invited to look at the discussion in (1975d) to see whether they can find anything there as clear as and more plausible than this.

Notice first that under this interpretation Putnam's approach involves a coherence theory of *truth* for reference statements; thus we are being invited to endorse a full-blown coherence theory of truth, or else to assign to reference-claims a second-class citizenship. This consequence will be uncomfortable only to those unhappy about idealistic implications; but there is a different problem that cuts across metaphysical allegiances, and which relates to an old conventionalist point due, once again, to Helmholtz.

Helmholtz observed that in geometry it is essential to distinguish between features of the form of representation and features of the information conveyed by such means.[15] If complexity and coherence are to play a role in semantics they will surely have to do so as features of content and not of the syntactic expressions which involve, as often as not, confusing, irrelevant notational factors. The point can be made on the basis of quite simple examples.

As we all know, we have in classical mechanics two familiar expressions for Newton's second law:

$$(*) \qquad F = ma$$

and

$$(**) \qquad F = m\frac{\partial^2 r}{\partial t^2} + m\omega \times (\omega \times r) + 2m\omega \times \frac{\partial r}{\partial t} + m\dot{\omega} \times r + m\,\frac{d^2 R}{dt^2}\,.$$

An uninformed person will look at these two formulas and say that they are very different: (*) is simple and (**) complex, (*) seems to involve only a few basic notions whereas (**) refers to several sorts of forces, 'fictitious' forces, possibly multiplying entities beyond necessity and thereby violating some methodological maxim. After reading or misreading Putnam he might be tempted to conclude that (*) tells us what force really is (what 'force' refers to) and that (**) must be eliminated on grounds of coherence.

But there is, I take it, no dispute over the fact that (*) and (**) are two different formulations of the very same law: (*) expresses it relative to inertial frames and (**) relative to frames rotating with angular velocity ω.[16] It follows that all the obvious differences between (*) and (**) must pertain not to what those expressions say, not to the law, but to the way in which they say it; therefore *not* to content or semantics but to syntax, and perhaps also to pragmatics. Now, as we are all fond of remembering, there was a fleeting moment of syntactic insanity in the development of logical positivism, when people wondered whether there was anything between syntax and its pragmatic link with empirical data. But they soon recovered and realized that semantics was there after all: Reichenbach told us about descriptive simplicity, Carnap and Popper about content, and with a sigh of relief we could once again believe that (*) and (**), whatever their differences, do say the same thing after all. Grant Quine all he wants about synonymy; surely it can't be part of the secret of 'Two Dogmas' that we should return to those troubled times in the pre-history of logical positivism?

And yet, several of Putnam's arguments seem to depend on the refusal to acknowledge this Helmholtzian distinction. Thus, he says that it is well known that "the demand of the special theory of relativity is that our reference system must be an inertial system" (1963, p. 115). But there is no such demand on reference systems either in the special theory or in classical physics; for it is now widely recognized that reference frames are representational artifacts intended to express the content of a theory in a manner suitable for computations; but when the goal is to represent it in a manner suitable for a better understanding of its content, the preferred form of expression is coordinate-free. One suspects that a good portion of Putnam's attack against Grünbaum is inspired by his Quinean reluctance to detach what pertains to forms of representations generally from what is intrinsic to the situation at hand.

This very attitude is displayed once again in Putnam's remarks concerning covariance. It is well known that Einstein's course towards the general theory was riddled with problems emerging from his insufficient appreciation of the role of coordinate systems in geometry and physics. When he finally achieved his covariant formulation of relativity he thought he had fulfilled a *desideratum* few other theories could satisfy; in particular, one that could not be satisfied by either classical physics or the special theory. In agreement with this standpoint Putnam claims that Einstein's insight on covariance was "a purely physical, not a logical insight, namely, the insight that the laws of nature can be written in a simple covariant form" (1963, p. 119). There was, in fact, no such insight. As Cartan explained in (1923), even Newton's theory can be given a simple covariant formulation. Wheeler noted not long ago that virtually every physical theory must allow for a covariant formulation since that requirement amounts to the condition that reference frames should play no more than a calculational role in the formulation of the theory.[17] When Einstein finally came to see this point (1918, p. 242) he still wanted to insist that a formulation of Newton's theory valid in all reference frames would have to be immensely complex.[18] Perhaps he was thinking of the increase in complexity in the step from (*) to (**), and then extrapolating to arbitrary frames. However this may be, simple covariant formulations of Newton's physics are now available in physics textbooks (see, e.g., Bradbury's *Theoretical Mechanics*, Section 2.21).

In view of all this, and in spite of the sketchiness of the coherence theory, one is entitled to conclude that it is intended as a doctrine which applies to syntactic phenomena, not to semantic noumena. If so, there is no reason to think it has anything to contribute to the problem of reference.

Let us now turn to the implicit-definition side of Putnam's account. We begin by recalling that two of the overriding goals of Putnam's picture of science, until recently anyhow, were to show that scientific claims are factual and that widely divergent scientific theories may talk about the same things and therefore conflict with each other.

The conventionalists have a relatively easy way to achieve these goals. They first distinguish a class *C* of statements that implicitly define the semantic force of the relevant theoretical expressions, and then they observe that conflicting theories must agree on *C* and disagree on something else, maybe on everything else. Putnam, and post-Quineans in general, cannot distinguish between *C* and the rest of a theory, so they are deprived of this strategy. Yet, they are as skeptical as the conventionalists were of traditional solutions to the problem of reference and, like them, they are often in sympathy with the idea that semantic force is fixed somehow, implicitly by theory.

Those post-Quineans inclined to pursue an implicit-definition semantics usually find themselves involved in the swamp of incommensurability, incommunicability and other inabilities that turn science into an idealist nightmare. The factuality of scientific knowledge becomes dubious, the possibility of radical conflict vanishes. Putnam is to be praised for sticking – or having stuck for so long – to those two realist hopes of factuality and intertheoretic reference. But his commitment to Quineanism has not made his job any easier; indeed, it may have made it impossible. For, consider how remarkable is the task that he has tried to accomplish.

The account of implicit definition that the conventionalists could understand was one that drained the defining conditions of all content – as Buridan's law demanded. It was because they wanted to have *some* statements with content that they were forced to draw something like the analytic-synthetic distinction and to separate the conventions in *C* from the rest of our background theories. As we just saw, Putnam cannot do that. His balancing act is therefore designed to give us a doctrine of reference that makes semantic force theory-dependent enough so that reference is fixed implicitly by theory, but not so theory-dependent that any particular claim will turn out to be true in virtue of that fact of semantics. Conventionalists certainly don't know how to do that trick. How exactly is it done?

At this point holists will wax eloquent on the black of convention, the white of fact and the grey of all theory; or about dictionaries and encyclopedias and so on. But all the prosaic conventionalist wants is an account of how a class of factual claims can fix reference implicitly. At times Putnam enumerates some of the sentences involved in the implicit characterization of reference (see, e.g., quotation on p. 13). Presumably there will be some non-empty class of all such claims, call it C. What do the sentences in C say? As far as I can tell, that whatever makes them true is such that they are true of it; and these claims are supposed to be factual, indeed, possibly false? The conventionalist can make no sense of this. If anyone can, the secret has been well kept.

Beyond the problem of how 'our' theory is to fix reference there is the further considerable problem of which particular theory is to be involved in this process. The question is *not* which theories we should use to decide whether an attribution of reference is true; the answer to this is, of course, the best theory around. The question is rather, what theory is involved in explaining what an attribution of reference *says*. There are those who don't see the distinction, but Putnam does. And he knows also that to choose our own theory would be improperly ethnocentric, and to choose a union of successive theories would be inconsistent. Hence our problem: what theory is to fix implicitly the common referent of Democritus', Bohr's and Feynman's 'atom'?

At times Putnam tells us that reference is not determined as what satisfies the current versions of our theories but as what satisfies them "optimally" (e.g., 1974b, p. 193) and his examples indicate that optimality is achieved by means of extensions or modifications of our theories, to be identified as science marches on to its Peircean omega-point. If so, reference is implicitly defined not by any theories around but by that of the ideal physicist in the end of time. In a way what we have here is the mirror image of Kripke's baptismal approach, what we might call a last-rites semantics in which reference is not fixed at the beginning but at the end, through a process which involves quasi-religious assumptions[19] and which takes place when it's too late to do anyone any good.

Barring Peircean miracles I assume that it is clear that implicit semantics cannot possibly perform the job Putnam wants done. If Euclid and Democritus were talking about the same things Riemann and Bohr were dealing with in their respective theories, if therefore almost everything we believe scientifically or otherwise may in fact be quite false, then reference cannot possibly be fixed by the condition that most or many of the relevant beliefs be true. What we need instead is a procedure that fixes reference not implicitly but independently of theory, not at the end of time but at the beginning. We need something like baptismal semantics.

II.b.2. *The Causal Theory*. About a decade ago Putnam started to develop a highly original and influential approach to semantics in terms of causality, according to which reference is fixed by procedures which, unlike those of implicit semantics, make no restrictions on how many of our beliefs must be true.

The main idea of his theory is that "the referent is fixed by the fact that [the language user] is causally linked to other individuals who were in a position to pick out the bearer of the name" (1973, p. 203). In due course this 'were' will become a 'were, are or will be'; for the causal link may stretch backwards to primeval baptisms, but also sideways and forward to experts. Let us first look at the link through ostensive definitions back to the original baptisms.

Regrettably, although Putnam is occasionally explicit on how reference is transmitted via ostensive definition (e.g., 1975a, p. 225) he never sets out to explain how reference emerges in this model. His repeated allusions to Kripke's doctrine of proper names suggests that he has in mind some generalization of that picture. He does say, for example, that as in the case of individuals and their proper names, expressions for kinds are attached to their referents via "introducing events" (e.g., 1973, p. 205); and the examples displayed often confirm this interpretation. Thus, we are told that physical magnitudes "are invariably discovered through their effects, and so the natural way to first single out a physical magnitude is as the magnitude responsible for certain effects" (1973, p. 202).[20] For example, Putnam's treatment of 'electricity' in (1973, IIA) evokes images of Ben Franklin holding his kite and shouting "I will henceforth use the word 'electricity' to name the unique and single entity, field, particle or process causally responsible for these painful proceedings I am currently witnessing" or words to that effect.

An introducing event such as the one we have just imagined has as its primary target a single physical object or process; for example, a particular energy exchange up there in the clouds. One thing the causal theory does very well in a variety of cases is to tell us how we come to have proper names for these things; Putnam wants it to tell us also how we come to have common names for the natural kinds to which they belong.

Now, I won't pretend to know what people mean when they talk about natural kinds; but I think I can mimic their language behavior so as to pass unnoticed among them. In this spirit of linguistic altruism I will report that the causal theory faces the problem of finding something in the introducing event that will distinguish those baptisms that create a proper name for the particular entity at hand, from those that create a common name for the (a?) natural kind to which that entity belongs.

What makes the difference, according to Putnam, is a complex of things, prominent among them a remarkable relation $same_{NK}$ (same-natural-kind; in the case of water, same liquid; in the case of gold, same metal; etc.). The theory is that given the speaker's interests and context, there is often or sometimes a unique set T of objects targeted in the introducing event(s) and a unique relation $same_{NK}$ such that the referent of the common name being introduced is the class of things that stand in the relation $same_{NK}$ to members of T. Now we have only two problems to solve: what is T and what is $same_{NK}$?

As we shall see, Putnam has little to say in reply to these questions. The reason why this appears not to affect his theory is that Putnam's presentation of it is accompanied by a wealth of imaginative and very appealing applications that lend to it a considerable *prima facie* plausibility. Yet, when these applications are closely examined, their intuitive plausibility vanishes as one notes that their adequacy entirely depends on assuming that the problem of fixing T and $same_{NK}$ has somehow been solved – even though virtually no indication is given as to how to solve it. One example will have to suffice as an illustration of this point.

In order to show that the extension of Archimedes' word 'chrysos' was the same as that of our 'gold' Putnam asks us to imagine that there is some object X such that (i) X is not gold, (ii) we can determine by means of tests that X isn't gold, but (iii) Archimedes couldn't tell X from real gold. And he argues as follows that, in spite of (iii), X was not in the extension of Archimedes' 'chrysos'.

Assume that Archimedes time-travels to the present bringing with him X and his language. After performing the tests to which we have alluded in (ii) "he would have been able to check the empirical regularity 'X behaves differently from the rest of the stuff I classify as 'chrysos' in several respects'. Eventually he would have concluded that 'X may not be gold'" (1975a, p. 237).

That X *may* not be gold Archimedes knew before his trip. The question is whether he has now any reason to think that X is not gold, and Putnam can't offer him any argument for this conclusion that does not beg the question. For, strictly speaking, all Archimedes could conclude after witnessing the indicated tests is that either X is not gold, or that the rest of the stuff he called 'chrysos' is not gold, or that maybe they are all gold after all, only of different kinds. There is simply no way to tell until we know what, for example, T is. And similar problems arise with Putnam's remaining examples.

One thing is clear: the target T of Putnam's introducing event couldn't possibly be a single entity, since Putnam wants to avoid the reduction of semantics to archeology which would immediately ensue. He emphasizes, for example, that the members of the extension of natural-kind words need not have a common hidden structure, and that they may share no more than superficial characteristics (1975a, p. 241). Or they may even have hidden structures but of different sorts. For example, he explains that 'jade' refers to two minerals "with two quite different structures" (1975a, p. 241). It follows that, if there is anything like an introduing event, its target T must be a class of things, what Putnam sometimes calls the "normal examples" (1975a, p. 243) or the "paradigms" (1975a, p. 245).

At this point the analogy between the causal theory for natural kinds and that for proper names begins to crumble. For it is clear that multiplying baptisms to increase the membership of T does not solve any problems; it merely brings us back to semantical archeology with many more things to dig for if we wish to know what we are talking about. We need a credible candidate for T and baptisms seem incapable of providing it. Putnam appears to consider two possible courses of action: (i) look for help in the old implicit-definition strategy; (ii) appeal to experts.[21]

The return of implicit semantics is adumbrated by a remark in the Introduction to Putnam's *Philosophical Papers* (1975e, p. x) where he notes that physical entities are "in a very complicated sense 'defined' by systems of laws," adding that the exact sense is discussed in (1973) and (1975a). But (1973) and (1975a) shed no light whatever on how systems of laws define anything; instead they explain the causal theory we are now trying to understand. If there is a link it must surely go the other way: the 'exact sense' in which the causal theory fixes reference might be given by fixing T and $same_{\mathrm{NK}}$ somehow implicitly by theory. This strategy might look promising until one becomes aware of the difficulties with implicit semantics discussed earlier.

From his seemingly inexhaustible bag of semantic tricks Putnam draws now his finest idea, the principle of division of linguistic labor. Its main point is that semantic force is partly a matter of what others in my linguistic community know. Thus even though – as we all know by now – Putnam cannot distinguish between an elm and a beech, whenever he uses the word 'elm' he refers only to elms and never to beeches. How does he succeed? Incautious readers might feel like saying: because someone baptized the right trees with the right names – or because the relation $same_{\text{Tree}}$ is doing its job full-time, regardless of whether Putnam knows it or not. Not really. The reason, we are now told, is that whenever a serious doubt arises concerning the appropriate referents Putnam can always call the neighborhood beech-expert who will solve his problem. More generally, the idea is that often natural-kind words are associated with criteria of recognition that only experts know about. The fact that we share a common language with them somehow allows that expert knowledge to play a role in the determination of the semantic force of our own words. This insightful idea, when properly worked out, is likely to contribute greatly to our understanding of language; but its ability to support Putnam's preferred picture of scientific knowledge is rather dubious.

To begin with, one is surprised to see Putnam offer his doctrine of experts not as a complement but as an alternative to the baptismal theory; for we are frequently told that experts *determine* reference.[22] If so, we have too many determinants of reference and their joint consistency becomes a problem.

Experts presumably use the best theories available to judge what $same_{\text{NK}}$ is; what they do about T is anybody's guess. In any case, it is clear that it should be a miracle if they should always agree with what Nature says about reference (through her choice of T and $same_{\text{NK}}$). Indeed, diachronically speaking, experts are bound to disagree with each other and therefore with Nature. I do not merely mean to say that at any given time there may be disagreement among experts; rather, that in the course of time, and unless scientific growth is converging or coming to an end, there is bound to be conflict among experts – if we believe, like Putnam, in intertheoretic reference.

The point is the same one we made for theories a few pages back: the experts now alive are hardly the wisest there will ever be. It would therefore be absurd to appoint them as reference-fixers for every technical word ever used, any more than we should use the $same_{\text{NK}}$ chosen by our theories as indistinguishable from the one chosen by Nature. If experts play a role, sanity demands that they should play roughly the same role at all stages of scientific development. At one point Putnam notes that we *might* want to use division of labor across time (1975a, p. 229) but, in fact, we *must* use it if experts have anything to do with fixing reference. My respect for Gell-Mann's opinions on the hidden structure of everything is virtually unbounded; but if I have to choose I would be a fool if I didn't grant more credence to the Mr. Spocks or R2–D2s of the future. Once again, the project is reduced to Peircean absurdity, for it is surely beyond the range of the believable that any of these scientists, or their expert successors in the blue yonder, will have the power to fix the reference of the words I am now using. If so, there would be one more reason to oppose the nuclear Holocaust: to insure that we have all been talking about something after all.

Neither baptisms not experts nor Nature nor coherence nor implicit definition have given Putnam what he had hoped for: believable intertheoretic reference and a believable doctrine of the *a priori*. That someone as resourceful as Putnam has tried and failed in so many ways should not cast shadows on the singer but on his song. Like so many others, Putnam has found reason to reject the traditional ways with meanings and he has moved away from them in directions first defined by Quine. He has, of course, diverged drastically from Quine on specific solutions; but his own initiatives are all grounded on the assumption that Quine's criticism of pre-Quinean semantics is correct. It is therefore of more than passing significance that on both of our topics, reference and the *a priori*, the broadly Quinean framework in which Putnam has placed himself has led to troublesome consequences.

On the topic of reference, it is remarkable that of the two major attempts to develop a conception of scientific knowledge intended to conform to Quinean standards, one of them, Kuhn's, tells us that disagreement on fundamentals is impossible, whereas the other, Putnam's, tells us that it *is* possible, but at the price of making knowledge of what different theories are talking about a virtual miracle.

On the matter of the *a priori*, the trouble might be displayed as follows. It would be hard to identify two more powerful *idées fixes* in analytic philosophy than these: that a good epistemology should not try to distinguish between matters of meaning and matters of facts, that a good epistemology should try to distinguish between virtually *a priori* matters and matters of fact. Quine uncovered the first intuition and was guided by it to reject the second one; the conventionalists acknowledged the second intuition and their account of it led them to disregard the first one. Like Russell before him – whom he resembles in so many respects – Putnam has allowed himself to be seduced by conflicting intuitions. Yet, the Quinean pull has deprived him of the semantic weapons with which the conventionalists had taken the *a priori* out of the idealist stronghold. As we saw, the meanings discovered by the semantic-realists and developed by the positivists proved sufficient to produce a theory of the *a priori* consistent with scientific realism. The effect of Quine's attack against those meanings might make one wonder if they weren't also necessary for that task. One may, indeed, wonder how essential those semantic tools were, as one watches Putnam, Goodman and their followers turning a friendly eye towards old idealist dogmas.

Indiana University and
University of California-Davis

NOTES

[1] *Logik Blomberg*, Kant's *Gesammelte Schriften* (Berlin: [Reimer]/De Gruyter [for the] Deutsche Akademie der Wissenschaften, 1900–) **24**[1], p. 41. (This edition is referred to below as Ak. Volume 24, Kant's *Vorlesungen*, is in two, continuously paginated volumes.) If we see a house from afar, for example, "we must necessarily have a representation of the parts, e.g., the window, etc. For if we did not see the parts we would not see the whole house either." *Pölitz Logik*, Ak **24**[2], pp. 510–511; see also *Wiener Logik*, Ak **24**[2], p. 841, *Logik Philippi*, Ak **24**[1], p. 410. In the first *Critique* Kant had written that "to analyze a concept [is] to become conscious of the manifold which I always think in it" (B 11/A 7) – cf. also *Logik Blomberg*, Ak **24**[1], p.131 – and in *Wiener Logik*, Ak **24**[2], p. 843, he says: "If we only knew what we know ... we would be astonished by the treasures contained in our knowledge."

[2] This 'nominal' version of analyticity is the one which inspired Frege's famous generalization in *Grundlagen* (*Grundlagen der Arithmetik*. Breslau: Koebner, 1884). For Kant analytic judgments are, in effect, those true in virtue of definitions and a handful of logical laws (identity, contradiction). For Frege, a judgment is analytic if it follows from definitions and *all* of logic. For none of them is analyticity identical with truth in virtue of meanings.

[3] e.g., Ak **4**, p. 266; also A 7/B 11.

[4] I realize that this interpretation is not orthodox. Its details are spelled out in my forthcoming *To the Vienna Station*. See also my 'Kant, Bolzano and the Emergence of Logicism,' *Journal of Philosophy* **79** (1982) 679–689.

[5] Thus, he asks, "what justifies the understanding to attribute to a subject *A* a predicate *B* which does not lie in the concept? Nothing, I say, but that the understanding *has* and *knows* the two concepts *A* and *B*. I think we must be in a position to judge about certain concepts merely because we have them ... "; and he adds: "this holds also in the case when the concepts are simple. But in this case the judgments which we make about them are certainly synthetic" (1837, p. 347). This conflicts with Kant's principle of synthetic judgments.

[6] I have done that in my 'From Geometry to Tolerance', forthcoming in the *Pittsburgh Studies in the Philosophy of Science* Series. In that paper I explain in what sense Carnap's syntax and Wittgenstein's grammar are direct descendants of the conventionalist tradition.

[7] For Russell see his (1899, pp. 699–700); for Frege (1899, p. 62).
[8] Cf. Russell's remark: "the meaning of fundamental terms cannot be defined but only suggested. If the suggestion does not evoke the right idea, there is nothing one can do" (1899, p. 702); and Frege's remark that when meanings are "logically simple one cannot give a real definition but must be satisfied with ruling out by means of hints the unwanted meanings . . . Hereby, of course, one always has to count on cooperative understanding, even guessing" (1899, p. 63).
[9] In Geach (1962), p. xi. Apart from the ambiguity of 'proposition' discussed below, Geach's formulation adds to the confusion through its reference to 'the' proposition in which *E* occurs. Geach regards this principle as obvious and says that he has used it to refute a large number of erroneous theories of reference.
[10] e.g. " . . . we only decide whether a body is rigid, its sides flat and its edges straight, by means of the very propositions whose factual correctness the examination is supposed to show" (1868, p. 39).
[11] Koenigsberger (1902–3), vol. I, p. 248.
[12] It would be a gross error to confuse this "decision to use certain words in certain ways" with what Grünbaum has called trivial semantic conventionalism. The latter doctrine assumes that we have a stock of meanings already available for reference, and that we arbitrarily decide to redistribute the association of words with such meanings. Here we are talking instead of a process through which we first identify or 'constitute' if you will, the meanings in question via the endorsement of the sentences in question. Nowhere is the realist Frege–Russell picture of meaning more misleading than in this particular context, since it deprives us of the possibility of distinguishing between this version of semantic conventionalism and its trivial counterpart.
[13] Although why people thought so, remains a mystery. The locus of the original refutation is Quine's 'Truth by Convention'. The doctrine refuted in that paper is one that Carnap regarded as incoherent: that we must try to justify our conventions. Putnam explicitly states as an objection to conventionalism that if there were such conventions, "I do not see how they could be *justified*" (1968, p. 188; his italics). And in his most recent treatment of the matter in (1981) he assumes that the positivists would like to justify their logical and other conventions, and (like Quine) deduces from this all sorts of awful things. Of course, the leading linguistic conventionalists (Carnap, Wittgenstein) would have regarded the idea that conventions call for justification – either by logic or by something else – as absurd. As for Quine's second try in 'Carnap and Logical Truth', I believe that Carnap's replies in the Schilpp (1963) volume provide sufficient refutation. A recent study by Geoffrey Hellman, 'Logical Truth by Linguistic Convention' (forthcoming) sheds a good deal of light on this matter.
[14] The side remarks are of the "one is not expected to give much of a reason for that kind of statement" sort (1962b, p. 240). The examples are not always very helpful. The paradigm case is, of course, Euclidean geometry – of which we are told that "before the development of non-Euclidean geometry . . . the best philosophical minds regarded [it] as virtually analytic" (sic) (1965b, p. 88). Next to Euclidean geometry, Putnam's favorite example appears to be "there are finitely many places in the Universe"; this, according to him, only became conceivable after Riemann (e.g. 1975c, p. xv; 1962b, p. 246). But one does not need to go far beyond the title of Koyré's *From the Closed World to the Infinite Universe* (Baltimore: Johns Hopkins Press, 1957) to realize that matters are a bit more complicated than Putnam thinks. Finally, on the issue of

inconceivability, consider the following: Let Oscar be someone holding a claim *C* in the particular way in which Putnam thinks one holds *a priori* claims. Let me assume that Oscar understands negation or that we can teach it to him without causing undue damage to his conceptual framework. Now, if Oscar believes *C* and understands negation, I dare conclude that he understands not-*C*. So, he understands not-*C* but cannot conceive it. What *does* this mean? Here is more food for thought: even though before Riemann it was "literally inconceivable" that there are finitely many places in the universe, it is *not* "literally inconceivable that a person could be a teakettle" (1975a, p. 244). Since – outside California anyhow – there are no pan-teakettlerist theories designed to broaden our conceptual horizons, one may safely conclude that Putnam's remarks on conceivability are best ignored in our search for his theory of the *a priori*.

[15] "The question which then forced itself upon me, and one which obviously belongs to the domain of the exact sciences, was at first only the following: how much of the propositions of geometry has an objectively valid sense? And how much, on the contrary, is only definition or the consequence of definitions or depends on the form of description? In my opinion this question is not to be answered all that simply" (Helmholtz, 1868, p. 39).

[16] *R* gives us the position of the origin of the rotating frame relative to an inertial frame; *r* the position of the moving particle relative to the rotating frame.

[17] In (1918) Einstein writes: "Were one to write Newton's gravitational mechanics in the form of absolutely covariant equations (four-dimensionally) one would be surely convinced that the principle (a) [the demand for covariance] excludes that theory not theoretically but practically" (p. 242). Cartan's (1923) would show him wrong. See also Havas (1964).

[18] See Wheeler (1973), p. 302.

[19] On the merits of Putnam's assumption of scientific convergence see Laudan (1981).

[20] From Democritus' atom to Maxwell's displacement current, Einstein's spacetime interval and Dirac's positron, the counterexamples to this claim are not hard to come by. There are, however, several cases where reference (that of 'Uranus', for example) appears to be fixed through a causal process, and this provides an excellent antidote to the strange things the early Kuhn used to say about incommensurability.

[21] A third course of action is to turn off the lights and bring in pragmatics. For example, Putnam says that $same_{NK}$ (and probably also *T*) will depend on the speaker's interests (1975a, p. 239). The darkness thus cast is not so great as to conceal inconsistencies. For example, bowing to common sense, Putnam concedes that Earthlings may, after all, refer to XYZ (Twin Earth 'water') when they say 'water', provided that they are using XYZ as they use real water. But then, one wonders why was Archimedes' object *X* not supposed to be in the extension of 'chrysos'.

[22] Thus we are told that the usage of experts "determines what many other people are referring to" (1974a, p. 274); that the reference of words like 'elm' "is fixed by the community, including the experts" (1975a, p. 265); that there is "social cooperation in the determination of reference" (1975a, p. 246) and that extension is "determined socially . . . owing to the division of linguistic labor" (1975a, p. 246).

REFERENCES

Bolzano, B. 1837. *Theory of Science*. Ed. and transl. by Rolf George. Berkeley: University of California Press, 1972.

Bradbury, T. C. 1968. *Theoretical Mechanics*. New York: Wiley.

Cartan, E. 1923. 'Sur les variétés à connexion affine et la théorie de la relativité généralisée,' *Ann. de l'Ecole Norm. Sup.* **40**, 325–412 and **41**, 1–25.

Einstein, A. 1918. 'Prinzipielles zur allgemeinen Relativitätstheorie,' *Annalen der Physik*, Ser. 4, **55**, 241–244.

Frege, G. 1899. Correspondence with Hilbert. In G. Gabriel *et al.*, *Wissenschaftlicher Briefwechsel*. Hamburg: Felix Meiner Verlag, 1976. (This is vol. 2 of his *Nachgelassene Schriften und Wissenschaftlicher Briefwechsel*, 1969–.)

Geach, P. T. 1962. *Reference and Generality*. Ithaca: Cornell University Press.

Glymour, C. 1980. *Theory and Evidence*. Princeton: Princeton University Press.

Grünbaum, A. 1972. *Philosophical Problems of Space and Time*. 2nd revised ed. *Boston Studies in the Philosophy of Science*, vol. 12. Dordrecht and Boston: D. Reidel.

Havas, P. 1964. 'Four-dimensional Formulations of Newtonian Mechanics and Their Relations to the Special and the General Theory of Relativity,' *Reviews of Modern Physics* **36**, 938–965.

Helmholtz, H. 1868. 'On the Facts Underlying Geometry.' In R. S. Cohen and Y. Elkana (eds.), *Epistemological Writings. The Paul Hertz/Moritz Schlick Centenary Edition of 1921*, newly transl. by M. Lowe. *Boston Studies in the Philosophy of Science*, vol. 37. Dordrecht and Boston: D. Reidel, 1977.

Koenigsberger, Leo. 1902–3. *Hermann von Helmholtz*. 3 vols. Braunschweig: Vieweg.

Laudan, L. 1981. 'A Confutation of Convergent Realism,' *Philosophy of Science* **48**, 19–49.

Mill, John Stuart. 1843. *A System of Logic*. London: Parker.

Putnam, H. 1959. 'Memo on Conventionalism.' In (1975b, I, pp. 206–214).

Putnam, H. 1962a. 'The Analytic and the Synthetic.' In (1975b, II, pp. 33–69).

Putnam, H. 1962b. 'It Ain't Necessarily So.' In (1975b, I, pp. 237–249).

Putnam, H. 1963. 'An Examination of Grünbaum's Philosophy of Geometry.' In (1975b, I, pp. 93–129).

Putnam, H. 1965a. 'A Philosopher Looks at Quantum Mechanics.' In (1975b, I, pp. 130–158).

Putnam, H. 1965b. 'Philosophy of Physics.' In (1975b, I, pp. 79–92).

Putnam, H. 1968. 'The Logic of Quantum Mechanics.' In (1975b, I, pp. 174–197).

Putnam, H. 1973. 'Explanation and Reference.' In (1975b, II, pp. 196–214).

Putnam, H. 1974a. 'Language and Reality.' In (1975b, II, pp. 272–290).

Putnam, H. 1974b. 'Reply to Gerald Massey.' In (1975b, II, pp. 192–195).

Putnam, H. 1975a. 'The Meaning of "Meaning".' In (1975b, II, pp. 215–271).

Putnam, H. 1975b. *Philosophical Papers*. 2 vols. Cambridge: Cambridge University Press. vol. **I**: *Mathematics, Matter and Method*; vol. **II**: *Mind, Language and Reality*.

Putnam, H. 1975c. 'Philosophy of Language and the Rest of Philosophy.' In (1975b, II, pp. vii–xvii).

Putnam, H. 1975d. 'The Refutation of Conventionalism.' In (1975b, II, pp. 153–191).

Putnam, H. 1975e. 'Science as Approximation to Truth.' In (1975b, I, pp. vii–xiv).
Putnam, H. 1981. *Reason, Truth and History*. Cambridge: Cambridge University Press.
Reichenbach, H. 1928. *The Philosophy of Space and Time*. New York: Dover, 1958.
Riehl, A. 1904. 'Helmholtz in seinem Verhältnis zu Kant,' *Kant Studien* 9, 261–285.
Russell, B. 1899. 'Sur les axiomes de la géométrie,' *Revue de Métaphysique et de Morale* 7, 684–707.
Schilpp, P. A. (ed.). 1963. *The Philosophy of Rudolf Carnap*. LaSalle, Ill.: Open Court.
Wheeler, J. A., Misner, C. W., Thorne, K. S. 1973. *Gravitation*. San Francisco: Freeman.

MORRIS N. EAGLE

THE EPISTEMOLOGICAL STATUS OF RECENT DEVELOPMENTS IN PSYCHOANALYTIC THEORY

I hope it is appropriate to begin with a personal note which is not entirely personal insofar as what I want to speak of is not only characteristic of Grünbaum the person but also permeates his work. I remember as a very young man being profoundly impressed by Unamuno's (1921) comment that a philosopher must also express his philosophy in "flesh and bone." I don't know whether or not I am remembering Unamuno's remarks accurately, but I know I was reacting against the sundering of one's articulated philosophy from the rest of one's lived life or, at best, the orthogonal relationship between the two and what greatly appealed to me in Unamuno's comments was the idea of unity, of living one's life as of a piece. I bring back these memories because, indeed, they are revivified by writing about Grünbaum. The qualities that are so apparent in his work – its clarity, its no-nonsense honesty, its attempt to fully understand another – also characterize Grünbaum as a person. He is also a man of great warmth, humor, and caring. In the relatively short time – since 1977 – I have known Adolf, I have always considered it one of my good fortunes that I can count him as my friend. I have also, again and again, been stimulated and enlightened by his thought and ideas. Therefore, I view with great pleasure the opportunity to contribute to this Festschrift for Grünbaum.

More than any other figure, Adolf Grünbaum has emerged as the contemporary philosopher and thinker who has subjected psychoanalytic claims to systematic probing, and comprehensive scrutiny. In the course of conducting this examination, he has thoroughly mastered the ideas and literature of Freudian psychoanalytic theory. When one considers that for most of his career Grünbaum has devoted himself to the philosophy of space and time, making seminal contributions in that area, and has only recently turned his attention to psychoanalytic theory, his mastery of the latter field is truly remarkable. Given the ignorance, vulgarity and crudeness of thought that has so often characterized debate in this area – on the part of both defenders and critics of psychoanalysis – the precision, care, and quality of Grünbaum's evaluations and criticisms are rare and exhilarating.

For the most part, Grünbaum has addressed himself to the foundational material of psychoanalytic theory, namely, Freud's writings. He has subjected

R. S. Cohen and L. Laudan (eds.), Physics, Philosophy and Psychoanalysis, 31–55.

certain basic claims and propositions contained in these writings to rigorous, logical examination. For example, he has demonstrated quite conclusively that because of the epistemologically contaminated status of clinical data derived from the psychoanalytic situation, such data have questionable probative value in the testing of psychoanalytic hypotheses (Grünbaum, 1982b). He has also shown that Freud's arguments for the "repression-aetiology of the psychoneuroses, and for the cardinal causal role of repressed ideation in committing parapraxes (slips) and in dreaming" are seriously flawed (Grünbaum, 1982a, p. 32).

Recently, there has been a great deal of ferment within the psychoanalytic community. A spate of articles and books have appeared which present basic reformulations of traditional psychoanalytic theory, some radical in nature, such as, for example, Kohut's (1971; 1977) self psychology and object relations theory.[1]

It has been suggested, in response to Grünbaum's devastating criticisms of aspects of Freudian theory, that while these criticisms may be appropriate to traditional psychoanalytic theory, they are not germane to the more recent 'progress' made in modifying traditional theory. For example, in a recent symposium on psychoanalysis in the *Journal of Philosophy*, Flax (1981) compares Grünbaum's criticisms of Freudian theory as epistemically akin to "throwing out physics because there are unresolved problems in Newton's theory" (p. 564). As Grünbaum (1981) notes, "this purported analogy suggests misleadingly that the epistemic difficulties which beset Freud's original formulations have been overcome by the much vaunted post-Freudian formulations of object relations theory" (p. 34). Grünbaum, of course, quite correctly rejects this contention. He notes that "the much vaunted post-Freudian versions have not remedied a single one of the methodological defects that I charged generically against the psychoanalytic method of clinical investigation" (1983, pp. 47–48). However, the purpose of Grünbaum's paper was to examine Freud's repression etiology argument, not to examine the methodological and epistemological status of post-Freudian formulations. One question I want to examine in this paper is whether it is, in fact, the case that recent modifications and formulations are better validated and on more secure epistemological foundations than traditional Freudian theory. Is it even possible to discern any increased awareness of methodological and epistemological issues in these recent reformulations? The other related purpose is to continue a discussion of the conceptual status of psychoanalytic theory begun earlier (Eagle, 1980a). It is fitting that this earlier work was initiated upon an invitation from Grünbaum to present a paper on psychoanalytic theory to the Center for the Philosophy of Science at the University of Pittsburgh.

The recent developments in psychoanalytic theory to which I limit myself in this paper center mainly on the work of the English object relations theorists (mainly Fairbairn and Guntrip), Mahler, and Kohut. This is not an arbitrary choice insofar as the work of these figures is frequently referred to in recent claims that progress, even breakthroughs, have been made in psychoanalytic theory and in recent descriptions of the 'widened scope' of psychoanalytic therapy. For the purposes of this paper, my interest is not so much in the substance and content as in the epistemological and conceptual status of recent reformulations of psychoanalytic theory.

As is the case for traditional psychoanalytic theory, recent developments do not constitute a monolithic entity. They range from formulations so vague and so suffused with jargon that there is serious question as to whether they contain any empirical meaning, to material of actual or potential empirical content capable of yielding testable hypotheses. I say potential as well as actual empirical content because what is often required as a first step in approaching this material is an unraveling of what is being said, a mining of the empirically meaningful statements from jargon. In reading much of this material, it is easy to be put off and discouraged by the ever present jargon and to decide, sometimes perhaps prematurely, that there is little or no empirical substance worth pursuing. And I should also make clear that when I speak of testable formulations and hypotheses, I mean it in the most liberal sense. That is, I do not necessarily refer to direct testability, but to the possibility of evaluating a claim in the light of all the available evidence.

Although recent developments are not unitary, there are certain general characteristics they tend to share. They all have in common either a rejection or some significant modification of Freudian instinct theory. Rejection of the central Freudian idea that all behavior is, either directly or indirectly, in the service of instinctual drive gratification is most explicit in Fairbairn's writings and also in Kohut's (1977) later work where he argues, much as Fairbairn does, that the peremptory pursuit of instinctual discharge is already a "disintegration product," a manifestation of a defective self. And in the writings of Mahler and her colleagues (1968; 1975) and Kernberg (1975; 1976), who do not explicitly reject Freudian instinct theory, one finds a relative and implicit de-emphasis and/or a more complex conception of the role of instinctual gratification in personality development. What replaces the uniquely central role of instinctual gratification – and this is a second characteristic of recent formulations – is an increasing emphasis upon object relations and the development of selfhood. The very terms 'object relations theory' and 'self psychology' to describe the theories, respectively, of Fairbairn and Kohut, is but one obvious manifestation of this shift in emphasis.

A third characteristic of recent formulations is a special emphasis on more severe pathology, focusing mainly on borderline conditions, narcissistic personality disorders, schizoid personalities (a term used mainly by the object relations theorists), and, in the case of Mahler, the severe childhood conditions of autism, childhood schizophrenia, and symbiotic psychosis. These forms of pathology are often referred to as 'pre-Oedipal' on the assumption that while the neuroses are linked mainly to Oedipal conflicts and traumas, the more severe conditions grow out of earlier, pre-Oedipal traumas and defects. Often accompanying this pre-Oedipal–Oedipal dichotomy is the parallel distinction between developmental arrest or defect on the one hand and intrapsychic conflict on the other (e.g., Stolorow and Lachmann, 1980). Also, specific claims are made regarding the etiology of these developmental arrests. For example, Kohut tells us that the mother's failure to provide emphatic mirroring to the infant and later parental failure to provide the child with opportunities for idealization of the parents are critical developmental determinants of failures in self-cohesiveness.

A fourth characteristic of recent developments, which pertains to psychoanalytic treatment, is a relative de-emphasis of interpretation and a correspondingly greater stress on the therapeutic role of the patient-analyst relationship. (Kernberg is a possible exception to this generalization.) For example, Kohut and his followers write about the therapeutic value of the therapist's empathic mirroring and about the importance of permitting so-called mirroring and idealizing transferences in narcissistic personality disorders, without offering interpretations of these transference reactions. And as another example, the object relations theorists have often described psychoanalytic therapy with schizoid patients in terms which suggest more a kind of re-parenting than an attempt to facilitate insight and make the unconscious conscious through interpretation (e.g., Guntrip, 1968).[2]

The fifth characteristic of recent formulations has to do more with the conceptual and epistemological status of psychoanalytic theory than with substantive modifications. I am referring to the recent tendency to reject as superflous and even harmful Freud's more abstract metapsychological theorizing and to argue that the core of psychoanalysis is its clinical theory, the language and concepts of which are presumably close to clinical observation and to the experiences of persons. This position has been most fully elaborated and articulated by Schafer (1976), Klein (1976), Gill (1976), and

Home (1966), but has also been indirectly associated with the object relations theorists and the writings of Kohut. While the latter have not written explicitly on the metapsychology versus clinical theory controversy, they have dealt with related issues. Thus, Guntrip (1968) has expounded on the concept of a "psychodynamic science," which he contrasts with the natural science approach. And Kohut (1959; 1971; 1977) has consistently taken the position that the use of "empathic introspection" and of "experience-near" concepts are defining characteristics of psychoanalysis. It should be noted here – and this will be more fully discussed later – that a common, if implicit, assumption in this area is that only empathically derived and experience-near concepts are appropriate for both an adequate description and explanation of phenomena having to do with selfhood and interpersonal relations. Or to put it conversely, that such phenomena 'require' a special set of empathically derived and experience-near explanatory concepts. The claim is that only a 'humanistic' methodology and explanatory theory can deal adequately with truly human phenomena. (The historical links of this position to an earlier *Verstehende* philosophy and to a programmatic dichotomy between *Geisteswissenchaften* and *Naturwissenschaften* should be apparent.)

The sixth and final characteristic I will describe is not necessarily intrinsic to recent developments, but it has certainly, in fact, been characteristic of them – namely, a tendency to engage in jargon and a concomitant failure to make even a minimal attempt to define one's terms clearly and to give them some clear empirical reference. Following their initial use, these terms and formulations then appear in subsequent papers by many different authors, with the tacit assumption that they are understood by both authors and readers. Indeed, if they are repeated often enough, they come to serve as explanatory concepts for other poorly understood phenomena. To offer just one example, Kohut and his followers 'explain' the alleged therapeutic repair of self defects by appealing to the process of "transmuting internalizations." But no systematic attempt is made to elucidate the very concept which is intended to serve a basic explanatory function. I want to add parenthetically that, as noted earlier, embedded in some of the jargon are some meaningful and perhaps even important ideas. It is because of this that I spoke of the need to unravel and clarify just what is being said.

Let me consider each of the above characteristics with the general intention of gauging the conceptual and epistemological status of recent modifications of traditional psychoanalytic theory. Do they represent advances and progress, as is often claimed in the literature? Or are they more accurately described as an expression of shifting fads and fashions? Do they succeed in escaping at least some of the sorts of criticisms leveled by Grünbaum against traditional theory? Do they fare better methodologically and epistemologically?

It will be recalled that the first two developments I referred to involve a rejection or de-emphasis of Freudian instinct theory and a concomitant stress on the development of selfhood and on so-called object relations. This position is most succinctly captured in Fairbairn's claim that "libido is primarily object-seeking (rather than pleasure-seeking, as in the classic theory)" (1952, p. 83). If one is not distracted by terms like 'libido' (here is a case where some unraveling of meaning is necessary), what Fairbairn is disputing here is the Freudian conception of the origin and nature of our interest in and involvement with objects (which includes both inanimate objects and other people). According to that conception, in order to develop an interest in objects, one is forced to overcome "primal hatred" of objects and an early and basic "primary narcissism" which makes one reluctant to invest interest and affect in objects (Freud, 1915). These early and basic tendencies are overcome only because objects are necessary for instinctual gratification.

A specific expression of this approach to the explanation of object relations is the Freudian account of infant-mother attachment which, as Bowlby (1969) notes, is an example of a 'secondary drive' theory. That is, the infant's attachment to mother is held to be secondary to or derived from her role in gratifying so-called primary drives. Thus, Freud (1940) writes that "love has its origins in attachment to the satisfied need for nourishment" (p. 188). In contrast to this view, Fairbairn is essentially arguing that involvement with objects and object-seeking are inborn and autonomous propensities, quite independent of hunger and other so-called primary drives.

All the evidence on human and infra-human behavior overwhelmingly supports the idea of a primary and autonomous attachment behavioral system, elements of which appear at birth or shortly after birth, and which is independent of the drives of hunger, sex, and aggression. The attachment system comprises such behavioral components as smiling, vocalizing, sucking, soothability, and readiness to respond to specific stimulus features. Furthermore, the classic Harlow (1958) study showed rather decisively that the infant monkey's attachment to its surrogate mother was not derived from or secondary to the latter's association with reduction of so-called primary drives (i.e., hunger, thirst). Rather such attachment seemed to be based on the autonomous need for what Harlow called "contact comfort." Infant monkeys were taken from their natural mothers at birth and were raised by artificial wire and terrycloth surrogate mothers. Harlow reasoned that if the infant monkey's attachment to its mother were secondarily derived from the association of mother with reduction of 'primary' drives, then the infant monkey would become attached to whichever surrogate mother reduced the primary drives of hunger and thirst. Instead the infant monkey developed an attachment to the (terrycloth) surrogate mother providing 'contact comfort' even when a different surrogate (wire) mother satisfied its so-called 'primary' drives. That tactile stimulation is also important for human infants is suggested by the finding that such stimulation of premature babies increased their weight gain significantly over controls (White and LaBarba, 1976).

As to whether infants have an early 'primal hatred' of objects and a narcissistic reluctance to respond to objects, all the evidence overwhelmingly refutes these ideas. Very young infants show selective preferences for novel visual and auditory stimuli (e.g., Friedman, Bruno, and Vietze, 1974); for one set of geometric features over another (e.g., McCall and Nelson, 1970; Ruff and Birch, 1974); for one set of colors over another (e.g., Fantz, 1958 and 1965; Fantz and Fagan, 1975); and for an optimal level of discrepancy from pre-existing stimuli (e.g., Kinney and Kagan, 1976). The young infant is capable of orienting visually to the source of a sound. Among other abilities he can recognize in one modality (vision) an object he has experienced in another modality (tactile). He can match events sharing the same temporal structure and can match intensities of a stimulus experienced in two different modalities. In addition, Emde and Robinson (1979) report that infants will interrupt feeding in order to look at a novel or interesting stimulus.

Now, what emerges from all the above research is that the Freudian conception of object relations and attachment in infancy is incorrect and that Fairbairn's intuitions regarding people's object-seeking propensities are supported by the available evidence. However, what is to be noted is that just about all the relevant evidence *comes from non-clinical rather than clinical sources. Further, there is nothing in the above account that would render Grünbaum's identification of the epistemological liabilities of clinical data inapplicable or that are even relevant to Grünbaum arguments*. Hence, it is just not clear which aspects of object relations theory critics such as Flax (1981) have in mind when they maintain that such developments have essentially rendered Grünbaum's criticisms outmoded.

The third general trend of recent developments I described above includes a special emphasis on presumably more severe, so-called pre-Oedipal, pathology and a description of the etiology of these conditions. Now I have little doubt that to a certain extent, this shifting emphasis is based on informally acquired clinical evidence. That is, it is entirely likely that many analysts and therapists found in listening to their patients that their major difficulties seemed to have to do not with Oedipal conflicts and anxieties, but with feelings of self fragility, with issues of dependency-independence and generally with issues of separation-individuation and autonomy (Mahler, 1968). But others, familiar with the same clinical evidence, have come to the very different conclusion that Oedipal issues are central in all pathology (e.g., Arlow and Brenner, 1964). Also, as Rangell (1980) has observed, through the years many analysts have treated patients very much like the ones described by Kohut, Kernberg and others without feeling the need to generate the special diagnostic categories of borderline conditions and narcissistic personality disorders. And finally, Gedo (1980) has noted that while the clinical descriptions contained in a casebook (Goldberg, 1978) prepared by followers of Kohut are replete with references to self defects and hardly mention Oedipally related conflicts, a similar casebook from analysts of the New York Psychoanalytic Institute (Firestein, 1978) breathes not a word about self defects but instead finds Oedipal conflicts central. Furthermore, these radical diagnostic differences occur despite the fact that the actual clinical phenomena (e.g., presenting symptomatology) are quite similar in the two casebooks.

Since observable behaviors (and reports of experience) are most often opaque as to the underlying unconscious conflicts or self defects they presumably reveal, it is apparent that the therapist's interpretation of these behaviors will play a major role in determining their significance. As Gedo's above example demonstrates, the analyst's theoretical persuasion will, in turn, be a critical determinant of how the behavior is interpreted, the meaning it is given. In addition, what especially complicates the issue for psychoanalysis is, as Grünbaum has shown, the epistemically contaminated status of the clinical data offered to support this or that theoretical interpretation. Patients often experience precisely the kind of 'insight' that corresponds to their therapist's theoretical orientation, Freudian patients reporting Freudian insights, Jungian patients reporting Jungian insights, and so on (Marmor, 1962).

Also Masling and Cohen (unpublished manuscript) point out in a recent paper, "from the very beginning of therapy, some content areas are reinforced explicitly and implicitly, and some not" (pp. 6–7). They report a series of studies which demonstrate that the patient's productions are markedly influenced by the therapist's subtle and most often unwitting reinforcements.

For example, in an analysis of a tape where Carl Rogers is the therapist, Murray (1956) showed that 68 client statements about independence were met by subtle approval, while none were disapproved. By contrast, 16 client statements about sex were met with disapproval while only 2 were approved. By the end of the therapy the client was concerned with independence and seemed unconcerned with sex. Truax (1966) essentially replicated Murray's findings.

Greenson (1967) reports an interaction with a patient which illustrates the same sort of reinforcement. Following is the description by Greenson (1967) of an interaction between him and a patient:

He had been a lifelong Republican (which I had known), and he had tried in recent months to adopt a more liberal point of view because he knew I was so inclined. I asked him how he knew I was a liberal and anti-Republican. He then told me that whenever he said anything favorable about a Republican politician, I always asked for associations. On the other hand, whenever he said anything hostile about a Republican, I remained silent, as though in agreement. Whenever he had a kind word for Roosevelt, I said nothing. Whenever he attacked Roosevelt, I would ask who did Roosevelt remind him of, as though I was out to prove that hating Roosevelt was infantile. I was taken aback because I had been completely unaware of this pattern. Yet, the moment the patient pointed it out, I had to agree that I had done precisely that, albeit unknowingly (p. 273).

Masling and Cohen conclude from their review of the above and similar reports and from "the mitotic splits between and within schools of psychotherapy" that "clinical evidence is an unsubstantial reed upon which to rest the foundation of a theoretical formulation" (p. 24).

Ostensibly supportive data are taken as confirmations of the very same theoretical position which, a broader view reveals, has contributed importantly to generating these data. That is, a consideration of the phenomenon of theory-congruent 'insights' is a strong indication that suggestion and compliance, however subtle and complex, are critical factors in generating these supposedly confirming data. Furthermore the more opaque and ambiguous productions, if interpreted in accord with one's theoretical bias, are also taken as confirmatory data. Therefore, if the productions are sufficiently opaque and ambiguous, each therapist, whatever his or her theoretical persuasion, can reach the verdict that they constitute confirmations.

What about the argument that all observations and facts are, as Popper (1962) and others have argued, theory-laden and therefore, that in this regard, the clinical data derived from psychoanalysis are no different from data obtained in other disciplines? In my view, this is not an adequate defense for a number of reasons. In other disciplines, the data are not as opaque, thus rendering the limits of interpretation of observations narrower. Observers of different theoretical orientations are more likely to agree on such basic matters as whether or not *x* occurred. For example, in a memory study, there will be universal agreement among observers as to whether or not a particular stimulus item was recalled, although these observers may have radically different conceptions of memory processes. By contrast, in the clinical situation, one is more likely to find lack of agreement at the basic level of the very nature of the data. Further, such disagreement is likely to be a function of the observer's theoretical preconception. Thus, as noted, where a follower of Kohut 'sees' self-defects in the patient's productions, an analyst from the New York Psychoanalytic Institute 'sees' Oedipal conflicts and a defensive retreat from and avoidance of such conflicts. And, to repeat, the very opaqueness of the data encourages radically different interpretations.

The defensive argument that all observations are theory-laden will not do for still another important reason. As von Eckardt (1981) has argued, in citing the work of Scheffler (1967) and Giere (1979), "the ideal of objective data is possible at least to this extent: that data relevant to a given theory *T* can be collected by someone whether or not he or she believes in *T* or, even, whether or not he or she has knowledge of *T*" (p. 572).[3] She then goes on to note Giere's point that a good scientific test "must be a statement that can reliably be determined to be true or false using methods that do not themselves presuppose that the hypothesis in question is true" (p. 95). But these are precisely the conditions violated by most clinical data – certainly those collected, interpreted, and evaluated by the therapist himself. This is true despite all the talk about therapists forming and testing hypotheses within the therapeutic session. When has a Freudian analyst tested the hypothesis that for his particular neurotic patient Oedipal conflicts are unimportant? When has a Kohutian analyst tested the hypothesis that a narcissistic patient has no significant self-defects or had adequate empathic mirroring as an infant? When has a Jungian analyst tested the hypothesis that his patient has no collective unconscious? Even if all these hypotheses could be tested, the fact is that typically such 'testing' in the clinical situation employs methods and entails assumptions which, precisely contrary to Giere's description, indeed, do themselves "presuppose that the hypothesis in question is true." The Oedipal conflicts, self-defects, and collective unconscious are all *assumed* to be true by their respective adherents and are employed to interpret ambiguous and opaque data.

Although the problem of generating biased data may also be present in non-clinical situations (e.g., Rosenthal, 1963), it is less pervasive and extensive. Most important, however, a recognition of the problem and the consequent implementation of standard safeguards (e.g., double-blind studies, control groups) represent effective means of dealing with the issue. Such safeguards are not present nor are they possible in the clinical situation.[4] The lack of these and other safeguards makes it difficult, if not impossible to select among various alternative interpretations.[5]

Many contemporary psychoanalytic writers do not seem to be either aware of or interested in these basic issues of reliability, hypothesis testing, or elementary rules of evidence. Consider as a striking case in point the claim, explicit and implicit in Kohut's writings, that lack of maternal empathic mirroring and later opportunities for idealization of parents in infancy and childhood are critical etiological determinants of self-defects and other difficulties associated with narcissistic personality disorders. As I have noted elsewhere (Eagle, in press), what is remarkable about this etiological claim is that it is based entirely on the production of adult patients in analytic treatment. There is not a single reference in the work of Kohut and his followers to developmental or longitudinal studies, to observations of children, to studies of mother-child interactions, or to any other related material. Furthermore, it appears that there is not even any recognition of the importance of this kind of evidence for etiological claims which are based on what presumably goes on in infancy or of the severe limitations of the clinical 'evidence' cited for these kinds of claims. The entire set of formulations, including the etiological claim, is then widely asserted and repeated in the literature and hailed as 'progress' in the development of psychoanalytic theorizing. In what sense and on what basis such shifts in theorizing represent progress is not at all clear.[6] In fact, I believe it can be truthfully said that as far as the evidential basis for and the nature of the inferential processes involved in recent etiological formulations are concerned, there is no evidence of any progress at all, perhaps even, some retrogression.

The *content* of certain formulations has, indeed, changed (e.g., from drive psychology to self psychology), but not their epistemological status. Hence, I believe that Grünbaum's criticisms, which were directed against certain aspects of Freudian theory, apply at least equally to more recent formulations ostensibly based on clinical data. Therefore, the argument that most, if not all, of Grünbaum's criticisms are not germane because they do not take into account recent 'progress' just does not hold any water.

A fourth characteristic of recent developments I noted above is a relative de-emphasis of interpretation and a correspondingly greater stress on the therapeutic role of the patient-therapist relationship in treatment. Insofar as in this shift of emphasis the truth value of interpretations becomes less of an issue and therefore, Freud's Tally argument less central, one could say that Grünbaum's criticisms are somewhat muted. That is, if one softens the claim that it is the truth of interpretations (their 'tallying with what is real') which is causally responsible for therapeutic progress (via insight), one will have become less subject to Grünbaum's criticisms of Freud's Tally argument. But, I will try to show that one escapes only *some* of Grünbaum's (and others') criticisms.

Some contemporary psychoanalytic theorists maintain that the patient-therapist relationship (whether described as the provision of empathic mirroring, a 'holding environment,' the 'real relationship,' re-parenting) is the primary therapeutic factor and that interpretations and accompanying insight are secondary and even superfluous. If one goes along with Bergin and Lambert (1978), one might want to argue that offering interpretations is merely a "congenial modus operandi" for the implementation of the really therapeutic process – the relationship.

Following this assumption, a number of questions immediately arises. For one, if interpretation and insight are dispensable, in what sense is the treatment distinctively psychoanalytic? If the patient-therapist relationship is the critical therapeutic variable, how does psychoanalytic treatment differ from the wide variety of therapies which have in common the attempted establishment of a (benevolent) relationship? On what basis can it rest its traditional claim that while other therapies are merely ameliorative, psychoanalytic treatment alone produces lasting modifications of personality organization? For an important rationale for this claim has centered on the role of insight in the alleged reorganization of personality ("making the unconscious conscious"; "where id was, there shall ego be"). With insight rendered dispensable and even superfluous, on what does the claim now rest, particularly in view of the fact that its central role is replaced by a factor common to most, if not all, therapies?

In the recent literature, a number of specific relationship factors have appeared as the leading 'candidates' to replace insight as the primary therapeutic vehicles. For example, Kohut (1977) refers to "transmuting internalizations," Volkan (1976) to identification with the analyst, Winnicott (1965) to the provision of a "holding environment" analogous to the milieu mother ideally provides for an infant, and so on. Common to all these concepts is the basic idea that the therapeutic relationship provides what was lacking in the patient's early experiences with his parents and thereby permits resumption of the developmental growth interrupted and aborted by early trauma. For example, according to Kohut, the therapeutic provision of empathic mirroring and the patient's use of the therapist as a "self object" (that is, someone who will help carry out psychological functions – such as regulation of self-esteem – one normally carries out for oneself) eventually facilitate the building of those psychic structures which were missing or defective (e.g., self-cohesiveness).

A wide host of questions are prompted by this formulation, including the question of the meaning of talk referring to building of psychic structures. But the point I want to focus on is the implicit assumption made by Kohut (and others) that if the provision of X to an adult patient in therapy is therapeutic, this is evidence that the parental failure to provide X during infancy or childhood is a critical etiological factor in the patient's pathology. For example, if the analyst's provision of empathic mirroring or if the patient's idealization of the analyst serve to repair self-defects (let us assume, for the sake of argument, that we know what this means), it follows, according to Kohut, that the *lack* of mirroring and opportunities for idealization in infancy and childhood were causative agents in producing these self-defects. It is then assumed that there is no further need for corroborating evidence from longitudinal-development studies.

Basing etiological hypotheses on clinical data is questionable for still additional reasons. Adult patients' reports of how they were treated in childhood are not only subject to omissions and distortions of memory, but also to suggestion and the rationalizing processes involved in the need to explain their illness and suffering. Given all these possible sources of distortion and selectivity, it is diffcult to see how such reports can serve as independent corroborative data for any etiological hypothesis. At best, they only tell us how an individual in a particular situation (with all its potential for suggestion, etc.) *now* 'remembers' his earlier experiences. After all, it is in the history of Freud's seduction hypothesis that one clearly sees the operation of distortion processes in early memories.

One can conclude that just as one must reject Freud's Tally argument, one must reject analogous current claims. That is, just as therapeutic success would not vouchsafe the validity or truth of Freud's etiological hypotheses, it does not vouchsafe the validity of current etiological hypotheses. The content has changed, but the nature of the claims and the form of the (incorrect) inferences are the same. Hence, again Grünbaum's criticisms apply equally to current developments in this area.

In the above discussion, we have been assuming, for the sake of pursuing the argument, that therapy is effective and successful. But in this area, the current psychoanalytic practice is no different from what it has always been. Claims for therapeutic efficacy are based entirely on case reports. There are no systematic controlled therapeutic outcome studies. Hence, despite all the talk about 'progress' and about the "broadened scope of psychoanalysis,"[7] the current claims regarding therapeutic efficacy rest on no more secure grounds than earlier claims. Again, it is difficult to see, despite claims to the contrary (Flax, 1981), how more recent developments escape the kinds of criticisms offered by Grünbaum.

The fifth characteristic of recent developments I have noted is the espousal of psychoanalytic clinical theory (and the rejection of its metapsychology) and of experience-near and empathically derived concepts.[8] The general argument is that the human phenomena with which psychoanalysis is concerned requires special concepts, methods, and kinds of understanding and explanation different from those characteristic of the natural sciences. The original formulation of the clinical theory versus metapsychology debate developed somewhat independently of recent substantive modifications of psychoanalytic theory. The original advocates of the clinical theory approach (e.g., Klein, 1976; Gill, 1976; Schafer, 1976) borrowed explicitly from the philosophical distinction between reasons and causes and argued that psychoanalysis was a discipline in which causes did not apply and only reasons were relevant.[9] However, a particular form of the argument has come to be associated with object relations theory and self psychology, in particular with the work of Guntrip (e.g., 1968) and Kohut (1959; 1971; 1977), respectively. As I have noted earlier, there seems to be an implicit assumption that while the language of abstract forces and energies may be appropriate to Freudian instinct theory, only empathically derived and experience-near concepts are suitable for the adequate description and explanation of phenomena having to do with selfhood and interpersonal relations.[10] As I have noted earlier, Guntrip has called for a "psychodynamic science" appropriate to the study of object relations and Kohut has specified empathic introspection as the identifying characteristic of the psychoanalytic method of achieving knowledge and understanding. Indeed, Kohut has taken the position that psychoanalytic formulations should limit themselves to empathically derived and experience-near concepts – features which, he makes clear, are characteristic of his self psychology in contrast to Freudian metapsychology.

Aside from the question of whether this is a legitimate position, is it even the case that Kohut limits his accounts to empathically derived and experience-near concepts? An examination of his writing will show that he does not. Instead, he employs abstract, metapsychological formulations as if they were experience-near concepts, all of which results in the abundant use of what Slap and Levine (1978) refer to as "hybrid concepts." Let me provide some examples taken from the Slap and Levine paper. In the first example, Kohut (1971) states that "the patient comprehends that the [pathological] condition is due to the fact that his self had temporarily become deprived of its cohesive narcissistic cathexis which had been uncontrollably siphoned into his actions" (p. 128). Assuming that one understands the terms employed, is narcissistic cathexis the kind of thing one can experience or comprehend being deprived of? Consider another example offered by Slap and Levine. "The presence of an unconscious fellatio fantasy in which swallowing the magical semen stands for the unachieved internalization and structure formation might well be assumed . . . . " (Kohut, 1971, p. 72). How can a fantasy stand for a metapsychological construct such as internalization or structure formation?

In the above examples, what is intermixed is the level of experience- and observation-near description and highly abstract, metapsychological experience- and observation-distant constructs as if they both referred to people's actual experiences and behavior. It is ironic that Kohut who so definitively espouses empathy and introspection as *the* distinctive hallmark of the psychoanalytic approach can so easily confuse what someone is experiencing or could be experiencing with an abstract metapsychological account or an etiological account.

Also, Kohut indulges in etiological speculations which could not possibly be based on empathy, that is, on an understanding of what another is (or was) experiencing or feeling. For example, even if one could empathically understand a patient's early experiences and even if one could assume that his current reports accurately describe those experiences, how can one *empathically* know the *etiological-causative role* of those experiences? Or, to take another example provided by Levine (1979), how can empathy permit one to make 'blind' diagnoses of the patient's parents and to understand their causative role in the patient's difficulties? Finally, although one may empathically resonate with another's feelings and fears, how does one empathically know about a patient's *structural defects*? How does one distinguish, empathically, someone's feelings and fantasies of being defective from the fact of having a structural defect (I'm assuming that we really know what we mean by the latter)? Is the latter the kind of knowledge one can ever gain empathically? Indeed, as Levine (1979) points out, for a therapist to claim empathic knowledge regarding a patient's self-defects may work to confirm the latter's fantasies about being defective and may also foreclose analytic exploration of such fantasies. To paraphrase an old saw: with empathy like this, one doesn't need misunderstanding.

It is apparent, then, from a careful reading of Kohut's work that his descriptions and formulations are no more experience-near nor empathically derived than traditional psychoanalytic formulations. Kohut's presents a straightforward environmental causal model of pathology in which defects and developmental arrests are brought about by early trauma (e.g., lack of maternal empathic mirroring). That is, Kohut essentially presents a clear-cut claim that there is a causal relationship between alleged parental behavior and the patient's current difficulties and pathology. This etiological-causal claim may or may not be correct (that is obviously an empirical question), but it is entirely unrelated to the issue of empathy. To clothe these formulations in the garment of empathy constitutes, I believe, a subtle claim for a kind of immunity from prosecution. That is, if one explicitly presents an etiological claim of the nature of *A* is causally related to *B*, then it is immediately apparent that in order for one's claim to merit serious attention, one must produce meaningful empirical evidence, preferably of a longitudinal-prospective kind. However, if one maintains that one's formulation is based on 'clinical evidence' empathically acquired, there is a tendency for others to be less critical and less demanding.

Perhaps it is easier for others to lose sight of the stark and startling fact that etiological claims concerning the effects of early empathic mirroring (or the lack of it) and of early idealization (or the lack of it) rest on no reliable and systematic evidence and do not include even a single longitudinal or follow-up study on the effects of these factors upon later development. But in this regard perhaps Kohut is no more culpable than many recent psychoanalytic writers who make similar assertions with equal lack of evidence.

Again, one must ask whether recent formulations are better validated or rest on more secure epistemological foundations and thereby escape the kinds of criticisms Grünbaum has directed to Freudian theory. It should be apparent from the foregoing that recent formulations do not escape such criticisms and that claims regarding the special use of experience-near and empathically derived concepts are without foundation.

The sixth and final characteristic of recent psychoanalytic writing I have noted is the extensive use of jargon and the failure to make some attempt to provide empirical referents for one's terms and concepts. Here one merely needs to provide some illustrative examples from writings which are representative of current developments. I begin with some examples by Kohut.

Kohut is particularly guilty of such jargon. We have already seen some examples in the discussion of his use of "hybrid concepts." An example each from his 1971 and 1977 books follows: " . . . the danger against which the ego defends itself by keeping the archaic grandiose self dissociated and/or in repression and the dedifferentiating influx of unneutralized narcissistic libido (toward which the threatened ego reacts with anxious excitement) and the intrusion of archaic images of a fragmented body self . . . " (1971, p. 152).

In discussing Mr. M, Kohut describes him as showing "a structural defect in the area of his goals, ideals, leading secondarily to an insufficient channelling of his exhibitionistic-grandiose-creative strivings toward well-integrated, firmly internalized goals. The absence of a sufficiently organized flow of grandiose-exhibitionistic libido toward a securely internalized set of ideals . . . " (1977, p. 123).

Kohut is certainly not alone in his use of jargon. Let me offer one or two examples from Mahler's (1968) writings. In describing the state of the young infant, Mahler writes that "the primordial energy reservoir that is vested in the undifferentiated 'ego-id' still contains an undifferentiated mixture of libido and aggression. As several authors have pointed out, the libidinal cathexis vested in symbiosis, by reinforcing the inborn instinctual stimulus barrier, protects the rudimentary ego from premature phase-unspecific strain – from stress traumata" (p. 9). Another example:

> Metapsychologically speaking, this seems to mean that, by the second month, the quasi-solid stimulus barrier (negative, because it is uncathected) – this autistic shell, which kept external stimuli out – begins to crack. Through the aforementioned cathectic shift toward the sensoriperceptive periphery, a protective, but also receptive and selective, positively cathected stimulus shield now begins to form and to envelop the symbiotic orbit of the mother-child dual unity . . . . This eventually highly selective boundary seems to contain not only the pre-ego self representations, but also the not yet differentiated, libidinally cathected symbiotic part objects, within the mother-infant symbiotic matrix (p. 15).

Finally, as an example of practically incomprehensible jargon, I offer a passage from a recent paper by Giovacchini (1981). In his discussion of whether one should conceptualize "the endopsychic structure that underlies the transitional phenomenon as a nurturing or functional modality" (p. 404), Giovacchini writes "what I am emphasizing is that the vicissitudes involved in the formation of this first mental construct have an important bearing on the formation of the postsymbiotic introjects which, in turn, serve as a source of cathexis that can lead to the stabilization of a function as well as the construction of self and object representations" (pp. 411–412). (For a general discussion of jargon in psychoanalytic writing, see Leites' (1971) *The New Ego*.

Would one want to say that the above kinds of formulations represent a degree of progress so evident as to vitiate and make inapplicable the kinds of criticisms to which Grünbaum subjects traditional psychoanalytic theory? Indeed, it seems to me that, on the contrary, Freud's writings and formulations for the most part, are paragons of precision and clarity compared to much current writing in so-called object relations theory and psychoanalytic self psychology. Indeed, it is the clarity of much of Freud's writing which, contrary to Popper's (1962) criticisms (see Grünbaum's [1979] comments on Popper's relatively uninformed evaluation of psychoanalytic theory), permits readier testing and even refutation of certain Freudian hypotheses. (For example, as we have seen earlier, Freud's 'secondary drive' explanation of infant-mother attachment is pretty much refuted or, at least, rendered highly improbable by the available evidence.)

In general, partly because of its older vintage, but also because of greater precision, certain propositions comprising Freudian theory have been subjected to empirical test (for example, see Fisher and Greenberg, 1978). For the most part, this is not the case with more recent formulations, in part, at least, because they are not presented in testable form. The *general* propositions relevant to object relations theory which have received extensive empirical support are, as noted earlier, those relating to our inborn object-seeking propensities and to an autonomous attachment system. Also, Mahler's (1968; and Mahler *et al.*, 1975) general claim that a significant, even central, dimension of psychological development is separation-individuation as well as the relationship posited between the child's independent exploratory behavior and the availability of a 'safe base' have received considerable empirical support (see Eagle, 1982). However, it is important to note a number of qualifications. One, in all cases, the empirical support is general and indirect (indeed, it must be ferretted out), is provided by extra-clinical sources, and is generated by research the context of which is not object relations theory, but an area quite independent of that theory. For example, the evidence supporting Fairbairn's speculations regarding our object-seeking propensities is derived from the general body of infant research rather than from specific attempts to test Fairbairn's formulations. (Indeed, it is likely that a large number of infant researchers are totally unfamiliar with Fairbairn's work.) Or, as another example, most animal research on the relationship between exploratory behavior and the availability of a 'safe base' does not even refer to Mahler's writings.

Having described and evaluated certain dominant features of recent developments in psychoanalytic theory, I want to make one additional point. There are certain familiar arguments one frequently comes across in discussions of the scientific status of psychoanalytic theory. Some of them go as follows (all taken from Flax, 1981): one cannot determine the scientific status of psychoanalysis because no one has successfully provided "an adequate account of the scientific process" (p. 563). "All data are 'epistemically contaminated'" (p. 563); mainly, physics "serves as a model for all science" (p. 564). Science would look very different if another discipline (e.g., biology) served as a model; psychoanalysis does "not fit into existing philosophies of science" but instead offers its own brand of "simultaneously empirical, intersubjective, hermeneutic, and emancipatory process and form of knowledge" (p. 569). My point is simply this: to a certain extent, complex discussions regarding the nature of science and whether or not psychoanalysis is a science distract one from attending to and recognizing the simple and stark fact that even from a most informal and common-sense point of view and even employing the most liberal and relaxed criteria, many of the claims and formulations I have described are either incoherent or without any evidential support. For the examples of jargon I gave the proper question is not "are they scientific?" (*whatever* one's conception of science), but do they have *any* coherent meaning? And as for certain etiological claims, such as those based *entirely* on interpretations of adult patient's reports, by even the loosest conception of what constitutes evidence, these claims are unsupported. It is really not a matter of some esoteric, technical conception of science. For ordinary planning and decisions in everyday life, one would insist on more secure evidence and a higher level of reasoning.

To return to the basic point of the paper: an examination of recent formulations in psychoanalytic object relations theory and self psychology makes it apparent that, contrary claims notwithstanding, Grünbaum's criticisms of Freudian theory are neither vitiated nor undone by these recent developments. In no way do current formulations somehow manage to weaken or even constitute a response to these criticisms. The clinical data generated by an object relations theory or self psychology approach are as epistemologically contaminated as data generated by the more traditional approach. There is a little, or perhaps even less, evidence available on therapeutic process and therapeutic outcome. And finally, the etiological claims made in more current formulations are perhaps even more logically and empirically flawed than Freud's etiological formulations.

York University

NOTES

[1] Perhaps more than in any other period in the history of psychoanalysis, one finds a great deal of writing examining the conceptual status of psychoanalytic theory. Interestingly enough, while in earlier periods advocates of psychoanalytic theory were intent on arguing its scientific status (e.g., Hook, 1959), the dominant recent trend within the psychoanalytic community has been to reject that claim and to identify psychoanalytic theory as a hermeneutic discipline (e.g., Schafer, 1976).

[2] One question that arises is to what extent treatment which radically downgrades interpretation, the importance of insight and of "making the unconscious conscious" can still legitimately be seen as psychoanalysis or even psychoanalytic (for a discussion of this issue, see Levine, 1979).

[3] This situation is approximated when data derived from the clinical situation are investigated (e.g., ratings made) by 'blind' judges external to that situation. For example, Sampson, Weiss and their colleagues at Mt. Zion Hospital in San Francisco report on a number of research projects in which data from tape recorded therapy sessions are rated on various dimensions by judges who are unfamiliar with the hypotheses being tested (Sampson and Weiss, 1977; Sampson *et al.*, 1977; Horowitz *et al.*, 1975).

[4] Although, as discussed in Note 3, data derived from the clinical situation can be treated in quasi-experimental fashion. For example, Luborsky (1967) has shown that the phenomenon of momentary forgetting in therapy is reliably linked to the subsequent emergence of certain conflictful themes. A quasi-experimental design is implemented by comparing patient verbalization segments following momentary forgetting with control segments (e.g., approximately the same place in the treatment session but no momentary forgetting). Similarly, Horowitz *et al.* (1975) found that the emergence of themes in treatment judged to be warded off by 'blind' judges are lawfully related to certain variables such as the therapist's neutrality. It is important to note that in both these studies (and other similar ones) it is researchers outside the clinical situation who rate and score the data and form and test hypotheses. It is also important to note that in these studies certain specific hypotheses concerning the *therapeutic process* rather than general psychoanalytic hypotheses are subjected to investigation. It is difficult to imagine that one could test in similar fashion certain general psychoanalytic propositions, such as whether identification with the aggressor helps resolve Oedipal conflicts in boys or whether lack of maternal mirroring produces self-defects. I make this point because it is precisely these latter sorts of inappropriate propositions on which clinical data are most frequently brought to bear.

[5] It seems to me that the failure to resolve these and other related problems (e.g., the elementary one of intersubjective reliability) has led to the current infatuation, within the psychoanalytic community, with the idea that psychoanalysis is properly classified

as a hermeneutic discipline and with the current emphasis on 'stories' and 'narratives' rather than on truth claims. If one cannot tackle successfully the problems of reliability and validity criteria for psychoanalytic accounts, I suppose the next best and most comforting step is to consider the issue irrelevant and to declare all accounts equally valid.

6 As I have noted above, progress *has* been made in certain areas – for example, in understanding the nature and basis of infant-mother attachment. And it may well be that something of the nature of what Kohut calls "empathic mothering" (assuming that this concept can be clarified and that its empirical referents can be made more explicit) will turn out to be an important factor in the normal development of certain aspects of personality. But if this does turn out to be the case, it will be established on the basis of reliable, extra-clinical data having to do with infant-mother interactions and, one would hope, generated by longitudinal, prospective studies. Furthermore, how all this would then be related to such issues as idealization, self-defects, self-cohesiveness, narcissistic disorders, archaic grandiosity and exhibitionism, "transmuting internalizations," etc., is, to put it mildly, far from clear.

7 Because there are more analysts (and more therapists of every persuasion) practising today than in the past, there is undoubtedly more competition for patients. If analysts limited themselves to patients who were, according to traditional criteria, ideal candidates for psychoanalysis, they would, very likely, soon find themselves in deep financial difficulty. The "broadened scope of psychoanalysis" comes along at a very convenient time insofar as it permits psychoanalytic work with a wider range of patients, thereby increasing the number of potential clients. I am suggesting that the changing nature of presenting symptoms of today's modal patient and the abundance of therapists available have far more to do with the "broadened scope of psychoanalysis" than theoretical and technical 'progress'.

8 This characterization applies mainly to the work of Kohut and Guntrip, not to Mahler's and certainly not to Kernberg's writings.

9 I will not review the arguments here (for a further discussion of this issue, see Eagle, 1980a).

10 The origin of this argument in the *Verstehende* philosophical movement and in the general philosophical distinction between *Naturwissenschaften* and *Geisteswissenshaften* will be recognized.

REFERENCES

Arlow, J. A. and Brenner, C. 1964. *Psychoanalytic Concepts and the Structural Theory*. New York: International Universities Press.

Bergin, A. E. and Lambert, M. J. 1978. 'The Evaluation of Therapeutic Outcomes.' In S. L. Garfield and A. E. Bergin (eds.), *Handbook of Psychotherapy and Behavior Change*. New York: John Wiley, pp. 139–189.

Bowlby, J. 1969. *Attachment and Loss*. Vol. I: *Attachment*. London: Hogarth Press.

Dilthey, W. 1961. *Meaning in History*. Ed. by H. P. Rickman. London: Allen and Unwin.

Eagle, M. 1980a. 'A Critical Examination of Motivational Explanation in Psychoanalysis,' *Psychoanalysis and Contemporary Thought* 3, 329–380. (Reprinted in Laudan, 1983).

Eagle, M. 1980b. 'Psychoanalytic Interpretations: Veridicality and Therapeutic Effectiveness,' *Noûs* 14, 405–425.

Eagle, M. 1982. 'Interests as Object Relations.' In J. Masling (ed.), *Empirical Studies in Analytic Theory*. Hillsdale, New Jersey: Laurence Earlbaum Associates. Also to appear in *Psychoanalysis and Contemporary Thought*.

Eagle, M. In press. 'Psychoanalysis and Modern Psychodynamic Theories.' In N. S. Endler and J. McV. Hunt (eds.), *Personality and the Behavior Disorders*. Revised ed. New York: John Wiley.

Eckardt, B. von. 1981. 'On Evaluating the Scientific Status of Psychoanalysis,' *Journal of Philosophy* 78, 570–572.

Emde, R. N. and Robinson, J. 1979. 'The First Two Months: Recent Research in Development Psychobiology.' In J. D. Noshpitz (ed.), *Basic Handbook of Child Psychiatry*, vol. I. New York: Basic Books.

Fairbairn, W. R. D. 1952. *Psychoanalytic Studies of the Personality*. London: Tavistock Publications.

Fantz, R. L. 1958. 'Pattern Vision in Young Infants,' *The Psychological Record* 8, 43–47.

Fantz, R. L. 1965. 'Visual Perception from Birth as Shown by Pattern Selectivity,' *Annals of the New York Acad. Sci.* 118, 793–814.

Fantz, R. L. and Fagan, J. F. 1975. 'Visual Attention to Size and Number of Pattern Details by Term and Pre-term Infants During the First Six Months,' *Child Development* 46, 3–18.

Firestein, S. 1978. *Termination in Psychoanalysis*. New York: International Universities Press.

Fisher, S. and Greenberg, R. P. 1978. *The Scientific Evaluation of Freud's Theories and Therapy*. New York: Basic Books.

Flax, J. 1981. 'Psychoanalysis and the Philosophy of Science: Critique or Resistance?' *Journal of Philosophy* 78, 561–569.

Freud, S. 1915. 'Instincts and Their Vicissitudes.' In *The Standard Edition of the Complete Psychological Works*, vol. 14. London: Hogarth Press, 1957.

Freud, S. 1940. 'An Outline of Psychoanalysis.' In *The Standard Edition of the Complete Psychological Works*, vol. 23. London: Hogarth Press, 1964.

Friedman, S., Bruno, L. A. and Vietze, T. 1974. 'Newborn Habituation to Visual Stimuli: A Sex Difference in Novelty Detection,' *Journal of Experimental Child Psychology* 18, 242–251.

Gedo, J. E. 1980. 'Reflections on Some Current Controversies in Psychoanalysis,' *Journal of the American Psychoanalytic Association* 28, 363–383.

Giere, R. 1979. *Understanding Scientific Reasoning*. New York: Holt, Rinehart and Winston.

Gill, M. M. 1976. 'Metapsychology is not Psychology.' In M. M. Gill and P. S. Holzman (eds.), *Psychology versus Metapsychology: Essays in Honor of George S. Klein*. New York: International Universities Press.

Giovacchini, P. L. 1981. 'Object Relations, Deficiency States, and the Acquisition of Psychic Structure.' In S. Tuttman *et al.* (eds.), *Object and Self: A Developmental Approach: Essays in Honor of Edith Jacobson*. New York: International Universities Press.

Goldberg, A. 1978. *The Psychology of the Self: A Casebook*. New York: International Universities Press.

Greenson, R. R. 1967. *The Technique and Practice of Psychoanalysis*, vol. I. New York: International Universities Press.

Grünbaum, A. 1977. 'How Scientific Is Psychoanalysis?' In R. Stern *et al.* (eds.), *Science and Psychotherapy*. New York: Haven Press.

Grünbaum, A. 1979. 'Is Freudian Psychoanalytic Theory Pseudo-scientific, by Karl Popper's Criterion of Demarcation?' *American Philosophical Quarterly* **16**, 131–141.

Grünbaum, A. 1980. 'Epistemological Liabilities of the Clinical Appraisal of Psychoanalytic Theory,' *Noûs* **14**, 307–385.

Grünbaum, A. 1981. 'Comments on Jane Flax's Paper.' Remarks delivered at meetings of the American Philosophical Association, Eastern Division, December 1981, Philadelphia, Pennsylvania. [See also Grünbaum, 1982c.]

Grünbaum, A. 1982a. 'Logical Foundations of Psychoanalytic Theory.' In W. K. Esler and H. Putnam (eds.), *Festschrift for Wolfgang Stegmüller*. Boston: D. Reidel.

Grünbaum, A. 1982b. 'Can Psychoanalytic Theory Be Cogently Tested "on the Couch"?' *Psychoanalysis and Contemporary Thought* **5**, 155–255, 311–436; and in a revised version ('The Foundations of Psychoanalysis') in Laudan (1983), pp. 143–309.

Grünbaum, A. 1983. 'Is Object Relations Theory Better Founded than Orthodox Psychoanalysis: Response to Jane Flax.' In Symposium on Psychoanalysis. *J. Phil.* **80**, 46–51.

Guntrip, H. 1968. *Schizoid Phenomena, Object Relations and the Self*. New York: International Universities Press.

Harlow, H. F. 1958. 'The Nature of Love,' *American Psychologist* **13**, 673–685.

Home, H. G. 1966. 'The Concept of Mind,' *International Journal of Psychoanalysis* **47**, 42–49.

Hook, S. (ed.) 1959. *Psychoanalysis, Scientific Method and Philosophy*. New York: New York University Press.

Horowitz, L. *et al.* 1975. 'On the Identification of Warded-off Contents: An Empirical and Methodological Contribution,' *Journal of Abnormal Psychology* **84**, 545–558.

Kanner, L. 1949. 'Problems of Nosology and Psychodynamics of Early Infantile Autism,' *American Journal of Orthopsychiatry* **19**, 416–426.

Kernberg, O. 1975. *Borderline Conditions and Pathological Narcissism*. New York: Jason Aronson.

Kernberg, O. 1976. *Object-Relations Theory and Clinical Psychoanalysis*. New York: Jason Aronson.

Kinney, D. G. and Kagan, J. 1976. 'Infant Attention to Auditory Discrepancy,' *Child Development* **47**, 155–164.

Klein, G. S. 1976. *Psychoanalytic Theory: An Exploration of Essentials*. New York: International Universities Press.

Klein, M. 1981. 'On Mahler's Autistic and Symbiotic Phases: An Exploration and Evaluation,' *Psychoanalysis and Contemporary Thought* **4**, 69–105.

Kohut, H. 1959. 'Introspection, Empathy and Psychoanalysis,' *Journal of the American Psychoanalytic Association* **7**, 459–483.

Kohut, H. 1971. *The Analysis of the Self*. New York: International Universities Press.

Kohut, H. 1977. *The Restoration of the Self*. New York: International Universities Press.

Laudan, L. (ed.). 1983. *Mind and Medicine: Problems of Explanation and Evaluation in Psychiatry and the Biomedical Sciences. Pittsburgh Series in Philosophy and History of Science*, vol. 8. Berkeley: University of California Press.

Leites, N. 1971. *The New Ego*. New York: Science House.

Levine, F. J. 1979. 'On the Clinical Application of Kohut's Psychology of the Self: Comments on Some Recently Published Case Studies,' *Journal of the Philadelphia Association for Psychoanalysis* 6, 1–19.

Luborsky, L. 1967. 'Momentary Forgetting during Psychotherapy and Psychoanalysis: A Theory and Research Method.' In R. R. Holt (ed.), *Motives and Thought: Essays in Honor of David Rapaport*. New York: International Universities Press.

Mahler, M. 1968. *On Human Symbiosis and the Vicissitudes of Individuation*. Vol. I: *Infantile Psychosis*. New York: International Universities Press.

Mahler, M., Bergman, A. and Pine, F. 1975. *The Psychological Birth of the Human Infant: Symbiosis and Individuation*. New York: Basic Books.

Marmor, J. 1962. 'Psychoanalytic Therapy as an Educational Process.' In J. Masserman (ed.), *Psychoanalytic Education*, vol. 5 of *Science and Psychoanalysis*. New York: Grune and Stratton.

Masling, J. and Cohen, I. 'Psychotherapy, Clinical Evidence, and the Self-Fulfilling Prophecy' (unpublished manuscript).

McCall, R. B. and Nelson, W. H. 1970. 'Complexity, Contours, and Areas as Determinants of Attention in Infants,' *Developmental Psychology* 3, 343–349.

Murray, E. J. 1956. 'A Content-Analysis Method for Studying Psychotherapy,' *Psychological Monographs* 70, no. 420 (entire issue).

Popper, K. R. 1962. *Conjectures and Refutations*. New York: Basic Books.

Rangell, L. 1980. 'Contemporary Issues in the Theory of Therapy.' In H. Blum (ed.), *Psychoanalytic Explorations of Technique: Discourse on the Theory of Therapy*. New York: International Universities Press.

Rosenthal, R. 1963. 'On the Social Psychology of the Psychological Experiment: The Experimenter's Hypothesis as Unintended Determinant of Experimental Results,' *American Scientist* 51, 268–283.

Ruff, H. A. and Birch, H. G. 1974. 'Infant Visual Fixation: The Effect of Concentricity, Curvilinearity, and Number of Directions,' *Journal of Experimental Child Psychology* 17, 460–473.

Sampson, H. and Weiss, J. 1977. 'Research on the Psychoanalytic Process: An Overview.' The Psychotherapy Research Group, Department of Psychiatry, Mt. Zion Hospital and Medical Centre, San Francisco, Bulletin No. 2.

Sampson, H., Weiss, J., and Gassner, S. 1977. 'Research on the Psychoanalytic Process II: A Comparison of Two Theories of How Previously Warded-off Contents Emerge in Psychoanalysis.' The Psychotherapy Research Group, Department of Psychiatry, Mt. Zion Hospital and Medical Center, San Francisco, Bulletin No. 3.

Schafer, R. 1976. *A New Language for Psychoanalysis*. New Haven: Yale University Press.

Scheffler, I. 1967. *Science and Subjectivity*. Indianapolis: Bobbs-Merrill.

Slap, J. W. and Levine, F. J. 1978. 'On Hybrid Concepts in Psychoanalysis,' *Psychoanalytic Quarterly* 67, 499–523.

Stern, D. N. 1980. 'The Early Development of Schemas of Self and Other, and of Various Experiences of Self with Other.' Paper presented at a symposium on 'Reflections on Self Psychology' at the Boston Psychoanalytic Society and Institute, Boston, Mass.

Stolorow, R. D. and Lachmann, F. M. 1980. *Psychoanalysis of Developmental Arrests: Theory and Treatment*. New York: International Universities Press.
Truax, C. 1966. 'Reinforcement and Nonreinforcement in Rogerian Psychotherapy,' *Journal of Abnormal Psychology* 71, 1–9.
Unamuno, M. de. 1921. *The Tragic Sense of Life*. London: Macmillan.
Volkan, W. 1976. *Primitive Internalized Object Relations: A Clinical Study of Schizoid, Borderline and Narcissistic Patients*. New York: International Universities Press.
White, J. L. and LaBarba, R. D. 1976. 'The Effects of Tactile and Kinesthetic Stimulation on Neonatal Development in the Premature Infant,' *Journal of Developmental Psychobiology* 9, 569–578.
Winnicott, D. W. 1965. *The Maturational Processes and the Facilitating Environment*. New York: International Universities Press.

CLARK GLYMOUR

THE THEORY OF YOUR DREAMS

The Interpretation of Dreams is often thought to be Freud's best book-length work. It was, indeed, Freud's first lengthy statement of a substantially original psychological theory. Freud wrote the book in the late 1890's and published it in 1900; it had a second edition in 1909, and thereafter many subsequent editions. By Freud's own account it was not well received by the scientific readers for whom it was intended, that is by physicians and academic psychologists, but it was something of a popular success. Freud's book found an audience among educated lay persons, and it was with this group that Freud found his most immediate and direct following. *The Interpretation of Dreams* is a long and complex book, and it is deeply revealing of Freud's procedures, his style of argument and his theoretical development.

The book begins with a thorough review of the scientific literature about dreams; but it is a review and not a criticism. Freud mentions some well-known features of memory in dreams – that what is dreamed of may often be long-forgotten memories, that the events recollected in dreams are often very unimportant, and that events that have recently occurred often figure in dreams. The important problem, for Freud, is what it is that determines the choice of the content of dreams. One suggestion, that the content of dreams is determined by external stimuli during sleep, Freud finds unsatisfactory – partly because such views do not present enough details about the connection between external stimuli and the contents of dreams, but also because Freud himself does not merely want *causes* of the dream content, he wants *motives* for that content; reasons. His choice of words suggests as much. Internal organic stimuli are another possible cause of the content of dreams; thus a 'dental stimulus' may bring about dreams of teeth falling out. This, at least, was a theory popular in medical circles in Freud's day. Freud's objection to these theories is an interesting one:[1]

> The obvious weakness of these attempted explanations, plausible though they are, lies in the fact that, without any other evidence, they can make successive hypotheses that this or that group of organic sensations enters or disappears from mental perception, till a constellation has been reached which affords an explanation of the dream (IV, p. 38).

In other words, this theory can produce explanations of any bit of dream

R. S. Cohen and L. Laudan (eds.), Physics, Philosophy and Psychoanalysis, 57–71.

content, and there is no straightforward way to test the hypotheses such explanations will involve. Freud, like most of us, was better at seeing the methodological weaknesses of theories advocated by others than he was at seeing similar weaknesses with his own views.

Freud also discussed a theory, due to Vold, which states that the content of dreams is determined by the position of the dreamer's limbs according to fairly definite rules – e.g., the actual position of the limb corresponds roughly to its position in the dream. Now Freud offers a curious but characteristic objection to this theory; he does not put forward any doubts as to whether it can be or has been tested, or that what it claims is true. He objects, rather, that such theories do not show how the specific content of the dream is determined: "I should be inclined to conclude from findings such as these that even the theory of somatic stimulation has not succeeded in completely doing away with the apparent absence of determination in the choice of what dream-images are produced" (IV, p. 9). This remark is characteristic of Freud: he regarded all events, psychological ones included, as determined, and thought it a great advantage of a theory or explanation that it succeeded in removing the appearance of chance. How Freud used this principle, and what validity it has, will concern us again later on.

Freud describes some of the commonly accepted psychological characteristics of the contents of dreams – their hallucinatory character, their nonsensicality, their 'craziness' in some cases, and so on. He mentions three theories – approaches really – to mental functioning in dreams. According to the first, all normal psychical activity occurs in sleep just as in a normal conscious state, but the state of sleep itself somehow produces the strange features of dreams. According to a second, dreams involve a lowering of mental activity – a loosening of connections, etc.; and according to still a third approach there are special mental capacities that are exercised in dreaming but which play no role in normal consciousness. Freud's own preference is clearly for the third kind of view, but he offers no arguments against the others. Finally, Freud turns to the literature on dream interpretation, which is, he points out, a popular literature, not a scientific one at all. Then, as now, 'interpreting' dreams was a popular pastime, and books on how to do it were common. Moreover, dream interpretation has a long history, so that there is nothing novel or modern in it. Freud distinguished between two approaches to the interpretation of dreams: in the holistic approach, the entire dream is replaced by some analogous, meaningful story. In the piecemeal approach, the dream is treated rather like a message in code, and individual pieces of the dream are decoded. Freud favors the latter procedure, which was one followed by popular dream books of the day. Such books gave keys for translating elements of the dream:

If I consult a 'dream book', I find that 'letter' must be translated by 'trouble' and 'funeral' by 'betrothal'. It then remains for me to link together the keywords which I have deciphered in this way and, once more, to transpose the result into the future tense. An interesting modification of the process of decoding, which to some extent corrects the purely mechanical character of its method of transposing, is to be found in the book written upon the interpretation of dreams by Artemidorus of Daldis. This method takes into account not only the content of the dream but also the character and circumstances of the dreamer; so that the same dream-element will have a different meaning for a rich man, a married man, or, let us say, an orator, from what it has for a poor man, a bachelor or a merchant (IV, p. 98–99).

Freud remarks in a footnote introduced in a later edition that Artemidorus used the associations of the dream-*interpreter* to the elements of the dream to determine the meanings of dream elements:

A thing in a dream means what it recalls to the mind – to the dream-interpreter's mind, it need hardly be said. An insuperable source of arbitrariness and uncertainty arises from the fact that the dream-element may recall *various* things to the interpreter's mind and may recall something different to different interpreters. The technique which I describe in the pages that follow differs in one essential respect from the ancient method: it imposes the task of interpretation upon the dreamer himself. It is not concerned with what occurs to the *interpreter* in connection with a particular element of the dream, but with what occurs to the *dreamer* (IV, p. 98n.).

And Freud tells us clearly what is wrong with popular dream interpretation:

It cannot be doubted for a moment that neither of the two popular procedures for interpreting dreams can be employed for a scientific treatment of the subject. The symbolic method is restricted in its application and incapable of being laid down on general lines. In the case of the decoding method everything depends on the trustworthiness of the 'key' – the dream-book, and of this we have no guarantee (IV, pp. 99–100).

Freud's introductory survey and criticisms of other views of dreams have given us fair criteria to judge his own theory by, for they are his criteria: a satisfactory dream theory ought to remove the appearance that the content of dreams has no determining causes, but it ought not to do so in such a way as to make its causal suppositions untested or untestable; a theory which proposes that dreams have a meaning should provide a general procedure for interpreting them, and a warrant which demonstrates the trustworthiness of that procedure. These are Freud's own demands on competing theories, and we must bear them in mind as we pursue *The Interpretation of Dreams*.

Freud tells us that he came to realize that dreams had significance, and how to interpret them, from the role they played in the associations of his patients. The method of dream interpretation was modeled on Freud's developing method for determining the causes of the symptoms of his neurotic patients; by now, 1900, that method had become what is known as 'free association.' The patient was simply to report everything that came into his mind, starting with some initial topic. In the case of dreams, the patient was to begin his association from any dream element. The important point was that the subject learn not to keep back any thought that should occur to him, however absurd, nonsensical, rude or improper it might be. He gives us no more definite account of his method, but illustrates it instead. The illustration is from one of his own dreams, which Freud recounts and which he interprets for us. He adopts this procedure, he tells us, because the only other dreams and associations available to him are from his patients, and they are suspect because his patients are mentally ill. Freud takes up the objection that such self-interpretations are liable to arbitrariness and are untrustworthy and he dismisses it in two sentences:

> No doubt I shall be met by doubts of the trustworthiness of 'self-analyses' of this kind; and I shall be told that they leave the door open to arbitrary conclusions. In my judgement the situation is in fact more favourable in the case of *self*-observation than in that of other people; at all events we may make the experiment and see how far self-analysis takes us with the interpretation of dreams (IV, p. 105).

Freud's reply to this sort of worry misses two points of complaint about his procedure. As an illustration of the method of free association, Freud's self-observations are confusing, as we shall see, because unlike cases in which a patient's associations are faithfully recorded by an analyst, in Freud's record we have no separation of association and interpretation. They are one. More importantly, the real question at issue is whether analyses of dreams, regardless of whether made by the dreamer or by another, are trustworthy. It is of no use to know that self-observation is more reliable than the observation of others if, in this context, both are exceedingly unreliable. The point is surely this: we know *before-hand* that, if a person is asked to associate his thoughts with elements of a dream, and report his associations, after a while we will be able to make up a cogent story, thought, fear, wish, or whatever from the resulting associations. We know that simply from our elementary psychological knowledge of people. There is nothing special about dreams in this regard; much the same could be done with rock formations or with blotches of ink or Thematic Apperception Tests or whatever. Unless he or she is enormously stupid, the person doing the associating can make up such a story from the elements of his or her associations. Indeed if the associater is clever, many such stories can be made up. Now the production of such associations from dream elements and the resulting stories will do nothing to establish that the *dreams* are expressing that story, that thought, anymore than the stories that result from people observing random ink blots or rock formations and then associating freely give evidence that the ink blots or the rock formations are expressing those stories. Accordingly, so far as Freud's aim is to establish that dreams express thoughts, and to determine what those thoughts are, his self-analysis is pointless; no matter how many dreams are analyzed, no 'experiment' has been conducted, or can be. We see, too, that the emphasis on the problem of self-observation as against observing others' associations is misplaced: the problem is with the inference from associations plus story to the conclusion that the story tells us what the dream expresses, and this inference is made neither more nor less difficult according to whether or not the associator and the person piecing the associations together are one and the same.

One might hope, in spite of the above, that when many dreams are collected from many persons, together with the connected associations and resulting stories, there will be some invariant, common features of the stories, the dreams, or their connections. If there are such common features, the question arises whether or not they are specific to associations made from dream elements or whether instead they are common as well to the stories made from associations starting from rock formations or ink-blots, etc. That can only be decided by collecting associations and stories from rock formations, ink-blots, etc., as well as from dreams. Then we would have something like an experiment, though a poor one, which might indicate that the associations and stories made from dream elements have peculiar features which make them different from the associations and stories that start with other subjects. That would, of course, still fall considerably short of demonstrating that the stories give the meaning that the dreams express. But even so, the experiment would be a poor one, and for a reason that also should make us wonder about Freud's procedure. The person, whoever he is, that is tying the associations together to make the story may very well be influenced in the kind of story he makes up by whether or not the associations begin with dreams on the one hand, or with rock formations on the other; the more so if he believes that the stories so made from dreams have special properties. In other words the dream interpreter's own views about what he is doing seem a relevant factor which may determine whether or not the dream stories and the rock formation stories do or do not have common features. Unless the dream interpreter's opinions on such matters could somehow be controlled, even the most dramatic differences in the two sorts of stories might be due solely to the interpreter's bias.

Freud's description of his method is defective in other, equally serious ways. There are no rules given for piecing associations together to make a story, and without them there seems little guarantee or even likelihood that from the same associations different interpreters would obtain the same thought or story, unless, of course, all of the interpreters were in the grip of some common theory. Again, Freud gives us no rules for stopping the dreamer's associations, or for insisting that he continue, or for coming back again and demanding new associations with an element of the dream previously subjected to association, and so on. It is pretty clear that in dealing with his patient's dreams, and with his own, Freud did all of these things. It should also be clear that without rules severely restricting when and how these things are to be done, the person conducting the interpretation may have it within his power, simply by stopping the dream associations when he likes, or insisting on their continuance when he likes, to obtain associations that will fit into a plausible story of whatever specific kind the interpreter should like to have.

The moral of all of this is that unless Freud provides much more detailed criteria than those I have described for interpreting dreams, his procedures are simply worthless. They tell us nothing. Let us see then how Freud proceeds. He illustrates his method by giving and interpreting one of his own dreams. The dream is as follows:

A large hall – numerous guests, whom we were receiving. – Among them was Irma. I at once took her on one side, as though to answer her letter and to reproach her for not having accepted my 'solution' yet. I said to her: "If you still get pains, it's really only your fault." She replied: "If you only knew what pains I've got now in my throat and stomach and abdomen – it's choking me" – I was alarmed and looked at her. She looked pale and puffy. I thought to myself that after all I must be missing some organic trouble. I took her to the window and looked down her throat, and she showed signs of recalcitrance, like women with artificial dentures. I thought to myself that there was really no need for her to do that. – She then opened her mouth properly and on the right I found a big white patch; at another place I saw extensive whitish grey scabs upon some remarkably curly structure which were evidently modelled on the turbinal bones of the nose. – I at once called in Dr. M. and he repeated the examination and confirmed it ... Dr. M. looked quite different from usual; he was very pale, he walked with a limp and his chin was clean-shaven ... My friend Otto was now standing beside her as well, and my friend Leopold was percussing her through her bodice and saying: "She has a dull area low down on the left." He also indicated that a portion of the skin on the left shoulder was infiltrated. (I noticed this, just as he did, in spite of her dress) . . . M. said: "There's no doubt it's an infection, but no matter; dysentery will supervene and the toxin will be eliminated." . . . We were directly aware, too, of the origin of the infection. Not long before, when she was feeling unwell, my friend Otto had given her an injection of a preparation of propyl, propyls . . . propionic acid . . . trimethylamin (and I saw before me the formula for this printed in heavy type) . . . Injections of that sort ought not to be made so thoughtlessly . . . And probably the syringe had not been clean (IV, p. 107).

Irma was actually a patient of Freud's who had relapsed; Freud had, the evening of the dream, written out an account of the case to give to Dr. M., as Freud says "in order to justify myself." M., Otto, and Leopold were all physician acquaintances of Freud's, and Otto had mentioned the case to Freud the previous day with, Freud thought, implied reproof. Given that background, Freud's interpretation of his own dream is an altogether plausible one, namely that it expresses his anxiety over the soundness of his own medical procedures, and that

The dream fulfilled certain wishes which were started in me by the events of the previous evening (the news given me by Otto and my writing out of the case history). The conclusion of the dream, that is to say, was that I was not responsible for the persistence of Irma's pains, but that Otto was. Otto had in fact annoyed me by his remarks about Irma's incomplete cure, and the dream gave me my revenge by throwing the reproach back on to him. The dream acquitted me of the responsibility for Irma's condition by showing that it was due to other factors – it produced a whole series of reasons. The dream represented a particular state of affairs as I should have wished it to be. *Thus its content was the fulfilment of a wish and its motive was a wish* (IV, p. 118–119).

As one reads the dream report over, Freud's interpretation of it certainly forms a plausible enough story. But our concern ought to be with how he came to the story, not just with whether the story itself is plausible. For Freud offers the dream as an illustration of his method, and the real conclusion that we are supposed to absorb is not that his dream of Irma meant such and such, but that his methods of dream interpretation are good ones. Now Freud gives us several pages of report of his associations with elements of the dream; though in fact what he gives us is not that at all, but something much more complex. For Freud's 'associations' include conclusions as to what dream elements mean or represent, queries regarding such conclusions, associations from such conclusions, that is, associations starting from elements of the interpretation of the dream, and, of course, associations with the dream elements themselves. Moreover, all of this is run together so that it is very difficult in many instances to determine just what is going on. Very little of what we get sounds like the free association procedure Freud has described for us.

Here are some passages:

> *I reproached Irma for not having accepted my solution; I said: "If you still get pains, it's your own fault."* I might have said this to her in waking life, and I may actually have done so. It was my view at that time (though I have since recognized it as a wrong one) that my task was fulfilled when I had informed a patient of the hidden meaning of his symptoms: I considered that I was not responsible for whether he accepted the solution or not – though this was what success depended on. I owe it to this mistake, which I have now fortunately corrected, that my life was made easier at a time when, in spite of all my inevitable ignorance, I was expected to produce therapeutic successes. – I noticed, however, that the words which I spoke to Irma in the dream showed that I was specially anxious not to be responsible for the pains which she still had. If they were her fault they could not be mine. Could it be that the purpose of the dream lay in this direction? (IV, pp. 108–109.)

The above sounds much more like a commentary on a bit of the dream and its connection with features of Freud's life than it does any bit of association. Sometimes, however, Freud sounds more like his own patient:

> *I took her to the window to look down her throat. She showed some recalcitrance, like women with false teeth. I thought to myself that really there was no need for her to do that.* I had never had any occasion to examine Irma's oral cavity. What happened in the dream reminded me of an examination I had carried out some time before of a governess: at a first glance she had seemed a picture of youthful beauty, but when it came to opening her mouth she had taken measures to conceal her plates. This led to recollections of other medical examinations and of little secrets revealed in the course of them – to the satisfaction of neither party (IV, p. 109).

In other places we get virtually no association with a dream element; instead we are simply told by Freud what it means, or we get associations with the interpretation. For example:

> *No matter.* This was intended as a consolation. It seemed to fit into the context as follows. The content of the preceding part of the dream had been that my patient's pains were due to a severe organic affection. I had a feeling that I was only trying in that way to shift the blame from myself. Psychological treatment could not be held responsible for the persistence of diphtheritic pains. Nevertheless I had a sense of awkwardness at having invented such a severe illness for Irma simply in order to clear myself. It looked so cruel. Thus I was in need of an assurance that all would be well in the end, and it seemed to me that to have put the consolation into the mouth precisely of Dr. M. had not been a bad choice. But here I was taking up a superior attitude towards the dream, and this itself required explanation (IV, p. 114).

And, finally, Freud stops certain chains of association:

> *In spite of her dress.* This was in any case only an interpolation. We naturally used to examine the children in the hospital undressed: and this would be a contrast to the manner in which adult female patients have to be examined. I remembered that it was said of a celebrated clinician that he never made a physical examination of his patients except through their clothes. Further than this I could not see. Frankly, I had no desire to penetrate more deeply at this point (IV, p. 113).

This is virtually all there is to Freud's characterization of his method. The method itself, so far as we can judge from Freud's description of it, is inadequate in every way to demonstrate the points Freud wishes to maintain, but his illustration of it is almost as unsatisfactory. Conclusions and haltings may, of course, be part of a subject's associations, but Freud's self-analysis still leaves one with little sense of how Freud applied his procedures to *others*.

What is striking about the first dream interpretation of the entire book, and the most detailed one, is that the interpretation offered is enormously plausible largely because it is an almost literal reading of the content of the dream, in which the blame for Irma's illness is placed with Otto, not Freud. But the *use* of the dream in Freud's argument is to get us to accept his method, which is then used to 'establish' any number of claims about dreams and their meaning. The method itself is worthless, for the reasons I have given earlier; the objections to it are obvious ones, and if they did not occur to Freud, they certainly would have occurred to many of his scientifically trained readers. It is hard to believe that they did not occur to Freud himself. The whole business seems the cheapest of rhetorical tricks: from the plausibility of the interpretation we are supposed to infer the reliability of the method at getting at the real meaning of dreams. But the Irma dream is one whose interpretation can be read almost on its face, and the elaborate 'analysis' Freud offers us contributes virtually nothing.

Freud's dream of Irma expresses a wish, and in the succeeding two chapters Freud attempts to convince the reader that in this regard the Irma dream is not only typical, but that without exception dreams universally express wishes. This seems a wholly implausible thesis: one thinks of dreams characterized by diffuse anxiety, and of nightmares. Freud, however, is not without resources; he introduces a distinction between what he calls the "manifest content" of the dream and the "latent content" of the dream. The former is the content of the dream as it is reported, the latter the content of the dream as it is interpreted. The two sorts of content, manifest and latent, may be entirely different, and while the manifest content of the dream may not express a wish, the latent content always does. The bulk of these chapters is devoted to establishing this proposition by giving us snippets of the interpretations of dreams which seem, on their face, not to express wishes at all, but which, when interpreted, do express wishes. Many of these dreams, Freud explains, were produced by his patients as evidence that his theory of dreams was in error, and he regards some of them as quite clever attempts at establishing that conclusion. But if his patients were convinced of his thesis by the interpretations Freud offers of their recalcitrant dreams, then they cannot have been very clever people. For by Freud's method every dream could be made out to be an expression of a wish just as every dream could, with almost equal ease, be made an expression of disgust or regret or fear or . . . The arguments are not better than the evidence for the reliability of the method of interpretation, and for that Freud has given us not a whit of evidence. The hypothesis that all dreams express wishes is simply one for which there is abundant negative evidence, and the introduction of the distinction between latent and manifest content is *nothing* more than the hypothesis that all of those dreams that appear to constitute counter-examples to the hypothesis have a hidden meaning which is the expression of a wish. Not a shred of evidence is presented for this supposition, since the dream interpretation procedure, or rather its results, constitutes no evidence at all, for reasons noted earlier. The distinction between manifest and latent content is then a perfectly *ad hoc* hypothesis; that is, an hypothesis introduced for the purpose of reconciling a theory with apparent counter-evidence, and without sustaining evidence of its own.

Freud's discussion does, however, make clearer certain of his devices. For one thing, we learn that he did not hesitate to reject associations, even those which led to an intelligible and plausible interpretation, if he disliked their direction. Thus:

> The patient, who was a young girl, began thus: "As you will remember, my sister has only one boy left now – Karl; she lost his elder brother, Otto, while I was still living with her. Otto was my favourite; I more or less brought him up. I'm fond of the little one too, but of course not nearly so fond as I was of the one who died. Last night, then, I dreamt that *I saw Karl lying before me dead. He was lying in his little coffin with his hands folded and with candles all round – in fact just like little Otto, whose death was such a blow to me*. Now tell me, what can that mean? You know me. Am I such a wicked person that I can wish my sister to lose the one child she still has? Or does the dream mean that I would rather Karl were dead than Otto whom I was so much fonder of?"
>
> I assured her that this last interpretation was out of the question (IV, p. 152).

Freud may have had good reason to believe that certain sorts of meanings were unlikely to be expressed in dreams. Of the example just quoted, Freud later remarks (IV, p. 248) that it was not a "typical" dream of death because the dreamer felt no grief in the dream, and therefore the wish the dream expressed was not a wish for her nephew's death, but something else. One can imagine that generalizations such as this one, which are rather independent of the free-association procedure, might be elaborated into a system of hypotheses about dreams, a system whose principles could be tested against one another. Something like that happened, I have suggested,[2] in the Rat Man Case with Freud's hypotheses about the aetiology of psychoneurosis. But, whatever one may imagine, in fact Freud seems never to do anything comparable with dreams. The point remains that Freud did not hesitate to cut short lines of association that were not leading in directions he thought appropriate.

Another of Freud's devices[3] is to confirm an interpretation by finding two or more elements of the dream which are independently associated with a key figure in the interpretation of the dream. Thus Freud reports the following dream told him by "A clever woman patient of mine . . . ":

"I wanted to give a supper-party, but I had nothing in the house but a little smoked salmon. I thought I would go out and buy something, but remembered then that it was Sunday afternoon and all the shops would be shut. Next I tried to ring up some caterers, but the telephone was out of order. So I had to abandon my wish to give a super-party" (IV, p. 147).

Freud rejected a first set of associations his patient had with the dream:

The associations which she had so far produced had not been sufficient to interpret the dream. [N.B. This is misleading; the patient's associations had suggested a meaning for the dream, but not one that Freud liked.] I pressed her for some more. After a short pause, such as would correspond to the overcoming of a resistance, she went on to tell me that the day before she had visited a woman friend of whom she confessed she felt jealous because her (my patient's) husband was constantly singing her praises. Fortunately this friend of hers is very skinny and thin and her husband admires a plumper figure. I asked her what she had talked about to her thin friend. Naturally, she replied, of that lady's wish to grow a little stouter. Her friend had enquired, too: "When are you going to ask us to another meal? You always feed one so well."

The meaning of the dream was now clear . . . 'What the dream was saying to you was that you were unable to give any supper-parties, and it was thus fulfilling your wish not to help your friend to grow plumper. The fact that what people eat at parties makes them stout had been brought home to you by your husband's decision not to accept any more invitations to supper in the interests of his plan to reduce his weight' (IV, p. 148).

All that was now lacking was some coincidence to confirm the solution. The smoked salmon in the dream had not yet been accounted for.

"How," I asked, "did you arrive at the salmon that came into your dream?" "Oh," she replied, "smoked salmon is my friend's favourite dish" (IV, p. 148).

Freud's story gives us evidence that the content of her dream made the woman think of the friend of whom she was jealous. That is what happens in the association: starting from the dream, she thinks of her friend. Why does she think of the friend? What about the dream makes her think of the friend? It would appear two things. First, the dream is about a failed attempt to give a supper-party and the previous day her friend had asked her when she would give another such party. And, second, the dream mentions smoked salmon which is her friend's favorite dish. The *apparent* causal links then are these:

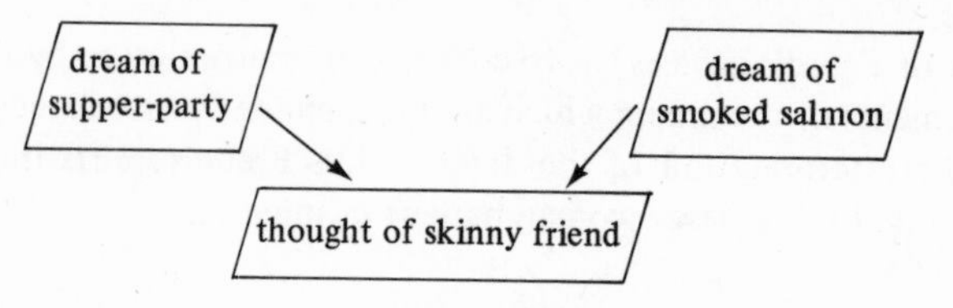

Now Freud transposes these causal relations evidenced by the patient's associations to give us:

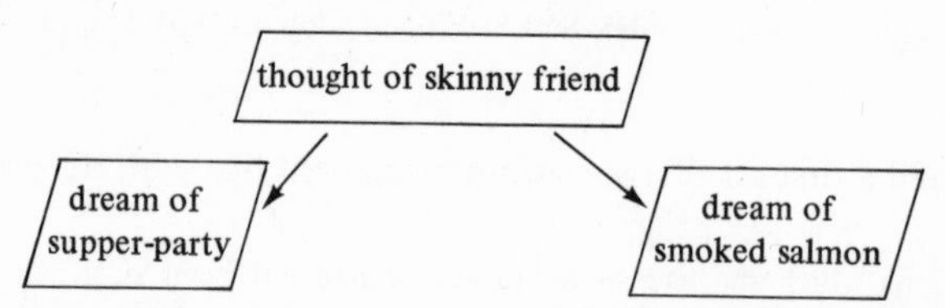

But what are the grounds for the transposition? Evidence for the first causal model is not necessarily evidence for the second. If we think *simply* in terms of the second causal picture, the fact that the dream both contains a failed attempt at a supper-party and mentions smoked salmon, and both of these elements lead to the remembrance of features of the patient's friend, seems an amazing coincidence that demands explanation. The best explanation seems to be that these elements of the dream have a common cause, and that cause has to do with a thought about the friend in question. But if we stand back for a moment we see that *this* coincidence is manufactured: one associates, at Freud's direction, until one thinks of something which has connections with several elements in one's dream; the several elements cause the common thought, not vice versa, and the coincidence requires no further explanation. The method of manufacture is all the explanation required. The *real* coincidence is that on the one hand, on one day the dreamer was visited by a friend, whose favorite dish is smoked salmon, and who had asked her when she, the dreamer, planned to have another supper-party, and, on the other hand, that night she dreamed of a supper-party and smoked salmon. Is *this* coincidence evidence of a causal connection between the encounter with her skinny friend in the dream? Perhaps it is, in the case at hand, for the real event and the dream are proximate in time and share a number of independent features, and one doubts that any other event so proximate would share these features. But, once again, Freud is not so much attempting to convince us of the correct interpretation of a specimen dream as to convince us of a *method*, and, once again, there is no reason to think the method trustworthy in general. As the span of time between the real event and the dream increases – and in many of Freud's applications it increases to years – the chances of coincidence increase immeasurably, and the inference to a causal connection becomes increasingly ill founded. The inference is all the worse when one is not sure whether the events recalled through association really occurred. That was Freud's actual situation at the close of the nineteenth century.

The Interpretation of Dreams formed a turning point in Freud's life and work: half a rotation from scientist towards mountebank. The book contrasts vividly with most of what Freud had published before. Freud's published work before the end of the 1890's seems to be honest, responsive to criticism, and sensitive to methodological and empirical difficulties. His writings from the period are nearly empty of the rhetorical trickery and evasiveness that mars his discussions of dreams. Freud's first book, on aphasia, is straightforwardly and coherently argued, pressing against the predominant accounts (e.g., Wernicke's[4]) of aphasic disorders the fact that certain predicted combinations of aphasias had not been found in the clinic. His collaboration with Breuer in *Studies on Hysteria* had produced an interesting body of cases and a rather sketchy theory to account for them, a sketch not very far removed from those common at the time in French psychiatric circles. Freud had produced and defended characterizations of two 'actual neuroses' – anxiety neurosis and neurasthenia – and hypotheses about their aetiology. He had offered an aetiology for hysteria which had clear empirical consequences, and he had proposed an account of the psychological mechanism of the disease. By my reading there is nothing intellectually dishonest in any of this. Why, then, the sudden near-chicanery and aloofness of *The Interpretation of Dreams*?

I believe that the dream book represents a complex and unwholesome resolution of an intellectual crisis that beset Freud in the very last years of the decade. Perhaps, as biographers claim, there was a personal crisis as well, in which case *The Interpretation of Dreams* may express a change of character, a kind of broken faith, a self-deception. But I will leave psychology aside, and make my conjecture from epistemology instead.

By 1897 Freud's attempts to develop an original and fundamental psychiatry were a shambles, and he knew it. His account of aphasias had been developed no further, and examples of the 'missing' combinations of aphasias were beginning to be found. His account of the sexual aetiology of 'actual neuroses' met with apparent counter-examples in the clinic, and was salvaged only by re-diagnosing the apparent counter-examples not as actual neurotics but as psychoneurotics – hysterics in particular. Freud's theory of the actual neuroses therefore depended for its legitimacy on the reliability of his developing methods of psychoanalysis. On the basis of the results of these methods – including free association and dream interpretation – Freud had published his theory of hysteria, a principal tenet of which was that every hysteric has as a child been seduced into precocious sexual activity. Yet by 1897 Freud was convinced that the seduction theory was founded on a tissue of falsehoods. In public he continued to defend his psychoanalytic methods; in private, we know from his correspondence with Fliess, he was consumed with misgivings. In print, he continued to claim as he had in *Studies on Hysteria* that his interventions, suggestions, and outright demands in therapy could have no contaminating effect on psychoanalytic conclusions. In letters he showed that he had every reason to abandon this smug conviction. Moreover, nothing in Freud's publications or letters from the period suggests that he was enjoying any considerable success as a therapist. Rather the reverse.

Faced with the evidence that the methods on which almost all of his work relied were in fact unreliable, Freud had many scientifically honorable courses of action available to him. He could have published his doubts and continued to use the same methods, reporting his results in company with caveats. He could have published his doubts and abandoned the subject. He could have attempted experimental inquiries into the effects of suggestion in his therapeutic sessions. He did none of these things, or others one might conceive. Instead he published *The Interpretation of Dreams* to justify by rhetorical devices the very methods he had every reason to distrust.

The Interpretation of Dreams reveals a kind of failure of Freud's character, a weakness of intellectual will. Yet it is not as though Freud abandoned reason with the publication of his dream book. In later years he could and did construct acute arguments, criticize and revise his own theories and behave as the savant he clearly was. What he could not do and did not do was to candidly and publicly face the issue of the warrant for believing in the reliability of his methods. Freud's 'tally argument', so ably discussed by Adolf Grünbaum (see Note 3), was a clever and ingenious counter-attack against criticisms of his methods, but it was not a serious treatment of the issue. A central tenet of Freud's argument was that only veridical psychoanalytic insight can produce enduring cures of psychoneurotics. He produced no empirical evidence for that thesis, I think because he had none to produce. In that respect, Freud's tally argument was but another rhetorical device. At the turn of the century Freud once and for all made his decision as to whether or not to think critically, rigorously, honestly, and publicly about the reliability of his methods. *The Interpretation of Dreams* was his answer to the public, and, perhaps, to himself.[5]

University of Pittsburgh

NOTES

[1] All references by volume and page number are to J. Strachey (ed.), *The Standard Edition of the Complete Psychological Works of Sigmund Freud*. 24 vols. (London: Hogarth Press and the Institute of Psychoanalysis, 1953–1974). The quotations from this edition cited in this paper are reprinted with the permission of George Allen and Unwin.

[2] Compare my 'Freud, Kepler, and the Clinical Evidence,' in R. Wollheim (ed.), *Freud* (New York: Doubleday, 1971), and Chapter 6 of *Theory and Evidence* (Princeton: Princeton University Press, 1980).

[3] For related assessments of this Freudian strategem, see Adolf Grünbaum, 'The Foundations of Psychoanalysis,' in L. Laudan (ed.), *Mind and Medicine: Problems of Explanation and Evaluation in Psychiatry and the Biomedical Sciences*, pp. 143–309. *Pittsburgh Series in Philosophy and History of Science*, vol. 8 (Berkeley: University of California Press, 1983). An earlier version of this paper, 'Can Psychoanalytic Theory Be Cogently Tested on the Couch?' appears in *Psychoanalysis and Contemporary Thought* 5 (1982), 155–255, 311–436.

[4] [See Carl Wernicke, 'The Symptom Complex of Aphasia: A Psychological Study on an Anatomical Basis' and the introductory essay by Norman Geschwind, 'The Work and Influence of Wernicke' in *Boston Studies in the Philosophy of Science*, vol. 4 (Dordrecht: D. Reidel, 1969), pp. 1–97 – Ed.]

[5] I wish to thank Adolf Grünbaum and Martha Harty for helpful conversions.

CARL G. HEMPEL

VALUATION AND OBJECTIVITY IN SCIENCE *

1. INTRODUCTION

The role of valuation in scientific research has been widely discussed in the methodological and philosophical literature. The interest in the problem stems to a large extent from the concern that value-dependence would jeopardize the objectivity of science. This concern is clearly reflected, for example, in Max Weber's influential writings on the subject.[1]

In my paper, I propose to consider principally some aspects of the problem which have come into prominence more recently. A discussion of these issues can contribute, I think, to clarifying the cognitive status of the methodology of science and, more generally, of epistemology.

2. VALUATION AS A MOTIVATING FACTOR IN SCIENTIFIC INQUIRY

The question of value-independence can be and has been raised concerning two quite different aspects of science, namely (1) the actual research behavior of scientists, and (2) the methodological standards for the critical appraisal and possibly the justification of scientific assertions and procedures.

There is no dispute about the important role that valuations of various kinds play in the first of these contexts. Moral norms, prudential considerations and personal idiosyncrasies clearly can influence a scientist's choice of a field and of problems to explore; they can also affect what methods of investigation are used, what others eschewed. Social and political values can lead to the deployment of strong research efforts in particular problem areas; they can also encourage the advocacy of ill-founded theories. And, of course, the decision of scientific investigators to adopt or to reject a given hypothesis or theory will, as a rule, be strongly influenced by their commitment to what might be called epistemic values or norms, as reflected in their adherence to certain methodological standards of procedure.

In these contexts, valuations are 'involved' in scientific research in the sense of constituting important motivational factors that affect the conduct of inquiry. Such factors must therefore be taken into account in efforts, such

R. S. Cohen and L. Laudan (eds.) Physics, Philosophy and Psychoanalysis, 73–100.

as those made in the psychology, the sociology, and the history of science, to *explain* scientific research behavior.

Explanations of this kind are scientific explanations. While they refer to certain values espoused by the scientists in question, they do not themselves posit any value judgments. Rather, they descriptively attribute to the scientific investigators a commitment to certain values and thus the disposition to act in accordance with them. The given research behavior is then explained as a particular manifestation of general preferential dispositions.

To explain why scientists took a certain step, such as adopting or rejecting a given theory, is neither to justify it as sound nor to exhibit it as unsound scientific procedure: speaking broadly and programmatically, the latter task calls for a critical appraisal of the theory in light of the available evidential and other systematic grounds that have a bearing on its acceptability.

Grünbaum expresses basically the same idea when he says that both warranted and unwarranted beliefs have psychological causes, and that the difference between them must be sought in the peculiar character of the causal factors underlying their adoption: "*a warrantedly held belief . . . is one to which a person gave assent in response to awareness of supporting evidence*. Assent in the face of awareness of a *lack* of supporting evidence is irrational, although there are indeed psychological causes in such cases for giving assent."[2]

Applying this general idea to a topical example, Grünbaum argues in lucid detail that criticisms of various features of psychoanalytic theory cannot be invalidated by contending, as has not infrequently been done, that the critics have a subconsciously motivated resistance to the ideas in question. For, first of all, this explanatory contention presupposes psychoanalytic theory and may therefore be question-begging; and, more importantly, "the invocation of purely psychological, extra-evidential explanations for *either* the rejection *or* the acceptance of the theory runs the risk of begging its validity, if only because *either attitude may well be prompted by relevant evidence*!"[3]

3. NORMATIVE VS. DESCRIPTIVE-NATURALISTIC CONSTRUALS OF METHODOLOGICAL PRINCIPLES

The familiar idea here invoked of critically appraising the warrant or the rationality of scientific claims assumes that there are clear objective criteria governing such appraisals. These criteria are usually thought of as expressible in terms of logical relations of confirmation or of disconfirmation between the claim in question and the available evidence, and possibly also in terms of certain other objective factors, to be mentioned soon.

It is this conception, I think, which has given rise to the question of objectivity and value-neutrality of science in its recent, philosophically intriguing form: to what extent, and, for what reasons, can scientific inquiry and scientific knowledge claims be characterized as subject to such objective methodological standards?

To the extent that such characterization is possible, proper scientific inquiry and its results may be said to be objective in the sense of being independent of idiosyncratic beliefs and attitudes on the part of the scientific investigators. It then is possible to qualify certain procedures – perhaps the deliberate falsification or the free invention of empirical evidence – as 'violations' of scientific canons, and to seek motivational explanations for them in terms of an overriding commitment to extra-scientific values, such as personal advancement, which conflict with the objective norms of proper scientific conduct.

In considering the question of objective standards for scientific inquiry, I will for convenience distinguish two extreme positions, to be called *methodological rationalism* and *methodological pragmatism*, or *naturalism*. These are ideal types, as it were. The views held by different thinkers in the field differ from those extremes in various ways, as will be seen later.

According to methodological rationalism, there are certain general norms to which all sound scientific claims have to conform. These are established largely on *a priori* grounds, by logical analysis and reconstruction of the rationale of the scientific search for knowledge. And they are expressible in precise terms, for example as purely logical characterizations of the relations between scientific hypotheses and evidence sentences that confirm or disconfirm them.

Methodological naturalism, on the other hand, holds that characterizations of proper scientific procedure must be formulated so as to reflect actual scientific practice rather than aprioristic preconceptions we may have about rational ways of establishing knowledge claims. Thomas Kuhn voices this view when he says that "existing theories of rationality are not quite right and . . . we must readjust or change them to explain why science works as it does. To suppose, instead, that we possess criteria of rationality which are independent of our understanding of the essentials of the scientific process is to open the door to cloud-cuckooland."[4]

Earlier, John Dewey had in a similar spirit rejected an aprioristic conception of methodology as "an affair of . . . fixed first principles . . . of what the Neo-scholastics call criteriology."[5]

Proponents of a pragmatist approach to methodology usually reject the conception that scientific inquiry is subject to standards that can be expressed in precise and fully objective terms.

I will now consider the two opposing views more closely and will try to show that there are stronger affinities between them than the controversies between their proponents might suggest.

4. NON-NATURALISTIC CONSTRUALS OF METHODOLOGICAL NORMS

The strongest and most influential efforts made in the past 50 years to establish methodological principles for empirical science in a rationalist vein were those of the analytic empiricists and kindred thinkers.

Their analytic endeavors no doubt drew encouragement from a tempting analogy between methodology and metamathematics. The latter discipline, too, does not aim at giving a descriptive account of the mathematical enterprise, but rather at formulating in precise terms certain objective standards for the soundness of mathematical claims and procedures.

Carnap's conception of the philosophy of science as the logical analysis of the language of science, and his and Popper's exclusion of psychological and sociological issues from the domain of epistemology reflect a broadly similar view of the methodology of science.

Metamathematics does not provide precise procedural rules for the solution of all mathematical problems. There is no general algorithm which will automatically lead to the discovery of significant new theorems or which, for any given formula of an axiomatized mathematical theory, will decide whether the formula is a theorem of that theory. But there is an algorithmic procedure which, for any given formula and any proposed proof of it, will decide whether the proof is valid and thus, whether the formula is a theorem of the system.

Similarly, a precise normative methodology of science cannot provide general procedural rules for the discovery of a new theory: such discovery, as has often been emphasized, requires creative scientific imagination; and so does even the discovery of feasible ways of testing a proposed theory. But it might well seem possible to formulate precise objective criteria which, for any proposed hypothesis H and evidence sentence E, determine whether or to what degree E confirms H; or perhaps to state purely comparative criteria determining which of two alternative theories is rationally preferable to the other in consideration of the available evidence *E* and possibly certain other objective factors.

This was indeed the basic conception underlying analytic-empiricist efforts to develop formal theories of confirmation or of logical probability.[6] Popper's concept of degree of corroboration of a hypothesis[7] reflects a similar formal bent.

The effort to explicate methodological concepts in precise logical terms is evident also in the attempts made by analytically oriented empiricists to characterize genuine empirical hypotheses in terms of verifiability or testability or confirmability, and in Popper's falsifiability criterion for scientific hypotheses. The same objective is illustrated by analytic models of scientific explanation, which impose certain logical conditions on the explanatory sentences and the sentences expressing what is to be explained.

The conditions thus set forth for empirical hypotheses and theories and for scientific explanations were in fact often put to normative-critical use; for example, in declaring the doctrine of neovitalism to be devoid of empirical content and to lack the status of an explanatory theory, or in rejecting the idea of explaining human actions on the basis of empathy or with the help of certain norms of rationality.[8]

The same kind of analytic approach has been used also to formulate methodological norms for scientific concept formation. This is hardly surprising since theory formation and concept formation in science are two faces of the same coin. Theories are formulated in terms of concepts. Concepts are characterized by the theories in which they function. This point is clearly reflected in the stepwise liberalization of the methodology of concept formation developed in the analytic tradition. It led from explicit definition to the introduction of concepts by reduction sentences and on to a holistic method by which an entire system of concepts to be employed in a theory is specified by formulating an axiomatized version of the theory and its intended interpretation.[9] In this process, theory formation and concept formation become inextricably fused.

Again, the methodological principles of concept formation were put to normative use, for example in the rejection as non-empirical or non-scientific of the idea of entelechy or of vital force which plays a central role in neovitalist doctrines.

The ideal, referred to earlier, of the objectivity of science would call for methodological norms which are objective in the sense that they determine unambiguous answers to problems of critical appraisal, so that different scientists applying them will agree in their verdicts. The criteria we have briefly considered are couched largely in the terms of logical theory; this bodes well for their objectivity.

But it must be noted that the criteria also make use of certain non-logical – more specifically: pragmatic – concepts, namely those of observation sentence and of observational term. For those criteria characterize the testability, the rational credibility, and cognate characteristics of a hypothesis by certain logical relations between the hypothesis and a body of evidence consisting of so-called observation sentences or basic sentences. These are taken to describe phenomena whose occurrence or non-occurrence can be established, with good intersubjective agreement, by means of direct observation. Similarly, the specification of scientific concepts by definition or reduction or by interpreted theoretical systems was taken to be ultimately effected with the help of so-called observational terms standing for directly and publicly observable characteristics of things or places or events.

It was just this intersubjective agreement here assumed in the use of observational terms and sentences that was seen as securing the objectivity of science at the evidential level. And methodological norms for the appraisal of scientific claims would then be objective and value-neutral since they called for precisely characterized logical relations between a hypothesis and a body of evidence that could be established with high intersubjective agreement by means of direct observation.

The criticism to which the notion of direct observability has been subjected in recent decades has necessitated considerable modifications in the analytic-empiricist construal of the evidential side of a critical appraisal, but this does not necessarily jeopardize the idea of the objectivity of science as characterized in Popper's remark that "the *objectivity* of scientific statements lies in the fact that they can be *inter-subjectively tested*".[10] I will revert to this issue later.

What is the cognitive status of methodological principles of the kind just considered? On what grounds are they propounded, and by what means can their adequacy by appraised? Let us consider first the views of some thinkers close to the rationalist position, especially Popper and Carnap.

5. NATURALISTIC AND VALUATIONAL FACETS OF POPPER'S METHODOLOGY

Karl Popper rejects the 'naturalistic' conception which views methodology as a study of the actual research behavior of scientists, arguing among other things that "what is to be called a 'science' and who is to be called a 'scientist' must always remain a matter of convention or decision".[11] As for his own methodology, which characterizes scientific hypotheses by their falsifiability, and which sees scientific progress in the transition to ever more highly corroborated and ever better testable theories, Popper holds that its principles are "*conventions*", which "might be described as the rules of the game of empirical science".[12]

In support of his methodology, Popper argues "that it is fruitful: that a great many points can be clarified and explained with its help," and that from the consequences of Popper's characterization of science "the scientist will be able to see how far it conforms to his intuitive idea of the goal of his endeavors".[13] He adds that the consequences of his definition enable us to detect inadequacies in older theories of knowledge. "It is by this method, if by any," Popper says, "that methodological conventions might be justified."[14]

Clearly, then, Popper's methodological conventions are not arbitrary: they are meant to meet certain justificatory requirements. Those I have just mentioned are rather unspecific; they could be applied also to a methodology of mathematics, for example. But Popper has more specific objectives in mind. As he tells us, his interest in methodology was stimulated by the thought that doctrines like astrology, Marxist theory of history, and Freud's and Adler's versions of psychoanalysis were unsatisfactory attempts at theorizing: they were protected against any possible empirical refutation by vagueness of formulation, by the use of face-saving conventionalist stratagems, or by being constructed so as to be totally untestable to begin with. In contrast, the general theory of relativity made far-reaching precise and specific predictions and thus laid itself open to severe testing and possible falsification.[15]

Popper considered these latter features as characteristic of genuine scientific theories and thus sought to construct a methodology that would systematically elaborate this conception, qualifying Einstein's and similar theories as scientific and excluding from the realm of science the unsatisfactory instances mentioned before. Popper's methodology therefore has a target: it is to exhibit the rationale of certain kinds of theories and theorizing which he judges to be scientific. Indeed, Popper notes that if we stipulate, as his methodology does, that science should "aim at *better and better testable* theories, then we arrive at a methodological principle . . . whose [unconscious] adoption in the past would rationally explain a great number of events in the history of science." "At the same time," he adds, the principle "gives us a statement of the task of science, telling us what should in science be regarded as *progress*."[16]

Thus, while Popper attributes to his methodological principles a prescriptive or normative character, he in effect assigns to them an empirical-explanatory facet as well. *This facet appears, not in the content of the norms, but in the justificatory claims adduced for them*; among them the claims that theories pre-analytically acknowledged as scientific are qualified as such by the methodology; that others, pre-analytically perceived as non-scientific are ruled out; and that important events in the actual history of science could be explained by the assumption that scientists in their professional research are disposed to conform to Popper's methodological norms.

The methodology of science as construed by Popper does therefore have a naturalistic facet in the sense that the justificatory claims made for it include empirical assertions. Indeed, Popper's methodology and similar ones have repeatedly been challenged on the ground that in important contexts scientists have not conformed to the stipulated methodological canons. I will not enter into those criticisms, however; my concern here is simply to note the naturalistic facet of Popper's methodology.

That methodology also has a valuational facet, as Popper, I think, would agree. His choice of methodological principles is prompted by his view that precise testability, large content, and the like, are characteristic features that properly scientific theories should possess. The valuation here involved is not moral or esthetic or prudential; it might rather be called an *epistemological* valuation, which, in the search for improved knowledge, assigns high value to susceptibility to severe tests and possible falsification.

A different epistemological valuation would be reflected in a set of methodological conventions that gives preference to a broad and suggestive but not very precise theory over a narrower, precise, and severely testable one.

Objectivity in the sense of intersubjective agreement among scientists on methodological decisions might well be preserved in spite of the valuational component – namely, to the extent that scientists *share* their epistemic values and the corresponding methodological commitments.

Both empirical information and epistemic valuations, then, are required for a justification or a critical appraisal of a methodology of the kind aimed at by Popper.[17] If his theory were formulated as a set of principles laid down strictly by convention to serve as the rules of a 'game of science' designed by Popper, it would have no methodological or epistemological interest. What lends it such interest is the fact that the rules are meant to afford an – assuredly idealized – account of a specific and very important set of procedures, namely, scientific research. It is by virtue of this claim that it has both the naturalistic and the valuational facets just indicated.

6. CARNAP ON THE EXPLICATION OF METHODOLOGICAL CONCEPTS

Similar considerations are applicable to Carnap's views on the analytic elaboration of methodological concepts and principles. Carnap has applied the procedure to diverse philosophical issues, among them those concerning the standards for a rational appraisal of the credibility of empirical hypotheses. This is the object of his theory of inductive probability, which Carnap presents as offering a precise characterization of the vague pre-analytic concept of the probability of a hypothesis.

Let us briefly consider the character of such precise characterizations and the grounds adduced in their support.

Carnap refers to conceptual clarification and refinement of the kind under discussion as *explication.*[18] He describes it as the replacement of a given, more or less inexact concept, the *explicandum*, by an exact one, the *explicatum*. He notes that the procedure is used also in science; for example, when the vague everyday concepts of hot and cold are replaced by a precise concept of temperature.

Explication plays an important role in analytic philosophy, where it has often been referred to as logical analysis or as rational reconstruction. All the accounts proposed by analytic empiricists for such notions as verification, falsification, confirmation, inductive reasoning, types of explanation, theoretical reduction, and the like are instances of explication, i.e., they propose explicit and precise reconstructions of vague concepts that play an important role in philosophical theories of knowledge.

Carnap lists four requirements which an adequate explication should satisfy:[19]

(1) "The explicatum must be *similar to the explicandum*" in the sense that in most cases in which the explicandum has so far been used, the explicatum applies as well; but some considerable deviations are permitted.

(2) The explicatum is to be characterized by rules of use which have "an *exact* form".

(3) "The explicatum is to be a *fruitful* concept" in the sense of permitting the formulation of an extensive system of laws or theoretical principles.

(4) "The explicatum should be as *simple* as possible."

The first of these requirements throws into relief what I called the descriptive facet of philosophical explication. A concern with descriptive fit is evident in the explicatory accounts that analytic empiricists offered of scientific testing, concept formation, explanation, theoretical reduction, and the like: these were formulated and often subsequently modified in consideration of actual scientific procedures. The complex system of explicatory definitions for empirical concepts that Carnap constructed in *The Logical Structure of the World*[20] incorporates a large amount of empirical knowledge, for example, about the structure of the color space. And scientific laws and theories are presupposed also in explications of scientific concepts along the lines of physicalism and logical behaviorism. Again, Carnap adduces certain modes of reasoning used in psychological research as showing that psychological concepts cannot generally be specified by means of reduction chains linking them to an observational vocabulary; he invokes this consideration to motivate a methodological conception which is quite close to holism, namely that of specifying a system of scientific concepts by means of partially interpreted postulates containing the concepts in question.[21]

In the case of Carnap's explicatory theory of inductive probability, the descriptive facet is more elusive.

By way of a rough characterization of the explicandum concept, $P(H, E)$, of the logical probability of hypothesis H relative to evidence E, Carnap states, among other things, that $P(H, E)$ is to represent the degree to which a person is rationally entitled to believe in H on the basis of E: and that it is also to be a fair betting quotient for a bet on H for someone whose entire evidence concerning H is E.[22]

Carnap's explicatum is formulated in terms of an axiomatized theory in which $P(H, E)$ can then be defined as a quantitative, purely logical relation between the sentences H and E; the axioms ensure that $P(H, E)$ has all the characteristics of a probability function.

The justification Carnap offers for his explication is, briefly, to the effect that the axioms of his theory of rational credibility reflect our intuitive judgments about rational belief, about types of bets it would be irrational to engage in, and the like.[23] As Carnap puts it, "the reasons to be given for accepting any axiom of inductive logic . . . are based upon our intuitive judgments concerning inductive validity, i.e., concerning inductive rationality of practical decisions (e.g., about bets). . . . The reasons are *a priori* [i.e.] independent both of universal synthetic principles about the world . . . and of specific past experiences."[24]

This argument by reference to our intuitive judgments concerning rational decisions seems to me to have a clear affinity to Goodman's view concerning the justification of rules of deductive or of inductive reasoning. Goodman holds that particular inferences are justified by their conformity with general rules of inference and that general rules are justified by their conformity with valid particular inferences. "The point is," he says, "that rules and particular inferences alike are justified by being brought into agreement with each other. *A rule is amended if it yields an inference we are unwilling to accept; an inference is rejected if it violates a rule we are unwilling to amend*."[25]

Now, when Carnap speaks of 'our' intuitive judgments concerning rationality, and when Goodman refers to particular inferences or to general rules which 'we' are unwilling to accept or to amend – who are 'we'? Surely, those intuitions and unwillingnesses are not meant to be just idiosyncratic; the idea is not: to everyone his own standards of rationality. The assumption must surely be that there is a body of widely *shared* intuitions and unwillingnesses, and that approximate conformity with them provides a justification for acknowledging as sound certain rules of deductive or inductive reasoning. Indeed, without such a body of shared ideas on sound reasoning, there would be no explicandum, and the question of an explicatory theory could not arise.

I think therefore that the grounds Carnap offers in support of his theory of rational inductive inference are not just *a priori*. To be sure, Carnap's formal theory of logical probability may be said to make no descriptive claims, but solely to provide, through its axioms, an "implicit definition," couched in purely logical terms, of its basic concept of logical probability. But the justificatory considerations adduced for accepting the theory are not simply *a priori*; for they make descriptive socio-psychological claims about shared intuitive judgments concerning the explicandum concept. Just as that concept is vague, so, admittedly, are those supporting claims. They do not specify, for one thing, exactly whose intuitions do, or are to, agree with the explication. But if, for example, some intuitive judgments adduced by Carnap were deemed counter-intuitive by a large proportion of scientists, mathematical statisticians, and decision-theorists, then surely they could not be invoked for justificatory purposes. Indeed, in a remark on just this point, Carnap acknowledges that scientists do not as a rule explicitly assign numerical degrees of credibility to hypotheses; but, he adds,

> it seems to me that they show, in their behavior, implicit use of these numerical values. For example, a physicist may sometimes bet on the result of a planned experiment; and, more important, his practical decisions with respect to his investment of money and effort in a research project show implicitly certain features . . . of his credibility function . . . If sufficient data about decisions of this kind made by scientists were known, then it would be possible to determine whether a proposed system of inductive logic is in agreement with these decisions.[26]

This passage comes very close to claiming that a theory of rational credibility should have the potential for providing at least an approximate descriptive and explanatory account of some aspects of the behavior of scientists on the basis of the degrees of rational credibility the theory assigns to scientific hypotheses.

The extent to which an explicatum meets Carnap's first requirement, demanding similarity to the explicans, will be constrained, however, by the three remaining requirements, which demand that the explicatum should function in a precise, comprehensive, and simple theory.[27]

It is particularly in these systematic requirements that the facet of epistemological valuation shows itself. Carnap presents some of them as general conditions of rationality. For example, he stipulates that the degree of rational credibility assigned to a hypothesis should be "dependent, not upon irrational factors like wishful or fearful thinking, but only on the totality of [the believer's] observational knowledge at the time ... "[28] This is a fundamental ideal of Carnap's theory, which implies that *P* must be a function of *H* and *E* alone. Another epistemological ideal is expressed in the area that the rational credibility of *H* on *E* should be definable exclusively in terms of purely logical attributes of *H* and *E*.

By reason of this restrictive requirement alone, Carnap's theory may be said to be adequate only to the critical appraisal of very simple kinds of hypotheses, but not to the complex considerations underlying the experimental testing of hypotheses or theories. The reason lies, briefly, in the Duhem–Quine argument, which has led to a holistic conception of scientific method and knowledge. The point of relevance to Carnap's view is that predictions of experimental findings cannot be deduced from the hypothesis under test alone, but only from the hypothesis taken in combination with an extensive system of other, previously accepted, hypotheses; broadly speaking, what evidence sentences are relevant to the hypothesis is determined by the entire theoretical system accepted at the time. If experimental findings conflict with theoretical predictions, some suitable adjustment has to be made, but not necessarily by rejecting the hypothesis ostensibly under test.

One consideration in choosing a suitable adjustment, it is often noted, is the desire to make a conservative change, one which changes the fundamental assumptions of the entire system as little as possible; another will be the concern to maintain or improve the simplicity and the systematic integration of the entire system. Thus, a rational decision as to whether the given hypothesis or another part of the system, or even the adverse experimental evidence itself, should count as discredited by a conflict between theory and new evidence will depend on considerations concerning the entire theoretical system and not only, as Carnap's requirement stipulates, on the hypothesis and the experimental evidence in question. The same difficulty faces, of course, the much narrower notions of verifiability and falsifiability.

It can be plausibly argued, however, that in various limited contexts of hypothesis-testing in science, such holistic considerations recede into the background, and the hypothesis is judged principally by the pertinent experimental evidence.[29]

Carnap's theory of rational credibility remains important as a carefully articulated explicatory model of the notion of rational credibility or of rational betting quotient for sufficiently simple testing or betting situations.

Whether adequate, more general and fully precise models can be constructed remains an open question. Thinkers favoring a pragmatist approach to the methodology of science have strong doubts on this score.

7. PRAGMATIST APPROACHES TO METHODOLOGY

Let us now take a glance at the pragmatist perspective on methodology, especially in the form it has been given by Thomas Kuhn.

Here, the emphasis on the descriptive facet of methodological claims is dominant and massive. An adequate methodological theory must be informed, on this view, by a close study of the history, sociology, and psychology of actual scientific research behavior. A proper descriptive and explanatory account of this kind can also, it is argued, provide methodological norms or standards of rationality for empirical inquiry.

Kuhn makes this basic assumption concerning the rational pursuit of knowledge: "Scientific behavior, taken as a whole, is the best example we have of rationality. Our view of what it is to be rational depends in significant ways, though of course not exclusively, on what we take to be the essential aspects of scientific behavior." Hence, "if history or any other empirical discipline leads us to believe that the development of science depends essentially on behavior that we have previously thought to be irrational, then we should conclude not that science is irrational, but that our notion of rationality needs adjustment here and there."[30]

As for the normative side of his methodology, Kuhn argues, briefly, as follows: "If I have a theory of how and why science works, it must necessarily have implications for the way in which scientists should behave if their enterprise is to flourish." Now, in the pursuit of their research, scientists behave in ways explored by descriptive and explanatory methodological studies. Those modes of scientific behavior have certain "essential functions," in particular what Kuhn calls the improvement of scientific knowledge. Hence, "in the absence of an alternate mode *that would serve similar functions*, scientists should behave essentially as they do if their concern is to improve scientific knowledge."[31]

It is for this reason that, in response to the question whether his methodological principles are to be read as descriptions or as prescriptions, Kuhn states that "they should be read in both ways at once".[32]

But the assignment of a prescriptive reading to a descriptive account of scientific research is not quite as straightforward as that. It presupposes epistemological idealization and valuation no less than does the formulation of a methodological theory by way of analytic explication. There are at least two reasons for this.

First, Kuhn's basic assumption that science is the best example we have of rationality expresses one broad epistemological valuation, a judgment as to what is to count as exemplary of the rational pursuit of knowledge. This valuation of scientific research as ranking highest on the rationality scale is posited and not further argued. Kuhn seems to suggest just this by his remark that he takes that judgment "not as a matter of fact, but rather of principle".[33]

Secondly, the behavior of scientists in the context of their professional work often shows facets that one would surely not regard as contributing to the improvement of scientific knowledge, but rather as interfering with it. Take, for example, the widespread intensive competition among specialists working in the same problem area and the familiar tendency it engenders to conceal their methods of approach and their unpublished results from one another. I very much doubt that Kuhn would want to see this kind of behavior included in a descriptive methodological account that may properly be given a prescriptive turn. And there are other features of actual scientific research behavior whose suitability for a prescriptive reading would require careful appraisal in light of prior epistemological values or conceptions as to what is "essential," as Kuhn puts it, to scientific progress. Some examples are the widespread practice of 'fudging' the evidence for a hypothesis and the less frequent outright faking of purported experiments and of experimental findings.

Kuhn seems to acknowledge this point when, in a passage quoted earlier in this section, he says that our notion of rationality depends on what we take to be the essential aspects of scientific behavior. The term 'essential' here surely refers to antecedently assumed – perhaps intuitively held – epistemological standards or values.

Thus, the justificatory grounds for methodological theories both of the naturalist and of the analytic-explicatory varieties have a descriptive facet and a facet reflecting epistemological valuation. Neither of the two construals of the methodology of science is purely *a priori*, and neither is purely descriptive.

8. 'DESIDERATA' AS IMPRECISE CONSTRAINTS ON SCIENTIFIC THEORY CHOICE

Despite the basic affinities we have considered, Kuhn's prescriptive methodology differs significantly from Carnap's or Popper's characterizations.

Analytic explicators aim at formulating precise general criteria for such contexts as the critical testing of hypotheses or the comparative appraisal of competing hypotheses or theories. Carnap based his rules of appraisal on his theory of rational credibility; Popper propounds rules for the game of science that are expressed in terms of precise concepts of falsifiability, corroboration, and the like.

Kuhn's pragmatist account of scientific research behavior, on the other hand, does not admit of a prescriptive reading in the form of a system of precise methodological rules.

This is clearly shown by Kuhn's characterization of the ways in which scientists appraise competing theories and eventually make a choice between them. Kuhn discusses this subject particularly for choices required in the context of a scientific revolution, when a paradigmatic theory that has long dominated research in its field is encountering mounting difficulties and is opposed by a new rival theory that has overcome some of those difficulties. Kuhn's ideas on the resolution of such conflicts are well known, and I will mention here only a few points that have an immediate bearing on the character of the prescriptive principles that might be gleaned from Kuhn's descriptive account.

Kuhn argues that the choice between competing theories is a matter left to the specialists in the field, whose appraisals of the merits of those theories are strongly influenced by certain shared preferences or values which have been molded in the course of their scientific training and their professional experiences. In particular, scientists widely agree in giving preference to theories exhibiting certain characteristics which have often been referred to in the methodological literature as "marks of a good hypothesis"; I will call them *desiderata* for short. Among them are the following: a theory should yield precise, preferably quantitative, predictions; it should be accurate in the sense that testable consequences derivable from it should be in good agreement with the results of experimental tests; it should be consistent both internally and with currently accepted theories in neighboring fields; it should have broad scope; it should predict phenomena that are novel in the sense of not having been known or taken into account when the theory was formulated; it should be simple; it should be fruitful.[34]

Kuhn reasons that while scientists are in general agreement concerning the importance of these features and attach great weight to them in deliberating about theory choice, the desiderata cannot be expressed in the form of precise rules of comparative evaluation which unambiguously single out one of two competing theories as the rationally preferable one. He arrives at this conclusion by arguing (A) that the individual desiderata are too vague to permit of explication in terms of precise criteria of accuracy, simplicity, scope, fruitfulness, etc., and (B) that even if precise criteria could be formulated for a comparison of two theories in regard to each of the desiderata, one of two competing theories might be superior to the other in regard to some of the desiderata, but inferior in regard to others: to permit overall comparison of the theories with regard to the totality of the desiderata, a further rule would therefore have to be constructed which, in effect, would assign different weights or priorities to the different desiderata. And again, Kuhn argues that it is not possible to formulate a precise and unambiguous rule of that kind which does sufficient justice to theory choice as actually practised in science.

A few words of amplification especially concerning (A). That the characterizations of the desiderata are vague is obvious. Some of them are also ambiguous. Kuhn notes, for example, that the requirement of simplicity favored the Copernican over the Ptolemaic theory if simplicity were judged by gross qualitative features, such as the number of circles required; but the two theories were substantially on a par if simplicity were judged by the ease of the computations required to predict the position of a planet.[35]

Or consider the desideratum of accuracy, which requires that "consequences deducible from a theory should be in demonstrated agreement with the results of existing experiments and observations."[36] Clearly, empirical data which a theory does not fit in this sense, should not count against the theory if the experiments yielding them were affected by factors not taken into account in the deduction, such as faults in the equipment or interference by disturbing outside factors. But a judgment as to whether such interference may have been present will depend on current theories as to what kinds of factors can affect the outcome of the given experiment. Thus, the desideratum of accuracy will have to be understood as requiring agreement between theoretically predicted findings and experimental data which, as judged by currently available information, are not vitiated by disturbing factors. This consideration leads to an extended form of holism: when a hypothesis conflicts with experimental evidence, then, as Duhem has pointed out, the conflict may be eliminated either by abandoning the hypothesis ostensibly under test, or by making changes elsewhere in the system of hypotheses accepted at the time; but, as just noted, there is also the possibility of rejecting the recalcitrant new evidence. It is not always the case that scientific theories are made to fit the observational or experimental data: often it is a well-established theory which determines whether given test findings can count as acceptable data.

This point is illustrated by the practice of eminent scientists. For example, the famous oil drop experiments by which Robert A. Millikan measured the charge of the electron yielded a number of instances in which the observed motion of the drops did not agree with Millikan's claim. He attributed the deviations to various possible disturbing factors and was in fact able, by screening some of these out, to obtain more uniform results. Yet, as Holton has shown by reference to Millikan's laboratory notes, there were observed cases of considerable deviation from the theoretically expected results, which Millikan did not include in his published data, assuming that something had gone wrong.[37]

The possibly disturbing factors mentioned by Millikan included fading of the battery that charged the condenser plates between which the oil drops were moving, partial evaporation and mass loss of an oil drop under observation, the observer's mistaking of dust particles for tiny oil drops, and several other possible occurrences. Broadly speaking, such sources of error could be checked and controlled by relevant knowledge that is quite independent of the hypothesis under test, which concerned the charge of the electron.

It might therefore seem reasonable to require that the attribution of adverse experimental findings to "disturbing factors" should never be based on the hypothesis under test, since otherwise any adverse evidence could be rejected simply on the ground that it conflicted with that hypothesis. Yet, this maxim is not generally adhered to in science. Thus, in certain deviant cases Millikan suggests, and offers some supporting reasons for, specific assumptions concerning the disturbing factors;[38] but there are other such cases recorded in his notebooks in which he simply comments "something wrong", "something the matter", "agreement poor", or the like.[39] And one might well argue that there was a good reason: in a large proportion of cases, the agreement between theoretically expectable and experimentally determined values was impressively close and gave grounds for the assumption that the hypothesis did exhibit a basic trait of nature.

These brief remarks were simply meant to illustrate that while in many cases, a judgment concerning the 'accuracy' of a theory may not pose great problems, it would be quite difficult to formulate a precise general criterion of accuracy of fit, which would take due account of the holistic character of scientific knowledge claims.[40]

The preceding considerations also have a bearing on Grünbaum's idea, mentioned earlier, that the warrant for a scientific claim lies in the evidence supporting it. A more detailed elaboration of this remark would call for consideration also of factors that are not evidence in the usual narrower sense; among them, I would think, features like the desiderata. Such a broad construal of the warrant of a hypothesis, however, leaves Grünbaum's point quite unaffected: a critic's hypothesized psychological resistance to some psychoanalytic doctrine is not a factor that has systematic relevance for the question whether the critic's objections are pertinent and well substantiated: indeed, as Grünbaum specifically notes in the passage quoted earlier, the critic's resistance might spring from an awareness of shortcomings in the systematic support that has been offered for the claims he is questioning.

As for the comments (A) and (B) outlined above concerning imprecision of the desiderata for good scientific theories, I think it of interest to note that quite similar views were expressed earlier by J. von Kries and by Ernest Nagel, and that Carnap agreed with them to some extent.[41]

Carnap (1950) expresses this view: "Inductive logic alone . . . cannot determine the best hypothesis on a given evidence, if the best hypothesis means that which good scientists would prefer. This preference is determined by factors of many different kinds, among them logical, methodological, and purely subjective factors" (p. 221). And he adds: "However, the task of inductive logic is not to represent all these factors, but only the logical ones; the methodological (practical, technological) and other nonlogical factors lie outside its scope" (p. 219). He then examines two among the factors mentioned by von Kries, which Carnap regards as purely logical, namely the extension and the variety of the confirming evidence for a hypothesis, and he sketches ways in which they might be given exact quantitative definitions; but he acknowledges that great difficulties remain for an attempt to include them into one precise quantitative concept of degree of confirmation (pp. 226ff.).

Against the idea that "in the logical analysis of science we should not make abstractions but deal with the actual procedures, observations, statements, etc., made by scientists", Carnap acknowledges that a pragmatic study of methodology is highly desirable (p. 217), but he warns that for the achievement of powerful results concerning sound decision-making, "the method which uses abstract schemata is the most efficient one" (p. 218).

Then he adds with characteristic candor that those who prefer to use powerful abstract methods are subject to "the ever present temptation to overschematize and oversimplify . . . ; the result may be a theory which is wonderful to look at in its exactness, symmetry, and formal elegance, and yet woefully inadequate for the tasks of application for which it is intended. (This is a warning directed at the author of this book by his critical superego.)" (p. 218.)

As we briefly noted, Carnap held that the question as to which of several hypotheses would be preferred by scientists on given evidence depended not only on logical characteristics of hypotheses and evidence, but also on methodological and on "purely subjective" factors. I do not know whether he thought it possible to offer precise explications of the relevant methodological considerations (which would require extra-logical concepts as well as logical ones), nor whether he would have considered a search for a general normative-explicatory account of comparative preferability of hypotheses as a promising project.

9. VALUATION, VAGUENESS, AND THE OBJECTIVITY OF SCIENCE

Desiderata of the kind we have considered have the character of epistemological norms or values. They do not enter into the content of scientific theories, but they serve as standards of acceptability or preferability for such theories; thus they function in the critical appraisal or in the justification of scientific claims. It is not to be wondered at that standards of evaluation are needed in this context: the problem of justifying theoretical claims can be intelligibly raised only to the extent that it is clear what objectives are to be achieved by accepting, or by according preference to, a theory.

Science is widely conceived as seeking to formulate an increasingly comprehensive, systematically organized, world view that is explanatory and predictive. It seems to me that the desiderata may best be viewed as attempts to articulate this conception somewhat more fully and explicitly. And if the goals of pure scientific research are indicated by the desiderata, then it is obviously rational, in choosing between two competing theories, to opt for the one which satisfies the desiderata better than its competitor.

The problem of formulating norms for the critical appraisal of theories may be regarded as a modern outgrowth of the classical problem of induction: it concerns the choice between competing comprehension theories rather than the adoption of simple generalizations and the grounds on which such adoption might be justified. And – disregarding for a moment the vagueness of the desiderata – the considerations sketched in the preceding paragraph might be viewed as *justifying* in a near-trivial way the choosing of theories in conformity with whatever constraints are imposed by the desiderata.

Note, however, that this kind of justification does not address at all what would be the central concern of the classical problem of induction, namely, the question whether there are any reasons to expect that a theory which, as judged by the desiderata, is preferable to its competitor, at a given time will continue to prove superior when faced with further, hitherto unexamined, occurrences in its domain.[42]

Since, at least so far, the desiderata can be formulated only vaguely, they do not unequivocally determine a choice between two theories. In particular they do not yield an algorithm which, in analogy to mathematical algorithms, effectively determines a unique solution for every problem within its domain. Indeed, the theory choices made by individual scientists committed to the desiderata are influenced also by factors that may differ from person to person, among them the scientists' individual construals of the desiderata as well as certain other factors which lie outside science and which may be more or less idiosyncratic and subjective.[43]

Does this jeopardize the objectivity of science? To be sure, idiosyncratic factors of the kinds just mentioned, as well as a variety of physical and sociocultural conditions, can affect individual choice behavior; they may all be relevantly invoked in explanatory accounts of decisions arrived at by particular investigators.

But that is equally true in cases where scientists seek to solve problems for which correct solutions can be characterized by precise, and perhaps even effective, criteria. A scientist working on a computational problem for which alternative methods of algorithmic solutions are available, may have a preference for, and may therefore employ, a particular one of these; a mathematical purist may after much effort produce an ingenious proof of a theorem in number theory which, unlike all previously available proofs, avoids any recourse to real or complex analysis. Consideration of such idiosyncratic factors is essential in explaining the mathematician's procedure; but it is irrelevant for a critical appraisal of the correctness of the computation or the validity of the proof. The criteria appropriate for the latter purpose are objective in the sense of making no reference whatever to individual preferences or values or to external circumstances.

In examining the objectivity of scientific inquiry, we will similarly have to ask whether, even in consideration of the new perspectives provided by pragmatist studies, scientific procedures including theory choice can still be characterized by standards that do not depend essentially on purely idiosyncratic individual factors.

I now think that plausible reasons can be offered in support of an objectivist but 'relaxed' rational reconstruction according to which proper scientific procedures are governed by methodological norms some of which are explicit and precise, while others – including very important ones – are vague. The requirements of deductive closure and of logical consistency for acceptable theories would be of the former kind; many other desiderata governing theory choice, of the latter.

A construal of this kind cannot, of course, claim to be a descriptive account of the practices actually observed by practitioners of an important socio-cultural pursuit broadly referred to as scientific research; the construal presupposes, as we saw other reconstructions must do, certain prior determinations, having the character of epistemic valuations, as to what peculiar features of that social enterprise are to count as characteristic of 'proper' science, as traits that make science scientific.

Let me briefly suggest some considerations that seem to me to favor a relaxed but objectivist construal of methodological principles.

To begin with – and this has to do with indicating important features of the explicandum – science is generally conceived as an objectivist enterprise where claims are subject to a critical appraisal in terms of standards that are not simply subjective and idiosyncratic; it is surely not regarded as a field in which 'everybody is entitled to his own opinion.'

Severe constraints are imposed by certain quite generally acknowledged norms. Among these are the demand for conformity with the standards of deductive logic, and the prohibition of logical inconsistencies: even though there is no general algorithmic test procedure for consistency, there is insistence on avoiding or in some way quarantining inconsistencies that may have been discovered. There are clear norms also for various methods of measurement and of testing statistical hypotheses.

And while norms like those represented by the desiderata are very much less explicit and precise, they surely do not license considerations that are idiosyncratic to some individual scientists as justificatory for theory choice.

The various desiderata can be said, it seems to me, to have an objectivist intent and to be amenable to discussion and possible further clarification.[44]

Empirical science, too, sometimes employs concepts which are characterized only vaguely, but whose application is not for that reason entirely arbitrary or a matter of purely subjective choice. Take, for example, the social status scale proposed by Chapin for rating American homes in regard to their socio-economic status.[45]

The total rating is based on appraisals of several component factors, each taken as a partial indicator of socio-economic status. Some of the factors are characterized quite precisely by criteria referring to the presence or absence in the living room of specified items such as hardwood floors, radios, etc., and their state of repair. On these, trained investigators will readily come to a good agreement in their ratings for a given home. Other component factors, however, are characterized much more vaguely; among them one that requires the investigator to express on a numerical rating scale his "general impression of good taste" shown in the living room. Here, the appraisal calls for ratings such as "Bizarre, clashing, inharmonious, or offensive (–4)" or "Attractive in a positive way, harmonious, quiet, and restful (+2)". On points of this kind, the 'reliability' of appraisals, as measured by the correlation between the ratings of different investigators, was expectably found to be lower than for items of the former kind. But it is reasonable to expect that by further training of the appraisers the reliability of their judgments could be enhanced. And the relevant training might well be effected with the help of paradigmatic examples rather than by means of fully explicit and precise criteria.

In a similar vein, Kuhn has stressed that, to a large extent, it is not by being taught rigorous definitions and rules that scientists learn how to apply their technical concepts and the formal apparatus of their theories to concrete problems in their field: they acquire that ability in considerable measure by being exposed to characteristic examples and by picking up other relevant clues in the course of their academic training and their professional experience.[46]

A relaxed objectivist construal of theory choice of the kind here adumbrated evidently falls far short of Carnap's rigorous conception of an explication (cf. Section 6 above). In particular, because of the vagueness of the desiderata, a relaxed account violates Carnap's requirement that the explicatum "be characterized by rules which have an *exact* form", and it does not satisfy well Carnap's condition that the explicatum be fruitful in the sense of permitting the construction of a precise and comprehensive theory. In these respects, a relaxed explication of rational theory choice bears no comparison to, say, Carnap's inductive logic as a precise explication of rational belief or of rational betting behavior. But it should be borne in mind that the virtues of that explication were achieved at the cost of strong idealization and simplification.

In some earlier papers,[47] I expressed the view that a methodological characterization of scientific theory choice as being essentially dependent on factors having the character of the desiderata does not warrant Kuhn's conception of scientific inquiry as being a rational procedure. I argued that a rational procedure consists in the pursuit of a specified goal in accordance with definite rules deliberately adopted on the ground that, in light of the available information, they offer the best prospects of attaining the goal. I concluded that insofar as scientific procedures are constrained only by considerations of the kind of the desiderata, they should be viewed as a-rational (though not as irrational), and I argued further that perhaps they might be qualified as latently functional practices in the sense of functionalist theories in the social sciences.[48]

That verdict would not apply, of course, to the many scientific procedures for which reasonably explicit and precise methodological standards can be formulated; some of these were mentioned earlier.

But in view of the considerations here outlined, it seems to me now that the characterization as a-rational or as latently functional does not do justice even to the broad process of theory choice; for it does not take sufficient account of the considerable role that precise and rule-governed reasoning does play in the critical appraisal of competing theories, which requires among other things a rigorous derivation of experimental implications and the performance of experimental tests that have to meet appropriate standards. But even the considerations adduced in appraising the satisfaction of the vaguer desiderata are typically perceived, it seems to me, as expressing not just individual taste or preference, but objective, if only roughly specified, traits of the competing theories. Thus, a less rigid construal of rationality may be indicated; the relaxed explication here adumbrated might be an attempt in this direction.

This conception leaves open the possibility that the methodology of science may gradually be formulated in terms of standards that are more explicit and precise than the desiderata we have considered. It also leaves room for the idea that the desiderata, which were taken here as reflecting the goal of scientific research or the idea of scientific progress, should be viewed, not as fixed once and for all, but as themselves subject to change in the course of the evolution of science.[49]

But I think it clear even at the present stage that scientific inquiry can be characterized by methodological principles which, while reflecting epistemological values, impose on scientific procedures and claims certain constraints of an objectively oriented, though partially vague, kind that preclude a view of science as an enterprise in which 'anything goes'.

University of Pittsburgh

NOTES

* This paper is based on work supported by the National Science Foundation under Grant No. SES–8025399.

[1] See, for example, Weber's essays, 'The Meaning of "Ethical Neutrality" in Sociology and Economics' and '"Objectivity" in Social Science and Social Policy' in Shils and Finch (1949). For a recent critical discussion of Weber's ideas, see Stegmüller (1979).

[2] Grünbaum (1972), p. 61; italics cited.

[3] Grünbaum (1980), p. 81; italics cited.

[4] Kuhn (1970b), p. 264.

[5] Dewey (1938), p. 9; italics cited.

[6] See, for example, the articles Hempel (1943) and (1945), which offer a logical analysis of the qualitative concept "*E* confirms *H*" for formalized languages of a simple kind; and Carnap's writings (1950; 1952; 1971a), which develop a comprehensive logical theory of the quantitative concept of degree of confirmation, or inductive probability.

[7] See, for example, Popper (1959), especially Appendix *IX; (1962), pp. 57–58; (1979), *passim.*

[8] See, for example, Hempel (1965), pp. 257–258 and pp. 469–472.

[9] Cf. Carnap (1956); Hempel (1973).

[10] Popper (1959), p. 44; italics cited.

[11] *Ibid.*, p. 52.

[12] *Ibid.*, p. 53; italics cited.

[13] *Ibid.*, p. 55.

[14] *Ibid.*

[15] See, for example, Popper (1962), pp. 33–39.

[16] Popper (1979), p. 356; italics and parentheses cited.

[17] I have here been concerned only with some general characteristics of Popper's methodology and have not considered his ideas of corroboration and of verisimilitude and other special issues. For a provocative discussion of the cognitive status of Popper's methodology, cf. Lakatos (1974) and Popper's reply (1974).

[18] See Carnap (1950), Chapter I.

[19] *Ibid.*, p. 7; italics cited.

[20] Carnap (1928).

[21] Carnap (1956), pp. 72–73.

[22] See, for example, Carnap (1963a), p. 967.

[23] Carnap (1963a); (1971a), pp. 13–16.

24 Carnap (1963a), pp. 978–979. The precise formulation of the relevant intuitive judgments and of their justificatory role is a rather subtle technical problem whose solution is based on results established by de Finetti, Kemeny, Shimony, and others. For details, see Kemeny (1963).
25 Goodman (1955), p. 67; italics cited.
26 Carnap (1963b), p. 990. See also Carnap's interesting remarks on p. 994 on differences and changes in intuitions concerning rational credibility.
27 In these respects, the explication of rational credibility by Carnap's precise theory of inductive probability is quite analogous to the explication of the concept of a correctly formed English sentence by a theory of English grammar. Such an explication must surely take account of the linguistic intuitions or dispositions of native speakers; but descriptive faithfulness has to be adjusted to the further objectives of constructing a grammar which is reasonably precise, simple, and general. For example, the idea that the conjunction of two well-formed sentences is again a well-formed sentence, and the idea that there must be an upper bound to the length of any well-formed sentence both have a certain intuitive appeal; but they are logically incompatible. Here, systematic-theoretical considerations will then decide which, if either, of the two is to be retained.
28 Carnap (1963a), p. 970; words in brackets supplied.
29 In this context, see Glymour's lucid and thought-provoking book (1980); it offers critical arguments against holism construed in the very comprehensive sense here adumbrated, and it propounds in detail a more restrictive account of the ways in which a hypothesis may be confirmed by empirical evidence.
30 Kuhn (1971), p. 144.
31 Kuhn (1970b), p. 237; italics cited.
32 *Ibid.*
33 Kuhn (1971), p. 144.
34 Kuhn (1977), pp. 321–322; also (1970a), pp. 205–206; (1970b), pp. 245–246. Van Fraassen (1980) discusses desirable characteristics of this kind under the heading of "virtues" of theories; see, for example, pp. 87–89.
35 Kuhn (1977), p. 324.
36 *Ibid.*, p. 321. The entire essay presents the ideas here referred to very suggestively. For other passages concerning those issues, see Kuhn (1970b), pp. 241, 245–246, 261–262; (1970a), pp. 199–200.
37 See Holton (1978). That essay deals with the fascinating controversy between Millikan and the physicist Felix Ehrenhaft. The latter had, in similar experiments, found a large number of cases that did not agree with Millikan's hypothesis and therefore rejected the latter. Millikan himself discusses the issue in his book (1917), especially Chapter VIII.
38 See Millikan (1917), pp. 165–172; also Holton (1978), p. 69.
39 Holton (1978), pp. 70–71.
40 Kuhn notes this point, for example, in (1977), p. 338.
41 Kries (1886), pp. 26, 29f; Nagel (1939), pp. 68–71; Carnap (1950), pp. 219–233.
42 Kuhn (1977); pp. 332–333 makes some remarks in a similar vein. It may be of interest to recall here that Reichenbach [cf. (1938), Section 43] proposed a principle of simplicity not just as a criterion for the *appraisal* of a given hypothesis, but as a rule for the inductive *discovery* of laws. Briefly, he argued that in the search for a law connecting several quantitative variables (such as temperature, pressure, and volume of

a gas), the method of always adopting the *simplest* hypothesis fitting the experimental data available at the time would, as the body of data grew, lead to a sequence of quantitative hypotheses which would mathematically converge on the law that actually connected the variables in question – provided there was such a law at all. In this justificatory argument for inductive reasoning to the simplest hypothesis, Reichenbach uses a rather precise characterization of the simplest curve, or surface, etc., through given data points. His method does not, however, take full account of the subtle and elusive questions concerning the simplicity of theories.

[43] Some such factors are briefly discussed in Kuhn (1977), p. 325; Laudan (1977) devotes a great deal of attention to considerations of this kind.

[44] It is broadly in this sense, I think, that Kuhn has recently characterized the application of the desiderata as a matter "of judgment, not taste" (1977, p. 337).

[45] Chapin (1935), Chapter XIX.

[46] See, for example, Kuhn (1974).

[47] Hempel (1979a; 1979b).

[48] Hempel (1979b), Section 8.

[49] Detailed arguments in favor of this conception have been offered by Laudan (1977); Kuhn (1977), too, allows for some changes of this kind; my discussion in (1979a), Section 6 deals briefly with this issue.

REFERENCES

Carnap, Rudolf. 1928. *Der Logische Aufbau der Welt*. Berlin-Schlachtensee: Weltkreis-Verlag. (English edition 1967, *The Logical Structure of the World and Pseudoproblems in Philosophy*. Berkeley: University of California Press.)

Carnap, Rudolf. 1950. *Logical Foundations of Probability*. Chicago: The University of Chicago Press.

Carnap, Rudolf. 1952. *The Continuum of Inductive Methods*. Chicago: The University of Chicago Press.

Carnap, Rudolf. 1956. 'The Methodological Character of Theoretical Concepts.' In Feigl, H. and M. Scriven (eds.), *Minnesota Studies in the Philosophy of Science*, vol. 1, pp. 38–76. Minneapolis: University of Minnesota Press.

Carnap, Rudolf. 1963a. 'My Basic Conceptions of Probability and Induction.' In Schilpp (1963), pp. 966–979.

Carnap, Rudolf. 1963b. 'Ernest Nagel on Induction.' In Schilpp (1963), pp. 989–995.

Carnap, Rudolf. 1971a. 'Inductive Logic and Rational Decisions.' In Carnap and Jeffrey (1971), pp. 5–31.

Carnap, Rudolf. 1971b. 'A Basic System of Inductive Logic, Part I.' In Carnap and Jeffrey (1971), pp. 35–165.

Carnap, Rudolf and Richard C. Jeffrey (eds.). 1971. *Studies in Inductive Logic and Probability*. Berkeley: University of California Press.

Chapin, F. Stuart. 1935. *Contemporary American Institutions*. New York and London: Harper and Brothers.

Dewey, John. 1938. *Logic: The Theory of Inquiry*. New York: Holt.

Fraassen, Bas C. van. 1980. *The Scientific Image*. Oxford: Clarendon Press.

Glymour, Clark N. 1980. *Theory and Evidence*. Princeton, N.J.: Princeton University Press.

Goodman, N. 1955. *Fact, Fiction and Forecast*. Cambridge, Mass.: Harvard University Press.

Grünbaum, Adolf. 1972. 'Free Will and Laws of Human Behavior.' In Feigl, H., K. Lehrer, and W. Sellars (eds.), *New Readings in Philosophical Analysis*, pp. 605–627. New York: Appleton-Century-Crofts.

Grünbaum, Adolf. 1980. 'The Role of Psychological Explanations of the Rejection or Acceptance of Scientific Theories,' *Transactions of the New York Academy of Sciences*, Series II, **19**, 75–90.

Hempel, Carl G. 1943. 'A Purely Syntactical Definition of Confirmation,' *Journal of Symbolic Logic* **8**, 122–143.

Hempel, Carl G. 1945. 'Studies in the Logic of Confirmation,' *Mind* **54**, 1–26, 97–121.

Hempel, Carl G. 1965. *Aspects of Scientific Explanation*. New York: The Free Press.

Hempel, Carl G. 1973. 'The Meaning of Theoretical Terms: A Critique of the Standard Empiricist Construal.' In Suppes, P. *et al.* (eds.), *Logic, Methodology and Philosophy of Science* **IV**, pp. 367–378. Amsterdam: North-Holland Publishing Company.

Hempel, Carl G. 1979a. 'Scientific Rationality: Analytic vs. Pragmatic Perspectives.' In Geraets, Th. F. (ed.), *Rationality Today/La rationalité aujourd'hui*, pp. 46–58. Ottawa: The University of Ottawa Press.

Hempel, Carl G. 1979b. 'Scientific Rationality: Normative vs. Descriptive Construals.' In Berghel, H., *et al.* (eds.), *Wittgenstein, the Vienna Circle, and Critical Rationalism*, pp. 291–301. Vienna: Hoelder-Pichler-Tempsky.

Holton, Gerald. 1978. 'Subelectrons, Presuppositions, and the Millikan-Ehrenhaft Dispute.' In Holton, G., *The Scientific Imagination: Case Studies*, pp. 25–83. Cambridge: Cambridge University Press.

Kemeny, John G. 1963. 'Carnap's Theory of Probability and Induction.' In Schilpp, pp. 711–738.

Kries, Johannes von. 1886. *Die Principien der Wahrscheinlichkeitsrechnung*. Freiburg: Akademische Verlagsbuchhandlung; 2nd ed. Tübingen: Mohr, 1927.

Kuhn, Thomas S. 1970a. *The Structure of Scientific Revolutions*. 2nd ed. Chicago: The University of Chicago Press.

Kuhn, Thomas S. 1970b. 'Reflections on My Critics.' In Lakatos, I. and A. Musgrave (eds.), *Criticism and the Growth of Knowledge*, pp. 231–278. Cambridge: Cambridge University Press.

Kuhn, Thomas S. 1971. 'Notes on Lakatos.' In Roger C. Buck and Robert S. Cohen (eds.), *PSA 1970, Proceedings of the 1970 Biennial Meeting, Philosophy of Science Association*, pp. 137–146. *Boston Studies in the Philosophy of Science*, vol. 8. Dordrecht: D. Reidel Publishing Company.

Kuhn, Thomas S. 1974. 'Second Thoughts on Paradigms.' In Suppe, F. (ed.), *The Structure of Scientific Theories*, pp. 459–482. Urbana: University of Illinois Press.

Kuhn, Thomas S. 1977. 'Objectivity, Value Judgment, and Theory Choice.' In Kuhn, T. S., *The Essential Tension*, pp. 320–339. Chicago: The University of Chicago Press.

Lakatos, Imre. 1974. 'Popper on Demarcation and Inductive.' In Schilpp (1974), pp. 241–273.

Laudan, Larry. 1977. *Progress and Its Problems*. Berkeley: University of California Press.

Millikan, Robert Andrews. 1917. *The Electron*. Chicago: The University of Chicago Press. Facsimile edition: Chicago: The University of Chicago Press, 1963.

Nagel, Ernest. 1939. *Principles of the Theory of Probability*. Chicago: The University of Chicago Press.
Popper, Karl R. 1959. *The Logic of Scientific Discovery*. London: Hutchinson.
Popper, Karl R. 1962. *Conjectures and Refutations*. New York: Basic Books.
Popper, Karl R. 1974. 'Lakatos on the Equal Status of Newton's and Freud's Theories.' In Schilpp (1974), pp. 999–1013.
Popper, Karl R. 1979. *Objective Knowledge*. Rev. ed. Oxford: Clarendon Press.
Quine, W. V. 1969. *Ontological Relativity and Other Essays*. New York and London: Columbia University Press.
Reichenbach, Hans. 1938. *Experience and Prediction*. Chicago: The University of Chicago Press.
Schilpp, Paul A. (ed.), 1963. *The Philosophy of Rudolf Carnap*. La Salle, Illinois: Open Court.
Schilpp, Paul A. (ed.). 1974. *The Philosophy of Karl Popper*. 2 vols. La Salle, Illinois: Open Court.
Shils, Edward A. and Henry A. Finch, translators and editors. 1949. *Max Weber on the Methodology of the Social Sciences*, Glencoe, Illinois: The Free Press.
Stegmüller, Wolfgang. 1979. 'Wertfreiheit, Interessen und Objectivität.' In W. Stegmüller, *Rationale Rekonstruktion von Wissenschaft und ihrem Wandel*, pp. 175–203. Stuttgart: P. Reclam.

ALLEN I. JANIS

SIMULTANEITY AND CONVENTIONALITY *

The role that conventionality plays in the standard definition of simultaneity in the special theory of relativity has been the subject of much debate. I should like to review certain aspects of that debate, and suggest an analogy between an argument thought by many to show the nonconventionality of standard synchrony and an earlier argument that was not generally thought to be damaging to the conventionalist point of view.

We may think about the simultaneity problem in the following context. Let A and B denote two distinct spatial locations fixed in an inertial frame F. Let E be an event that occurs at B. The problem is to decide what event at A is simultaneous (relative to F) with E. Equivalently, one can ask how to synchronize (ideal) clocks at rest at A and B.

Einstein's approach to the problem (Einstein, 1905, pp. 38–40 of the English translation; however, note Scribner, 1963, for correction of an error in the translation) was equivalent to the following treatment of an idealized thought experiment. Suppose that a ray of light (traveling in vacuum) leaves A at time t_1 (as measured by the clock at rest there), chosen so that the ray's arrival at B coincides with E. Let this ray be instantaneously reflected back to A, arriving at time t_2. Then standard synchrony is defined by saying that E is simultaneous with the event at A whose time coordinate is $(t_1 + t_2)/2$. This definition amounts to requiring that the speed of the ray from A to B is the same as for the return trip from B to A.

The view that the choice of standard synchrony is a convention, rather than being singled out by facts about the physical universe, has been argued particularly by Reichenbach (see, for example, Reichenbach, 1958, pp. 123–135) and Grünbaum (see, for example, Grünbaum, 1973, pp. 342–368). Briefly stated, these arguments assert that the only nonconventional basis for claiming that two distinct events are not simultaneous would be the possibility of a causal influence connecting the events. If the universe were Newtonian, the possibility of arbitrarily fast causal influences would single out a unique event at A as being simultaneous with E. Given, however, that no causal influence can propagate with a speed greater than that of light in vacuum, any event at A whose time coordinate is in the open interval between t_1 and t_2 could be defined to be simultaneous with E. In terms of

R. S. Cohen and L. Laudan (eds.), Physics, Philosophy and Psychoanalysis, 101–110.

Reichenbach's ϵ-notation, the time coordinate t_E of the event at A that is simultaneous with E would be given by $t_E = t_1 + \epsilon (t_2 - t_1)$, where the choice of any particular ϵ in the range $0 < \epsilon < 1$ is a matter of convention; standard synchrony, of course, corresponds to the choice $\epsilon = 1/2$.

In opposition to this conventionalist point of view, many claims of purportedly convention-free ways (or, at least, ways free of nontrivial conventions) of choosing a particular value of ϵ have been made and debated in the literature. I should like to comment on certain aspects of this debate, without attempting a comprehensive review of the subject.

Let me begin with what I shall call the simplicity argument. According to this argument, the choice $\epsilon = 1/2$ should be made on the basis of simplicity. Only that choice makes the speed of light isotropic. Indeed, one can define simultaneity (as, in fact, Einstein did) by requiring equal transit times for light in opposite directions, and never mention ϵ at all. Thus only the relation of equality is used, and the question of choosing a particular value for a parameter (or introducing a particular number without referring to an arbitrary parameter) never arises.

The counterargument is that, since the equality of the one-way speeds of light is itself a convention, this choice does not simplify the postulational basis of the theory. In rejecting the simplicity argument, Grünbaum has written (Grünbaum, 1973, p. 356):

> On the contrary, it is the postulated fact that light is the fastest signal which assures that *each one* of the permissible values of ϵ will be equally compatible with all possible matters of fact which are independent of how we decide to set the clock at [B]. Thus, the value $\epsilon = 1/2$ is not simpler than the other values in the inductive sense of assuming less in order to account for our observational data, but only in the descriptive sense of providing a *symbolically* simpler representation of the theory explaining these data.

In effect, the conventionalist argument is that the whole range of possible values of ϵ is implicit in the physics of the situation, even though it is possible to frame the definition of standard synchrony in ways that tend to conceal this fact.

Many of the arguments against the conventionalist view have been based on particular physical methods of establishing simultaneity (or, equivalently, measuring the one-way speed of light). Salmon (1977) discusses a number of such methods, and argues that each involves a nontrivial convention. For example, one of the methods he discusses makes use of the law of conservation of linear momentum to conclude that two particles of equal mass, situated initially at the midpoint between A and B and separated by an explosion, must arrive at A and B simultaneously. But, as Salmon points out (Salmon, 1977, p. 273), "The confirmation – indeed, the very enunciation – of this law requires the use of the concept of one-way velocity." Since one-way velocities cannot be determined without synchronized clocks at the two ends of the spatial interval that is traversed (or something equivalent to such synchronization), it is circular to use conservation of momentum to define simultaneity.

It seems to me that this whole class of arguments based on the use of physical phenomena can be approached in the following way. Note first of all that these arguments describe phenomena in ways that are compatible with the standard formulations of physics; to do otherwise would be to assert at the outset that standard physics is untenable. (I should perhaps emphasize here that, as part of the conceptual framework of this essay, I assume the correctness of the special theory of relativity, including at least the possibility of standard synchrony. Challenges to special relativity are beyond the scope of this work.) These standard formulations, in turn, are compatible with (and may even require, as the conservation-of-momentum example illustrates) standard synchrony. Since the claim of these arguments is that they lead to a unique choice of synchrony, it thus follows that this choice must be the standard one. It becomes instructive, then, to imagine what would happen if these arguments and their associated phenomena were to be described from the point of view of a nonstandard synchrony.

Since such a description might involve more than two fixed spatial locations, let me explain precisely what I mean by nonstandard synchrony in a more general context. Let us start with a standard Minkowskian coordinate system (x, y, z, t) associated with the inertial frame F. Consider an arbitrary family of nonintersecting, spacelike hyperplanes, and let the members of the family be labeled by a parameter t'. We may now use this parameter in defining simultaneity; that is, events will be defined to be simultaneous if and only if they lie in the same hyperplane t' = constant. This definition of simultaneity contains the one stated previously in terms of Reichenbach's ϵ: for any two fixed spatial locations A and B (i.e., having fixed values of x, y, and z), and any event E at B, the event at A having the same value of t' as E will satisfy the Reichenbach definition of simultaneity for some value of ϵ in the range $0 < \epsilon < 1$; that ϵ is in this range is guaranteed by the spacelike nature of the hyperplanes. If the hyperplanes t' = constant form the same family as the hyperplanes t = constant, the synchrony is the standard one; otherwise the synchrony is nonstandard.

Let me make two observations concerning this definition of nonstandard synchrony. The first is that the hyperplanes of simultaneity, $t' =$ constant, corresponding to a nonstandard synchrony are the same hyperplanes that would give standard synchrony in a suitably chosen inertial frame different from F. It should be emphasized, however, that the question of conventionality of simultaneity being discussed in this essay is posed solely from the point of view of any one inertial frame F. The second observation is that for some purposes a more general definition of nonstandard synchrony might be useful, one in which the hypersurfaces of simultaneity need not be hyperplanes. This greater generality is not needed here, however.

Consider, now, any of the arguments based on particular physical phenomena, and imagine that a complete description of everything that occurs is given in terms of standard synchrony. It is clear that an alternative description could be given in terms of an arbitrary nonstandard synchrony; the only question is whether such a description violates any basic (nonconventional) postulate of physical theory. An example of how a basic postulate might be violated would be if, instead of using a nonstandard synchrony in the sense I have defined, one were to choose a synchrony corresponding to an ϵ outside the range $0 < \epsilon < 1$. Such a description would permit a time-labeling in which effects precede their causes. By contrast, it seems to me that any appeal to physical laws to rule out a nonstandard synchrony is necessarily doomed to failure. Rather than the laws ruling out nonstandard synchrony, the description in terms of nonstandard synchrony indicates how the laws should be reformulated if such a synchrony is adopted. As an example, consider the conservation-of-momentum argument cited earlier. With nonstandard synchrony, the correct formulation of this law would imply that two particles of equal mass and zero total momentum would not move with equal speeds in opposite directions, but rather with the speeds that one would discover by transforming equal speeds with standard synchrony to the given nonstandard synchrony. Other examples may be more subtle, but the conclusion should still hold that the argument cannot show standard synchrony to be nonconventional.

To look at this from another point of view, recall that nonstandard synchrony in an inertial frame F assigns events the same time coordinates (except possibly for trivial changes) that they would be assigned by observers using standard synchrony in a different inertial frame. Thus nonstandard synchrony cannot yield time assignments that are inconsistent with the events themselves, but only possibly inconsistent with the interpretation of these events that is used in F. If this interpretation is based on physical laws that are discovered and formulated within the framework of standard synchrony, then any resulting argument that nonstandard synchrony is not allowed will be circular.

Let me give one further example of how thinking about a particular set of phenomena in terms of nonstandard synchrony can show that what may seem at first sight to be a convincing argument against the conventionalist point of view must, in fact, be invalid. The argument runs as follows. A ray of light travels from *A* to *B* and back to *A*. The experimenter has the option of inserting a block of glass in the path of the light ray during either (or both) of the one-way trips. Thus, with a single clock at *A*, four round-trip times can be measured, corresponding to the ray traveling solely through vacuum on both legs of the trip, through glass only on the first leg, through glass only on the return leg, and through glass on both legs. It is a straightforward matter to write an equation expressing each round-trip time in terms of the known distance between *A* and *B* and the two unknown one-way velocities on the two legs of the trip; this equation is easily seen to be linear in the reciprocals of the velocities. By use of the four measured round-trip times, one thus obtains four linear equations for the reciprocals of the four unknown one-way velocities (the velocities with or without glass on each leg of the trip). One then finds the four one-way velocities, without use of conventions, by solving these equations.

Let us begin the analysis of this argument by considering a clock at *B*. Suppose first that this clock is in standard synchrony with the one at *A*. It is clear that the one-way velocities found with the aid of the clocks at *A* and *B* will be consistent with the round-trip times measured at *A*, and consequently must be solutions of the set of equations. Suppose now that the clocks at *A* and *B* are put into nonstandard synchrony. It is still clear that the resulting one-way velocities are consistent with the round-trip times measured at *A*, for any round-trip time will still be the sum of the two corresponding one-way times. Thus this new set of one-way velocities must also satisfy the equations. Thus, by thinking in terms of a nonstandard synchrony, it is clear that the argument cannot possibly single out a unique synchrony. Furthermore, since we have a case of four linear equations for four unknowns admitting an infinitude of solutions, it is clear that the equations must not be independent; this is easily seen to be the case.

An entirely different sort of argument against the conventionalist point of view has been given by Malament (1977), who argues that standard synchrony is the only simultaneity relation that can be defined, relative to a given inertial frame, from the relation of (symmetric) causal connectibility. He makes this argument by proving two propositions. The first of these shows that standard synchrony is so definable. The second proposition establishes the uniqueness claim, provided that "minimal, seemingly innocuous conditions are imposed" (Malament, 1977, p. 297). It is on this second proposition that I should like to focus.

Let me begin by mentioning that not everyone would agree that unique definability is necessary for a simultaneity relation to be nonconventional. It might be argued, on the contrary, that a relation is nonconventional if it is definable in terms of structures that are intrinsic to Minkowski space-time, even in the absence of uniqueness. Taking this point of view, Spirtes (1981, Ch. VI, Sec. F) has argued that Malament's first proposition shows that standard synchrony is nonconventional, the import of Malament's second proposition being that nonstandard synchronies are conventional if symmetric causal connectibility is the only intrinsic relation in Minkowski space-time. If, on the other hand, space-time is temporally oriented, then Spirtes shows the existence of infinitely many nonconventional simultaneity relations. I prefer a different view of conventionality. If faced with more than one 'nonconventional' relation, on what basis can one of them be singled out for use other than by making a conventional choice? In such a situation, I would say that the choice of standard (or some other) synchrony was conventional. It thus seems to me that any argument asserting that Malament's results demonstrate the nonconventionality of standard synchrony in an unoriented space-time must include his second proposition as an essential ingredient.

Let me now give some definitions that will facilitate the statement of Malament's second proposition. Let $p = (p_0, p_1, p_2, p_3)$ and $q = (q_0, q_1, q_2, q_3)$ be points of R^4. The Minkowski inner product on R^4 is defined by $(p, q) = p_0q_0 - p_1q_1 - p_2q_2 - p_3q_3$, and the norm on R^4 is defined by $|p| = (p, p)$. The two-place relation κ of (symmetric) causal connectibility is defined by $p \kappa q \equiv |p - q| \geqslant 0$. Thus, causal connectibility of two events means that a signal traveling with a speed not exceeding that of light (in vacuum) can pass between them. Let O be the time-like line of some inertial observer, and let Sim_O be the standard simultaneity relation relative to O. A bijective map $f\colon R^4 \longrightarrow R^4$ is called an O causal automorphism iff for all points p and q in R^4 it satisfies the two conditions $p \kappa q$ iff $f(p) \kappa f(q)$ and $p \in O$ iff $f(p) \in O$. Finally, a two-place relation S on R^4 is said to be implicitly definable from κ and O iff for all O causal automorphisms f and all points p and q of R^4 it satisfies the condition $S(p, q)$ iff $S(f(p), f(q))$.

In terms of these definitions, Malament (1977, pp. 297–8) states his second proposition as follows: "Suppose S is a two-place relation on R^4

where (i) S is (even just) implicitly definable from κ and O; (ii) S is an equivalence relation; (iii) S is non-trivial in the sense that there exist points $p \in O$ and $q \notin O$ such that $S(p, q)$. Then S is either Sim_O or the universal relation (which holds of all points)."

Recall the definition of simultaneity I gave earlier in terms of space-like hyperplanes t' = constant. It is clear that any of the standard or nonstandard simultaneity relations so defined satisfy conditions (ii) and (iii) of Malament's second proposition. Thus it is condition (i) that singles standard synchrony out from among this whole class of possible synchronies. That it does so is not surprising, since the family of hyperplanes corresponding to standard synchrony is invariant (in the sense that each hyperplane of the family is mapped either onto itself or onto another member of the family) under the class of O causal automorphisms, whereas that corresponding to any of the nonstandard synchronies is not.

Consider now the following way of defining a nonstandard synchrony in an inertial frame F. First use the relation of causal connectibility to define standard synchrony relative to an observer O at rest in F. This definition, whose possibility is guaranteed by Malament's first proposition, amounts to a definition of the family of hyperplanes that I have called the t = constant hyperplanes: to each $p \in O$ there corresponds exactly one member of the family, consisting of all q such that $\mathrm{Sim}_O\,(p, q)$. The hyperplanes t' = constant corresponding to a nonstandard synchrony can now be defined by specifying their orientation relative to the t = constant hyperplanes. This orientation can be specified by three numbers, which could, for example, be the three components of the velocity relative to F of an inertial frame in which the t' = constant hyperplanes provide standard synchrony. (In one spatial dimension, a single number suffices, which is equivalent to Reichenbach's ϵ. The additional numbers required in more dimensions correspond to a direction-dependent ϵ.) If we introduce the three dimensionless parameters β_x, β_y, and β_z as the ratios of these velocity components to the speed of light in vacuum, then the only thing needed to define a nonstandard synchrony beyond what is needed to define standard synchrony is to specify the numerical values of three parameters, each of which is restricted to the doubly open interval $(-1, 1)$. (In one spatial dimension, the single parameter β that is needed can be related to ϵ by $\beta = 2\epsilon - 1$, at least if appropriate trivial conventions are adopted.)

It is of interest to see another way of defining nonstandard synchrony from standard synchrony. In a frame F' in which the hyperplanes t' = constant provide standard synchrony, the original world-line O corresponds to an observer in uniform motion. The relation Sim_O thus is a nonstandard relation in F'. Similarly, in the original frame F one could choose a world-line O' corresponding to an observer at rest in F', and Malament's definition of $\mathrm{Sim}_{O'}$ would yield a nonstandard relation in F. Thus we have two equivalent ways of defining the nonstandard synchrony in F by using Malament's definition of standard synchrony. In the first, we start with standard synchrony relative to O and then specify the parameters that yield the nonstandard synchrony in terms of the hyperplanes $t' =$ constant. In the second, we start by specifying the parameters that yield the world-line O' and then use Malament's definition of standard synchrony relative to O'. Since O' can be defined in terms of the same parameters used to define the $t' =$ constant family of hyperplanes, it is clear that these two approaches incorporate the same ingredients.

If a particular nonstandard synchrony has been defined by this second approach, then a proof of its uniqueness can be given analogous to that of Malament's second proposition. We define O' causal automorphisms and implicit definability from κ and O' by substituting O' for O in the definitions of O causal automorphisms and implicit definability from κ and O, respectively. If we then substitute O' for O in Malament's conditions (i) and (iii), the resulting proposition can be proved by making the same substitution in the appropriate steps of Malament's proof.

The situation can perhaps be more simply described as follows. Malament has shown that if a simultaneity relation is implicitly definable from κ and O (and if it satisfies his conditions (ii) and (iii) and is not the universal relation), then it must be the relation Sim_O. But from the point of view of an inertial frame in which O does not represent rest, Sim_O is a nonstandard simultaneity relation. Thus the only ingredients necessary to define a nonstandard synchrony are those necessary to define Sim_O (viz κ and O) plus a set of numbers that specify O' relative to O.

It thus appears to me that if one wishes to use Malament's propositions to argue that the unique choice of standard synchrony is not conventional, it must be on the basis that it is an artificial complication to introduce the parameters that specify a particular O', for it is only their introduction that is needed to specify a particular nonstandard synchrony. Such an argument is reminiscent of what I earlier referred to as the simplicity argument. The argument there was that the introduction of ϵ was an artificial complication. A definition equivalent to that for $\epsilon = 1/2$ could be given without mentioning ϵ, or any particular number, at all. The argument in the present case is likewise that a definition equivalent to that for $\epsilon = 1/2$ (which corresponds to $\beta = 0$)

can be given without mentioning ϵ (or β), or any particular number, at all. (I refer here to a single parameter since the context of the earlier discussions involved two fixed spatial locations, which define a single spatial direction. Both arguments are easily generalized to more dimensions.)

What I should like to suggest, then, is that requiring implicit definability from κ and O (as opposed to implicit definability from κ and O', where the value of a parameter would have to be given in order to specify which O' is meant) is not very different from requiring equal speeds of light in opposite directions (as opposed to unequal speeds, where the value of a parameter would have to be given in order to specify which choice of unequal speeds is meant). If this suggestion is valid, then it seems to me that an argument to the effect that Malament's results show the nonconventionality of standard simultaneity should carry neither more nor less weight than the earlier simplicity argument: those that found the simplicity argument unconvincing should be similarly unconvinced by an argument based on Malament's results; those that are persuaded by Malament's results should have been similarly persuaded by the simplicity argument. If not, wherein lies the essential difference?

University of Pittsburgh

NOTE

* My thinking about the general subject of this essay has benefited greatly from many conversations with Adolf Grünbaum and Wesley Salmon; any errors, however, are original with me. It is a particular pleasure to dedicate this essay, with admiration and affection, to Adolf Grünbaum in honor of his 60th birthday.

REFERENCES

Einstein, A. 1905. 'Zur Elektrodynamik bewegter Körper,' *Annalen der Physik* 17, 891–921. English translation in *The Principle of Relativity*, pp. 35–65. New York: Dover, 1952.

Grünbaum, A. 1973. *Philosophical Problems of Space and Time*. 2nd enlarged ed. *Boston Studies in the Philosophy of Science*, vol. 12. Dordrecht, Boston: D. Reidel.

Malament, D. 1977. 'Causal Theories of Time and the Conventionality of Simultaneity,' *Noûs* 11, 293–300.

Reichenbach, H. 1958. *The Philosophy of Space and Time*. New York: Dover.

Salmon, W. 1977. 'The Philosophical Significance of the One-Way Speed of Light,' *Noûs* 11, 253–292.

Scribner, C. 1963. 'Mistranslation of a Passage in Einstein's Original Paper on Relativity,' *American Journal of Physics* 31, 398.
Spirtes, P. 1981. 'Conventionalism and the Philosophy of Henri Poincaré.' Ph. D. Dissertation, University of Pittsburgh.

LARRY LAUDAN

THE DEMISE OF THE DEMARCATION PROBLEM *

1. INTRODUCTION

We live in a society which sets great store by science. Scientific 'experts' play a privileged role in many of our institutions, ranging from the courts of law to the corridors of power. At a more fundamental level, most of us strive to shape our beliefs about the natural world in the 'scientific' image. If scientists say that continents move or that the universe is billions of years old, we generally believe them, however counter-intuitive and implausible their claims might appear to be. Equally, we tend to acquiesce in what scientists tell us not to believe. If, for instance, scientists say that Velikovsky was a crank, that the biblical creation story is hokum, that UFOs do not exist, or that acupuncture is ineffective, then we generally make the scientist's contempt for these things our own, reserving for them those social sanctions and disapprobations which are the just deserts of quacks, charlatans and con-men. In sum, much of our intellectual life, and increasingly large portions of our social and political life, rest on the assumption that we (or, if not we ourselves, then someone whom we trust in these matters) can tell the difference between science and its counterfeit.

For a variety of historical and logical reasons, some going back more than two millennia, that 'someone' to whom we turn to find out the difference usually happens to be the philosopher. Indeed, it would not be going too far to say that, for a very long time, philosophers have been regarded as the gatekeepers to the scientific estate. They are the ones who are supposed to be able to tell the difference between real science and pseudo-science. In the familiar academic scheme of things, it is specifically the theorists of knowledge and the philosophers of science who are charged with arbitrating and legitimating the claims of any sect to 'scientific' status. It is small wonder, under the circumstances, that the question of the nature of science has loomed so large in Western philosophy. From Plato to Popper, philosophers have sought to identify those epistemic features which mark off science from other sorts of belief and activity.

Nonetheless, it seems pretty clear that philosophy has largely failed to deliver the relevant goods. Whatever the specific strengths and deficiencies of

R. S. Cohen and L. Laudan (eds.), Physics, Philosophy and Psychoanalysis, 111–127.

the numerous well-known efforts at demarcation (several of which will be discussed below), it is probably fair to say that there is no demarcation line between science and non-science, or between science and pseudo-science, which would win assent from a majority of philosophers. Nor is there one which *should* win acceptance from philosophers or anyone else; but more of that below.

What lessons are we to draw from the recurrent failure of philosophy to detect the epistemic traits which mark science off from other systems of belief? That failure might conceivably be due simply to our impoverished philosophical imagination; it is conceivable, after all, that science really is *sui generis*, and that we philosophers have just not yet hit on its characteristic features. Alternatively, it may just be that there are no epistemic features which all and only the disciplines we accept as 'scientific' share in common. My aim in this paper is to make a brief excursion into the history of the science/non-science demarcation in order to see what light it might shed on the contemporary viability of the quest for a demarcation device.

2. THE OLD DEMARCATIONIST TRADITION

As far back as the time of Parmenides, Western philosophers thought it important to distinguish knowledge (*episteme*) from mere opinion (*doxa*), reality from appearance, truth from error. By the time of Aristotle, these epistemic concerns came to be focussed on the question of the nature of *scientific* knowledge. In his highly influential *Posterior Analytics*, Aristotle described at length what was involved in having scientific knowledge of something. To be scientific, he said, one must deal with causes, one must use logical demonstrations, and one must identify the universals which 'inhere' in the particulars of sense. But above all, to have science one must have *apodictic certainty*. It is this last feature which, for Aristotle, most clearly distinguished the scientific way of knowing. What separates the sciences from other kinds of beliefs is the infallibility of their foundations and, thanks to that infallibility, the incorrigibility of their constituent theories. The first principles of nature are directly intuited from sense; everything else worthy of the name of science follows demonstrably from these first principles. What characterizes the whole enterprise is a degree of certainty which distinguishes it most crucially from mere opinion.

But Aristotle sometimes offered a second demarcation criterion, orthogonal to this one between science and opinion. Specifically, he distinguished between know-how (the sort of knowledge which the craftsman and the engineer possess) and what we might call 'know-why' or demonstrative understanding (which the scientist alone possesses). A shipbuilder, for instance, knows how to form pieces of wood together so as to make a seaworthy vessel; but he does not have, and has no need for, a syllogistic, causal demonstration based on the primary principles or first causes of things. Thus, he needs to know that wood, when properly sealed, floats; but he need not be able to show by virtue of what principles and causes wood has this property of buoyancy. By contrast, the scientist is concerned with what Aristotle calls the "reasoned fact"; until he can show why a thing behaves as its does by tracing its causes back to first principles, he has no scientific knowledge of the thing.

Coming out of Aristotle's work, then, is a pair of demarcation criteria. Science is distinguished from opinion and superstition by the certainty of its principles; it is marked off from the crafts by its comprehension of first causes. This set of contrasts comes to dominate discussions of the nature of science throughout the later Middle Ages and the Renaissance, and thus to provide a crucial backdrop to the re-examination of these issues in the seventeenth century.

It is instructive to see how this approach worked in practice. One of the most revealing examples is provided by pre-modern astronomy. By the time of Ptolemy, mathematical astronomers had largely abandoned the (Aristotelian) tradition of seeking to derive an account of planetary motion from the causes or essences of the planetary material. As Duhem and others have shown in great detail,[1] many astronomers sought simply to correlate planetary motions, independently of any causal assumptions about the essence or first principles of the heavens. Straightaway, this turned them from scientists into craftsmen.[2] To make matters worse, astronomers used a technique of *post hoc* testing of their theories. Rather than deriving their models from directly-intuited first principles, they offered hypothetical constructions of planetary motions and positions and then compared the predictions drawn from their models with the observed positions of the heavenly bodies. This mode of theory testing is, of course, highly fallible and non-demonstrative; and it was known at the time to be so. The central point for our purposes is that, by abandoning a demonstrative method based on necessary first principles, the astronomers were indulging in mere opinion rather than knowledge, putting themselves well beyond the scientific pale. Through virtually the whole of the Middle Ages, and indeed up until the beginning of the seventeenth century, the predominant view of mathematical astronomy was that, for the reasons indicated, it did not qualify as a 'science'.

(It is worth noting in passing that much of the furor caused by the astronomical work of Copernicus and Kepler was a result of the fact that they were claiming to make astronomy 'scientific' again.)

More generally, the seventeenth century brought a very deep shift in demarcationist sensibilities. To make a long and fascinating story unconscionably brief, we can say that most seventeenth century thinkers accepted Aristotle's first demarcation criterion (viz., between infallible science and fallible opinion), but rejected his second (between know-how and understanding). For instance, if we look to the work of Galileo, Huygens or Newton, we see a refusal to prefer know-why to know-how; indeed, all three were prepared to regard as entirely scientific, systems of belief which laid no claim to an understanding grounded in primary causes or essences. Thus Galileo claimed to know little or nothing about the underlying causes responsible for the free fall of bodies, and in his own science of kinematics he steadfastly refused to speculate about such matters. But Galileo believed that he could still sustain his claim to be developing a 'science of motion' because the results he reached were, so he claimed, infallible and demonstrative. Similarly, Newton in *Principia* was not indifferent to causal explanation, and freely admitted that he would like to know the causes of gravitational phenomena; but he was emphatic that, even without a knowledge of the causes of gravity, one can engage in a sophisticated and *scientific* account of the gravitational behavior of the heavenly bodies. As with Galileo, Newton regarded his non-causal account as 'scientifical' because of the (avowed) certainty of its conclusions. As Newton told his readers over and again, he did not engage in hypotheses and speculations: he purported to be deriving his theories directly from the phenomena. Here again, the infallibility of results, rather than their derivability from first causes, comes to be the single touchstone of scientific status.

Despite the divergence of approach among thinkers of the seventeenth and eighteenth centuries, there is widespread agreement that scientific knowledge is apodictically certain. And this consensus cuts across most of the usual epistemological divides of the period. For instance, Bacon, Locke, Leibniz, Descartes, Newton and Kant are in accord about this way of characterizing science.[3] They may disagree about how precisely to certify the certainty of knowledge, but none quarrels with the claim that science and infallible knowledge are co-terminous.

As I have shown elsewhere,[4] this influential account finally and decisively came unraveled in the nineteenth century with the emergence and eventual triumph of a *fallibilistic* perspective in epistemology. Once one accepts, as most thinkers had by the mid-nineteenth century, that science offers no apodictic certainty, that all scientific theories are corrigible and may be subject to serious emendation, then it is no longer viable to attempt to distinguish science from non-science by assimilating that distinction to the difference between knowledge and opinion. Indeed, the unambiguous implication of fallibilism is that there is no difference between knowledge and opinion: within a fallibilist framework, scientific belief turns out to be just a species of the genus opinion. Several nineteenth century philosophers of science tried to take some of the sting out of this *volte-face* by suggesting that scientific opinions were more probable or more reliable than non-scientific ones; but even they conceded that it was no longer possible to make infallibility the hallmark of scientific knowledge.

With certainty no longer available as the demarcation tool, nineteenth century philosophers and scientists quickly forged other tools to do the job. Thinkers as diverse as Comte, Bain, Jevons, Helmholtz and Mach (to name only a few) began to insist that what really marks science off from everything else is its *methodology*. There was, they maintained, something called 'the scientific method'; even if that method was not fool-proof (the acceptance of fallibilism demanded that concession), it was at least a better technique for testing empirical claims than any other. And if it did make mistakes, it was sufficiently self-corrective that it would soon discover them and put them right. As one writer remarked a few years later: "if science lead us astray, more science will set us straight".[5] One need hardly add that the nineteenth century did not invent the idea of a logic of scientific inquiry; that dates back at least to Aristotle. But the new insistence in this period is on a fallible method which, for all its fallibility, is nonetheless superior to its non-scientific rivals.

This effort to mark science off from other things required one to show two things. First, that the various activities regarded as science utilized essentially the same repertoire of methods (hence the importance in the period of the so-called thesis of the 'unity of method'); secondly, the epistemic credentials of this method had to be established. At first blush, this program of identifying science with a certain technique of inquiry is not a silly one; indeed, it still persists in some respectable circles even in our time. But the nineteenth century could not begin to deliver on the two requirements just mentioned because there was no agreement about what the scientific method was. Some took it to be the canons of inductive reasoning sketched out by Herschel and Mill. Others insisted that the basic methodological principle of science was that its theories must be restricted to observable entities (the nineteenth century requirement of '*vera causa*').[6] Still others, like Whewell and Peirce, rejected the search for *verae causae* altogether and argued that the only decisive methodological test of a theory involved its ability successfully to make surprising predictions.[7] Absent agreement on what 'the scientific method' amounted to, demarcationists were scarcely in a position to argue persuasively that what individuated science was its method.

This approach was further embarrassed by a notorious set of ambiguities surrounding several of its key components. Specifically, many of the methodological rules proposed were much too ambiguous for one to tell when they were being followed and when breached. Thus, such common methodological rules as "avoid *ad hoc* hypotheses", "postulate simple theories", "feign no hypotheses", and "eschew theoretical entities" involved complex conceptions which neither scientists nor philosophers of the period were willing to explicate. To exacerbate matters still further, what most philosophers of science of the period offered up as an account of 'the scientific method' bore little resemblance to the methods actually used by working scientists, a point made with devastating clarity by Pierre Duhem in 1908.[8]

As one can see, the situation by the late nineteenth century was more than a little ironic. At precisely that juncture when science was beginning to have a decisive impact on the lives and institutions of Western man, at precisely that time when 'scientism' (i.e., the belief that science and science alone has the answers to all our answerable questions) was gaining ground, in exactly that quarter century when scientists were doing battle in earnest with all manner of 'pseudo-scientists' (e.g., homeopathic physicians, spiritualists, phrenologists, biblical geologists), scientists and philosophers found themselves empty-handed. Except at the rhetorical level, there was no longer any consensus about what separated science from anything else.

Surprisingly (or, if one is cynically inclined, quite expectedly), the absence of a plausible demarcation criterion did not stop *fin de siècle* scientists and philosophers from haranguing against what they regarded as pseudo-scientific nonsense (any more than their present-day counterparts are hampered by a similar lack of consensus); but it did make their protestations less compelling than their confident denunciations of 'quackery' might otherwise suggest. It is true, of course, that there was still much talk about 'the scientific method'; and doubtless many hoped that the methods of science could play the demarcationist role formerly assigned to certainty. But, leaving aside the fact that agreement was lacking about precisely what the scientific method was, there was no very good reason as yet to prefer any one of the proposed 'scientific methods' to any purportedly 'non-scientific' ones, since no one had managed to show either that any of the candidate 'scientific methods' qualified them as 'knowledge' (in the traditional sense of the term) or, even more minimally, that those methods were epistemically superior to their rivals.

3. A METAPHILOSOPHICAL INTERLUDE

Before we move to consider and to assess some familiar demarcationist proposals from our own epoch, we need to engage briefly in certain metaphilosophical preliminaries. Specifically, we should ask three central questions: (1) What conditions of adequacy should a proposed demarcation criterion satisfy? (2) Is the criterion under consideration offering necessary or sufficient conditions, or both, for scientific status? (3) What actions or judgments are implied by the claim that a certain belief or activity is 'scientific' or 'unscientific'?

(1) Early in the history of thought it was inevitable that characterizations of 'science' and 'knowledge' would be largely stipulative and *a priori*. After all, until as late as the seventeenth century, there were few developed examples of empirical sciences which one could point to or whose properties one could study; under such circumstances, where one is working largely *ab initio*, one can be uncompromisingly legislative about how a term like 'science' or 'knowledge' will be used. But as the sciences developed and prospered, philosophers began to see the task of formulating a demarcation criterion as no longer a purely stipulative undertaking. Any proposed dividing line between science and non-science would have to be (at least in part) explicative and thus sensitive to existing patterns of usage. Accordingly, if one were today to offer a definition of 'science' which classified (say) the major theories of physics and chemistry as non-scientific, one would thereby have failed to reconstruct some paradigmatic cases of the use of the term. Where Plato or Aristotle need not have worried if some or even most of the intellectual activities of their time failed to satisfy their respective definitions of 'science', it is inconceivable that we would find a demarcation criterion satisfactory which relegated to unscientific status a large number of the activities we consider scientific or which admitted as sciences activities which seem to us decidedly unscientific. In other words, the quest for a latter-day demarcation criterion involves an attempt to render explicit those shared but largely implicit sorting mechanisms whereby most of us can agree about paradigmatic cases of the scientific and the non-scientific. (And it seems to me that there is a large measure of agreement at this paradigmatic level, even

allowing for the existence of plenty of controversial problem cases.) A failure to do justice to these implicit sortings would be a grave drawback for any demarcation criterion.

But we expect more than this of a *philosophically* significant demarcation criterion between science and non-science. Minimally, we expect a demarcation criterion to identify the *epistemic* or *methodological* features which mark off scientific beliefs from unscientific ones. We want to know what, if anything, is special about the knowledge claims and the modes of inquiry of the sciences. Because there are doubtless many respects in which science differs from non-science (e.g., scientists may make larger salaries, or know more mathematics than non-scientists), we must insist that any philosophically interesting demarcative device must distinguish scientific and non-scientific matters in a way which exhibits a surer epistemic warrant or evidential ground for science than for non-science. If it should happen that there is no such warrant, then the demarcation between science and non-science would turn out to be of little or no philosophic significance.

Minimally, then, a philosophical demarcation criterion must be an adequate explication of our ordinary ways of partitioning science from non-science and it must exhibit epistemically significant differences between science and non-science. Additionally, as we have noted before, the criterion must have sufficient precision that we can tell whether various activities and beliefs whose status we are investigating do or do not satisfy it; otherwise it is no better than no criterion at all.

(2) What will the formal structure of a demarcation criterion have to look like if it is to accomplish the tasks for which it is designed? Ideally, it would specify a set of individually necessary and jointly sufficient conditions for deciding whether an activity or set of statements is scientific or unscientific. As is well known, it has not proved easy to produce a set of necessary and sufficient conditions for science. Would something less ambitious do the job? It seems unlikely. Suppose, for instance, that someone offers us a characterization which purports to be a necessary (but not sufficient) condition for scientific status. Such a condition, if acceptable, would allow us to identify certain activities as decidedly unscientific, but it would not help 'fix our beliefs', because it would not specify which systems actually were scientific. We would have to say things like: "Well, physics *might* be a science (assuming it fulfills the stated necessary conditions), but then again it *might* not, since necessary but not sufficient conditions for the application of a term do not warrant application of the term." If, like Popper, we want to be able to answer the question, "when should a theory be ranked

as scientific?",[9] then merely necessary conditions will never permit us to answer it.

For different reasons, merely sufficient conditions are equally inadequate. If we are only told: "satisfy these conditions and you will be scientific", we are left with no machinery for determining that a certain activity or statement is *unscientific*. The fact that (say) astrology failed to satisfy a set of *merely sufficient* conditions for scientific status would leave it in a kind of epistemic, twilight zone – possibly scientific, possibly not. Here again, we cannot construct the relevant partitioning. Hence, if (in the spirit of Popper) we "wish to distinguish between science and pseudo-science",[10] sufficient conditions are inadequate. The importance of these seemingly abstract matters can be brought home by considering some real-life examples. Recent legislation in several American states mandates the teaching of 'creation science' alongside evolutionary theory in high school science classes. Opponents of this legislation have argued that evolutionary theory is authentic science, while creation science is not science at all. Such a judgment, and we are apt to make parallel ones all the time, would *not* be warranted by any demarcation criterion which gave only necessary *or* only sufficient conditions for scientific status. Without conditions which are both necessary and sufficient, we are never in a position to say "*this* is scientific: but *that* is unscientific". A demarcation criterion which fails to provide both sorts of conditions simply will not perform the tasks expected of it.

(3) Closely related to this point is a broader question of the purposes behind the formulation of a demarcation criterion. No one can look at the history of debates between scientists and 'pseudo-scientists' without realizing that demarcation criteria are typically used as *machines de guerre* in a polemical battle between rival camps. Indeed, many of those most closely associated with the demarcation issue have evidently had hidden (and sometimes not so hidden) agendas of various sorts. It is well known, for instance, that Aristotle was concerned to embarrass the practitioners of Hippocratic medicine; and it is notorious that the logical positivists wanted to repudiate metaphysics and that Popper was out to 'get' Marx and Freud. In every case, they used a demarcation criterion of their own devising as the discrediting device.

Precisely because a demarcation criterion will typically assert the epistemic superiority of science over non-science, the formulation of such a criterion will result in the sorting of beliefs into such categories as 'sound' and 'unsound', 'respectable' and 'cranky', or 'reasonable' and 'unreasonable'. Philosophers should not shirk from the formulation of a demarcation criterion merely because it has these judgmental implications associated with it. Quite the reverse, philosophy at its best should tell us what is reasonable to believe and what is not. But the value-loaded character of the term 'science' (and its cognates) in our culture should make us realize that the labelling of a certain activity as 'scientific' or 'unscientific' has social and political ramifications which go well beyond the taxonomic task of sorting beliefs into two piles. Although the cleaver that makes the cut may be largely epistemic in character, it has consequences which are decidedly non-epistemic. Precisely because a demarcation criterion will serve as a rationale for taking a number of *practical* actions which may well have far-reaching moral, social and economic consequences, it would be wise to insist that the arguments in favor of any demarcation criterion we intend to take seriously should be especially compelling.

With these preliminaries out of the way, we can turn to an examination of the recent history of demarcation.

4. THE NEW DEMARCATIONIST TRADITION

As we have seen, there was ample reason by 1900 to conclude that neither certainty nor generation according to a privileged set of methodological rules was adequate to denominate science. It should thus come as no surprise that philosophers of the 1920's and 1930's added a couple of new wrinkles to the problem. As is well known, several prominent members of the *Wiener Kreis* took a syntactic or logical approach to the matter. If, the logical positivists apparently reasoned, epistemology and methodology are incapable of distinguishing the scientific from the non-scientific, then perhaps the theory of meaning will do the job. A statement, they suggested, was scientific just in case it had a determinate meaning; and meaningful statements were those which could be exhaustively verified. As Popper once observed, the positivists thought that "verifiability, meaningfulness, and scientific character all coincide."[11]

Despite its many reformulations during the late 1920's and 1930's verificationism enjoyed mixed fortunes as a theory of meaning.[12] But as a would-be demarcation between the scientific and the non-scientific, it was a disaster. Not only are many statements in the sciences not open to exhaustive verification (e.g., all universal laws), but the vast majority of non-scientific and pseudo-scientific systems of belief have verifiable constituents. Consider, for instance, the thesis that the Earth is flat. To subscribe to such a belief in the twentieth century would be the height of folly. Yet such a statement is verifiable in the sense that we can specify a class of possible observations which would verify it. Indeed, every belief which has ever been rejected as a part of science because it was 'falsified' is (at least partially) verifiable. Because verifiable, it is thus (according to the 'mature positivists'' criterion) both meaningful and scientific.

A second familiar approach from the same period is Karl Popper's 'falsificationist' criterion, which fares no better. Apart from the fact that it leaves ambiguous the scientific status of virtually every singular existential statement, however well supported (e.g., the claim that there are atoms, that there is a planet closer to the sun than the Earth, that there is a missing link), it has the untoward consequence of countenancing as 'scientific' every crank claim which makes ascertainably false assertions. Thus flat Earthers, biblical creationists, proponents of laetrile or orgone boxes, Uri Geller devotees, Bermuda Triangulators, circle squarers, Lysenkoists, charioteers of the gods, *perpetuum mobile* builders, Big Foot searchers, Loch Nessians, faith healers, polywater dabblers, Rosicrucians, the-world-is-about-to-enders, primal screamers, water diviners, magicians, and astrologers all turn out to be scientific on Popper's criterion – just so long as they are prepared to indicate some observation, however improbable, which (if it came to pass) would cause them to change their minds.

One might respond to such criticisms by saying that scientific status is a matter of degree rather than kind. Sciences such as physics and chemistry have a high degree of testability, it might be said, while the systems we regard as pseudo-scientific are far less open to empirical scrutiny. Acute technical difficulties confront this suggestion, for the only articulated theory of degrees of testability (Popper's) makes it impossible to compare the degrees of testability of two distinct theories *except when one entails the other*. Since (one hopes!) no 'scientific' theory entails any 'pseudo-scientific' one, the relevant comparisons cannot be made. But even if this problem could be overcome, and if it were possible for us to conclude (say) that the general theory of relativity was more testable (and thus by definition more scientific) than astrology, it would not follow that astrology was any less worthy of belief than relativity – for testability is a semantic rather than an epistemic notion, which entails nothing whatever about belief-worthiness.

It is worth pausing for a moment to ponder the importance of this difference. I said before that the shift from the older to the newer demarcationist orientation could be described as a move from epistemic to syntactic and semantic strategies. In fact, the shift is even more significant than that way of describing the transition suggests. The central concern of the older tradition had been to identify those ideas or theories which were worthy of belief. To judge a statement to be scientific was to make a *retrospective* judgment about how that statement had stood up to empirical scrutiny. With the positivists and Popper, however, this retrospective element drops out altogether. Scientific status, on their analysis, is not a matter of evidential support or belief-worthiness, for all sorts of ill-founded claims are testable and thus scientific on the new view.

The failure of the newer demarcationist tradition to insist on the necessity of retrospective evidential assessments for determining scientific status goes some considerable way to undermining the practical utility of the demarcationist enterprise, precisely because most of the 'cranky' beliefs about which one might incline to be dismissive turn out to be 'scientific' according to falsificationist or (partial) verificationist criteria. The older demarcationist tradition, concerned with actual epistemic warrant rather than potential epistemic scrutability, would never have countenanced such an undemanding sense of the 'scientific'. More to the point, the new tradition has had to pay a hefty price for its scaled-down expectations. Unwilling to link scientific status to any evidential warrant, twentieth century demarcationists have been forced into characterizing the ideologies they oppose (whether Marxism, psychoanalysis or creationism) as untestable in principle. Very occasionally, that label is appropriate. But more often than not, the views in question can be tested, have been tested, and have failed those tests. But such failures cannot impugn their (new) scientific status: quite the reverse, *by virtue of failing the epistemic tests to which they are subjected, these views guarantee that they satisfy the relevant semantic criteria for scientific status*! The new demarcationism thus reveals itself as a largely toothless wonder, which serves neither to explicate the paradigmatic usages of 'scientific' (and its cognates) nor to perform the critical stable-cleaning chores for which it was originally intended.

For these, and a host of other reasons familiar in the philosophical literature, neither verificationism nor falsificationism offers much promise of drawing a useful distinction between the scientific and the non-scientific.

Are there other plausible candidates for explicating the distinction? Several seem to be waiting in the wings. One might suggest, for instance, that scientific claims are well tested, whereas non-scientific ones are not. Alternatively (an approach taken by Thagard),[13] one might maintain that scientific knowledge is unique in exhibiting progress or growth. Some have suggested that scientific theories alone make surprising predictions which turn out to be true. One might even go in the pragmatic direction and maintain that science is the sole repository of useful and reliable knowledge. Or, finally, one might propose that science is the only form of intellectual system-building which proceeds cumulatively, with later views embracing earlier ones, or at least retaining those earlier views as limiting cases.[14]

It can readily be shown that none of these suggestions can be a necessary and sufficient condition for something to count as 'science', at least not as that term is customarily used. And in most cases, these are not even plausible as necessary conditions. Let me sketch out some of the reasons why these proposals are so unpromising. Take the requirement of well-testedness. Unfortunately, we have no viable over-arching account of the circumstances under which a claim may be regarded as well tested. But even if we did, is it plausible to suggest that all the assertions in science texts (let alone science journals) have been well tested and that none of the assertions in such conventionally non-scientific fields as literary theory, carpentry or football strategy are well tested? When a scientist presents a conjecture which has not yet been tested and is such that we are not yet sure what would count as a robust test of it, has that scientist ceased doing science when he discusses his conjecture? On the other side of the divide, is anyone prepared to say that we have no convincing evidence for such 'non-scientific' claims as that "Bacon did not write the plays attributed to Shakespeare", that "a mitre joint is stronger than a flush joint", or that "off-side kicks are not usually fumbled"? Indeed, are we not entitled to say that all these claims are much better supported by the evidence than many of the 'scientific' assumptions of (say) cosmology or psychology?

The reason for this divergence is simple to see. Many, perhaps most, parts of science are highly speculative compared with many non-scientific disciplines. There seems good reason, given from the historical record, to suppose that most scientific theories are false; under the circumstances, how plausible can be the claim that science is the repository of all and only reliable or well-confirmed theories?

Similarly, cognitive progress is not unique to the 'sciences'. Many disciplines (e.g., literary criticism, military strategy, and perhaps even philosophy) can claim to know more about their respective domains than they did 50 or 100 years ago. By contrast, we can point to several 'sciences' which, during certain periods of their history, exhibited little or no progress.[15] Continuous, or even sporadic, cognitive growth seems neither a necessary nor a sufficient condition for the activities we regard as scientific. Finally, consider the requirement of cumulative theory transitions as a demarcation criterion. As several authors[16] have shown, this will not do even as a necessary condition for marking off scientific knowledge, since many scientific theories – even those in the so-called 'mature sciences' – do not contain their predecessors, not even as limiting cases.

I will not pretend to be able to prove that there is no conceivable philosophical reconstruction of our intuitive distinction between the scientific and the non-scientific. I do believe, though, that we are warranted in saying that none of the criteria which have been offered thus far promises to explicate the distinction.

But we can go further than this, for we have learned enough about what passes for science in our culture to be able to say quite confidently that it is not all cut from the same epistemic cloth. Some scientific theories are well tested; some are not. Some branches of science are presently showing high rates of growth; others are not. Some scientific theories have made a host of successful predictions of surprising phenomena; some have made few if any such predictions. Some scientific hypotheses are *ad hoc*; others are not. Some have achieved a 'consilience of inductions'; others have not. (Similar remarks could be made about several non-scientific theories and disciplines.) *The evident epistemic heterogeneity of the activities and beliefs customarily regarded as scientific should alert us to the probable futility of seeking an epistemic version of a demarcation criterion*. Where, even after detailed analysis, there appear to be no epistemic invariants, one is well advised not to take their existence for granted. But to say as much is in effect to say that the problem of demarcation – the very problem which Popper labelled 'the central problem of epistemology' – is spurious, for that problem *presupposes* the existence of just such invariants.

In asserting that the problem of demarcation between science and non-science is a pseudo-problem (at least as far as philosophy is concerned), I am manifestly not denying that there are crucial epistemic and methodological questions to be raised about knowledge claims, whether we classify them as scientific or not. Nor, to belabor the obvious, am I saying that we are never entitled to argue that a certain piece of science is epistemically warranted and that a certain piece of pseudo-science is not. It remains as important as it ever was to ask questions like: When is a claim well confirmed? When can we regard a theory as well tested? What characterizes cognitive progress? But once we have answers to such questions (and we are still a long way from that happy state!), there will be little left to inquire into which is epistemically significant.

One final point needs to be stressed. In arguing that it remains important to retain a distinction between reliable and unreliable knowledge, I am not trying to resurrect the science/non-science demarcation under a new guise.[17] However we eventually settle the question of reliable knowledge, the class of statements falling under that rubric will include much that is not commonly regarded as 'scientific' and it will exclude much that is generally considered 'scientific'. This, too, follows from the epistemic heterogeneity of the sciences.

5. CONCLUSION

Through certain vagaries of history, some of which I have alluded to here, we have managed to conflate two quite distinct questions: What makes a belief well founded (or heuristically fertile)? And what makes a belief scientific? The first set of questions is philosophically interesting and possibly even tractable; the second question is both uninteresting and, judging by its checkered past, intractable. If we would stand up and be counted on the side of reason, we ought to drop terms like 'pseudo-science' and 'unscientific' from our vocabulary; they are just hollow phrases which do only emotive work for us. As such, they are more suited to the rhetoric of politicians and Scottish sociologists of knowledge than to that of empirical researchers.[18] Insofar as our concern is to protect ourselves and our fellows from the cardinal sin of believing what we wish were so rather than what there is substantial evidence for (and surely that is what most forms of 'quackery' come down to), then our focus should be squarely on the empirical and conceptual credentials for claims about the world. The 'scientific' status of those claims is altogether irrelevant.

University of Pittsburgh

NOTES

* I am grateful to NSF and NEH for support of this research. I have profited enormously from the comments of Adolf Grünbaum, Ken Alpern and Andrew Lugg on an earlier version of this paper.

1 See especially his *To Save The Phenomena* (Chicago: University of Chicago Press, 1969).

2 This shifting in orientation is often credited to the emerging emphasis on the continuity of the crafts and the sciences and to Baconian-like efforts to make science 'useful'. But such an analysis surely confuses agnosticism about first causes – which is what really lay behind the instrumentalism of medieval and Renaissance astronomy – with a utilitarian desire to be practical.

[3] For much of the supporting evidence for this claim, see the early chapters of Laudan, *Science and Hypothesis* (Dordrecht: D. Reidel, 1981).
[4] See especially Chapter 8 of *Science and Hypothesis.*
[5] E. V. Davis, writing in 1914.
[6] See the discussions of this concept by Kavaloski, Hodge, and R. Laudan.
[7] For an account of the history of the concept of surprising predictions, see Laudan, *Science and Hypothesis*, Chapters 8 and 10.
[8] See Duhem's classic *Aim and Structure of Physical Theory* (New York: Atheneum, 1962).
[9] Karl Popper, *Conjectures and Refutations* (London: Routledge and Kegan Paul, 1963), p. 33.
[10] *Ibid.*
[11] *Ibid.*, p. 40.
[12] For a very brief historical account, see C. G. Hempel's classic, 'Problems and Changes in the Empiricist Criterion of Meaning,' *Revue Internationale de Philosophie* 11 (1950), 41–63.
[13] See, for instance, Paul Thagard, 'Resemblance, Correlation and Pseudo-Science,' in M. Hanen *et al., Science, Pseudo-Science and Society* (Waterloo, Ont.: W. Laurier University Press, 1980), pp. 17–28.
[14] For proponents of this cumulative view, see Popper, *Conjectures and Refutations*; Hilary Putnam, *Meaning and the Moral Sciences* (London: Routledge and Kegan Paul, 1978); Władysław Krajewski, *Correspondence Principle and Growth of Science* (Dordrecht, Boston: D. Reidel, 1977); Heinz Post, 'Correspondence, Invariance and Heuristics,' *Studies in History and Philosophy of Science* 2 (1971), 213–55; and L. Szumilewicz, 'Incommensurability and the Rationality of Science,' *Brit. Jour. Phil. Sci.* 28 (1977), 348ff.
[15] Likely tentative candidates: acoustics from 1750 to 1780; human anatomy from 1900 to 1920; kinematic astronomy from 1200 to 1500; rational mechanics from 1910 to 1940.
[16] See, among others: T. S. Kuhn, *Structure of Scientific Revolutions* (Chicago: University of Chicago Press, 1962); A. Grünbaum, 'Can a Theory Answer More Questions than One of Its Rivals?', *Brit. Journ. Phil. Sci.* 27 (1976), 1–23; L. Laudan, 'Two Dogmas of Methodology,' *Philosophy of Science* 43 (1976), 467–72; L. Laudan, 'A Confutation of Convergent Realism,' *Philosophy of Science* 48 (1981), 19–49.
[17] In an excellent study ['Theories of Demarcation Between Science and Metaphysics,' in *Problems in the Philosophy of Science* (Amsterdam: North-Holland, 1968), 40ff)], William Bartley has similarly argued that the (Popperian) demarcation problem is not a central problem of the philosophy of science. Bartley's chief reason for devaluing the importance of a demarcation criterion is his conviction that it is less important whether a system is empirical or testable than whether a system is 'criticizable'. Since he thinks many non-empirical systems are nonetheless open to criticism, he argues that the demarcation between science and non-science is less important than the distinction between the revisable and the non-revisable. I applaud Bartley's insistence that the empirical/non-empirical (or, what is for a Popperian the same thing, the scientific/non-scientific) distinction is not central; but I am not convinced, as Bartley is, that we should assign pride of place to the revisable/non-revisable dichotomy. Being willing to change one's mind is a commendable trait, but it is not clear to me that such revisability addresses the central *epistemic* question of the well-foundedness of our beliefs.

18 I cannot resist this swipe at the efforts of the so-called Edinburgh school to recast the sociology of knowledge in what they imagine to be the 'scientific image'. For a typical example of the failure of that group to realize the fuzziness of the notion of the 'scientific', see David Bloor's *Knowledge and Social Imagery* (London: Routledge and Kegan Paul, 1976), and my criticism of it, 'The Pseudo-Science of Science?' *Phil. Soc. Sci.* **11** (1981), 173–198.

PHILIP L. QUINN

GRÜNBAUM ON DETERMINISM AND THE MORAL LIFE

Adolf Grünbaum is probably best known to professional philosophers for his technical work on philosophical problems of space and time and, more recently, on philosophical problems of psychoanalysis. However, his philosophical concerns extend to less technical questions of more general interest. Because other contributors to this volume have chosen to focus their discussions on Grünbaum's technical work in philosophy of science, I shall try to round out the picture a bit by considering what Grünbaum has to say about one such question of general interest. I have chosen to concentrate on Grünbaum's treatment of determinism. Although he has not written a great deal on this subject, what he has written has been widely reprinted and translated, and has thus presumably been widely influential.[1] In this brief paper, I cannot discuss in depth all aspects of Grünbaum's treatment of determinism. I shall therefore attend only to his response on behalf of determinism to the challenge posed by what he takes to be the argument from morality. I must omit consideration of his no less interesting discussions of such topics as determinism and predictability, and determinism and quantum physics. In defense of this selection of topics, I shall say only that it seems to me that it is the argument from morality, or some moral worry akin to it, which perturbs most people who find determinism a threatening doctrine.

MATTERS METHODOLOGICAL

Simply stated, determinism with respect to human actions is the thesis that, for every human action *A* that occurs, there is some condition *C* such that it obtains prior to *A*'s occurrence, its obtaining does not entail the occurrence of any other action of *A*'s doer, and its obtaining is a causally sufficient condition for *A*'s occurrence. Since the scope of the thesis is restricted to human actions, it does not conflict with the otherwise plausible claim that some microphysical events in the quantum domain have no causally sufficient conditions. Nor does the thesis entail fatalism with respect to human actions, for it does not entail that any causally sufficient condition for the occurrence of any action obtains as a matter of logical necessity.

Many empiricists who are sympathetic to determinism hold that causation

R. S. Cohen and L. Laudan (eds.), Physics, Philosophy and Psychoanalysis, 129–151.

is to be understood or analyzed in terms of subsumption under laws. Among them is Grünbaum. He says:

> The deterministic conception of *human behavior* is inspired by the view that man is an integral part and product of nature and that his behavior can reasonably be held to exhibit scientifically ascertainable regularities just as any other *macroscopic* sector of nature (Feigl *et al.*, 1972, p. 606).

Such philosophers would probably wish to strengthen the deterministic thesis to the claim that, for every human action *A* that occurs, there is some condition *C* such that it obtains prior to *A*'s occurrence, its obtaining does not entail the occurrence of any other action of *A*'s doer, and there is some true law-like proposition which entails that whenever a condition of the type of *C* obtains an action of the type of *A* occurs. Besides being more cumbersome than the earlier formulation, this version of the thesis has an important liability in the present context. While it is still unsettled how causation is to be analyzed, the question of whether there are true singular causal statements that are not instances of general regularities remains open. But even if there are, the weaker thesis of determinism with respect to human actions would, if plausible, still present a serious challenge to many beliefs and practices associated with common sense morality. In order not to burden the defender of determinism with excess baggage by way of unnecessary assumptions, I shall henceforth conduct the discussion on the hypothesis that it is the weaker thesis which is at issue. Those who care to do so should find it simple enough to formulate parallel arguments relevant to the stronger thesis.

Inquiry should begin with a due sense of modesty about what philosophical argument is likely to contribute to resolving the question of whether the thesis of determinism with respect to human actions is true. On account of the logical form of the thesis, which begins with a universal quantifier followed by an existential quantifier, it is not likely that it is, strictly speaking, either falsifiable or verifiable on the basis of examination of particular cases. A critic who attacks the thesis by claiming to be introspectively aware that some of his own actions, at any rate, have no causally sufficient conditions of the sort specified by the thesis is committed to the claim that his introspective powers extend to perceiving reliably the absence of causal conditions, and not merely to reliably not perceiving their presence. But in the light of what modern science tells us about neurophysiological determinants of behavior, which are never objects of introspective awareness, and about mechanisms of repression, which sometimes block behavioral determinants from introspective awareness, this claim to robust introspective powers is bound to seem unwarranted and unconvincing. Similarly someone who defends the thesis cannot hope to prove it by an exhaustive enumeration of positive instances. As long as all the human actions that will ever occur have not been examined, the possibility remains that some as yet unexamined action will permanently frustrate the scientific search for its causally sufficient conditions, no matter how many successes scientists have had in finding causally sufficient conditions for other actions they have examined. This is not to deny, of course, that successful causal explanations of some actions or classes of action would corroborate, or at least enhance the plausibility of the deterministic thesis. Thus, when Grünbaum writes

> I wish to emphasize, however, that the categorical truth of that deterministic assertion can be established inductively *not* by logical analysis alone but requires the working psychologist's empirical discovery of specific causal laws (Feigl *et al.*, 1972, p. 608),

I would merely add that, even on the most utopian assumptions about the future of psychology, what will be inductively established is not the categorical truth of the deterministic assertion but only some high degree of empirical warrant for it.

While we await the future of psychology, utopian or otherwise, what is there left for philosophers to do? For defenders of determinism like Grünbaum, there is the task of defeating objections. Many people claim to know things about human moral life which they suppose to be severe anomalies for determinism with respect to human actions, either because they are outright inconsistent with such determinism or because they would be inexplicable if determinism were true. A defense of determinism in response to such allegations can proceed in several ways. It may be denied that people know the things about moral life they take to be anomalous data for determinism; or it may be denied that the data people have count as anomalous for determinism. Grünbaum is aware that both these tactical options are open to him. He says:

> For I shall maintain that in important respects the data are *not* what they are alleged to be. And insofar as they are, I shall argue that they are not evidence against determinism. Nay, I shall claim that in part, these data are first rendered intelligible by determinism (Feigl *et al.*, 1972, p. 608).

The particular moral phenomena Grünbaum considers include the assignment of responsibility for action, the rationale for inflicting punishment, and moral attitudes such as remorse and regret. To the extent that he can make good on the claim that such phenomena provide us with no data that are anomalous for determinism with respect to human actions, Grünbaum will have defeated or defused certain objections to determinism. The ball will have been returned to the court of the opponent of determinism.

In the remainder of this paper, I shall re-examine some of the moral phenomena Grünbaum considers. I shall argue that they have aspects which seem anomalous for determinism and that Grünbaum's discussion does not succeed in dispelling this appearance of anomaly. If I succeed in my task, the ball will be back in the determinist's court, and further play can continue from that point.

Of course, it should be realized that at the present time the whole game of defending determinism with respect to human actions is philosophically optional, rather than mandatory. After all, it is not as if psychologists and social scientists had succeeded in finding causally sufficient conditions for vast numbers of human actions or in subsuming numerous kinds of human action under well-corroborated causal laws. So it is not as if determinism were an essential element in a current scientific paradigm which would crumble into revolutionary chaos unless the appearance of anomaly were removed. One could, I suppose, hold off from pressing objections against determinism or offering rebuttals to such objections until such time as it became urgent to do so because a successful science of human behavior based on deterministic assumptions was in serious intellectual conflict with alleged moral data. Still, there is something to be said for considering the objections to determinism with respect to human actions even in the absence of a respectable deterministic science of human action. If bad objections to determinism might block certain lines of scientific inquiry, then it would be well to defeat them promptly. Though optional, the game may nonetheless be pragmatically desirable.

AGENT OPACITY

According to Grünbaum, a determinist's rationale for the practice of assigning responsibility for actions to agents rests on a distinction between voluntary behavior and behavior which literally occurs under compulsion. He stipulates that behavior occurs under compulsion, in the somewhat technical sense he requires for the purposes of his argument, if

> ... we are literally being physically restrained from without in implementing the desires which we have upon reacting to the total stimulus situation in our environment and are physically made to perform a different act instead (Feigl *et al.*, 1972, p. 610).

Thus, when a stronger man forces my finger to press a button, I am literally compelled to blow up the bridge; but when an armed thief offers me the choice of my money or my life, I am not literally compelled to hand over my money. Of course bits of behavior that occur under compulsion are hardly actions in the ordinary sense: to behave under compulsion is to be a patient rather than an agent. Grünbaum explicitly takes notice of this fact, but he does not make much of it since the difficulty is merely terminological (Feigl *et al.*, 1972, p. 611). The important point is that the determinist is not committed to the view that all human behavior occurs under compulsion. Instead, the determinist holds that even those bits of behavior which are voluntary actions are causally determined, presumably in part by the agent's desires rather than independent of or contrary to them. And so, for the determinist, to assign responsibility for a voluntary action to an agent is to assert that the agent's desires played the right kind of efficacious role in the causal generation of that action. Assessments of the extent to which an agent is to be held accountable for the voluntary actions for which he is responsible will then introduce familiar complications having to do with excusing or mitigating circumstances. But complications and borderline cases should not be allowed to obscure, or distract attention from, the fact that the determinist can distinguish between bits of behavior which literally occur under compulsion and for which the agent is not to be assigned any responsibility and bits of behavior which are voluntary actions and for which the agent is to be assigned some responsibility. The voluntary actions are those bits of behavior generated in part by causal chains that pass, so to speak, through the agent by way of efficacious desires.

As far as I can tell, the determinist's story about the grounds for assigning responsibility for actions to agents is fine as far as it goes. If our practice of assigning responsibility contained no other feature needing explanation than the difference in the way we treat behavior occurring under compulsion and voluntary action, then the story would go far enough. But is the determinist's story complete as it stands? I think not.

Consider typical circumstances under which inquiries aimed at assigning responsibility are undertaken. A state of affairs that engages our interests, presumably in virtue of some positive or negative value it had, is found to obtain. A search for its causal antecedents is begun. Causal chains are traced back to one or more agents, if possible, and responsibility for producing the state of affairs is assigned to those agents, if there are any. Except in unusual circumstances, no attempt is then made to trace causal chains back, through the agents, in order to assign responsibility to any antecedent causally sufficient conditions for the agents' actions of the sort determinists presume to obtain. A bridge collapses, for example. What caused this to happen? The explosion of a time bomb was, in the circumstances, a necessary part of a causally sufficient condition. What caused the explosion of the time bomb? In the circumstances, a necessary part of a causally sufficient condition was Smith's placing the bomb on the bridge and setting the timing mechanism. Why did Smith do this? Smith's desire to destroy some cars was, in the circumstances, a necessary part of a causally sufficient condition for his action. Why did Smith have such a desire? Well, in the circumstances, a necessary part of a causally sufficient condition for his having that desire was an unfortunate experience he had involving his father's car on his first date. And so on.

The story raises two questions that cry out for answers. First, why not assign responsibility for the collapse of the bridge to the bomb and terminate the causal inquiry forthwith? In other words, why does the assignment of responsibility reach behind the bomb in order to latch onto Smith? And, second, why not assign responsibility for the collapse of the bridge to Smith's father's car and decline to end the causal inquiry at any earlier stage? In other words, why does the assignment of responsibility not reach behind Smith in order to latch onto his father's car? Considered as concrete objects involved in the causal genesis of the bridge's collapse, the bomb, Smith, and Smith's father's car are pretty much on a par. Each is a constituent in some event which in the circumstances is a necessary and non-redundant part or conjunct of some condition causally sufficient for the collapse of the bridge. In the absence of some villain, Jones, who has hypnotized Smith, why then do we single out poor Smith and assign responsibility for the collapse of the bridge to him and not to the bomb or the car? Is there some good reason, either in principle or in practice, for singling out for special treatment from among several concrete individual constituents of events that are necessary parts of causally sufficient conditions for the collapse of the bridge, the ones that happen to be human agents? We do this as a matter of course and take the practice of so doing to be justified, generally speaking, though it may lead to uncomfortable results in some applications. But what, if anything, underlies and accounts for the rationality of the practice?

It should be noted that the determinist's opponent has a straightforward answer to this question. Smith is singled out for special treatment because his action is distinguished from certain other macroscopic events which contribute causally to the bridge's collapse, such as the bomb's explosion or the car's behavior, in virtue of lacking an antecedent causally sufficient condition. His action, so to speak, originates the tip of a branch in the causal tree whose trunk produces the collapse of the bridge. A general principle governing assignments of responsibility on this view would be that an individual has some responsibility for an event only if the individual is a constituent of some event which *both* is either identical to, or a necessary part of a causally sufficient condition for, the event in question *and* occurs despite the absence of any causally sufficient condition for its occurrence. Obviously this condition, though necessary, would not be sufficient for responsibility. If cosmic history began with an uncaused first event such as the Big Bang, then it would be an event which is a necessary part of a causally sufficient condition for many later events and which occurs despite the absence of a causally sufficient condition for its occurrence, yet its physical constituents, whatever they might be, would not be responsible for those later events. On the assumption that the bomb's explosion is not triggered by amplification of some indeterministic quantum mechanical process, the bomb is not responsible for the collapse because its explosion has a causally sufficient condition and no other event of which it is a constituent is either identical to the collapse or a necessary part of a causally sufficient condition for the collapse. And, for similar reasons, the car is not responsible for the collapse. On this view, assignments of responsibility reach back to the tips of branches of causal trees, if there are any, and go no farther. If the view were correct, it would render intelligible the features of our practice of assigning responsibility I have been highlighting. Can the determinist give an account that does at least as well?

I think the determinist can tell a fairly plausible story to explain why it is no part of our practice to allow the assignment of responsibility to home in on the bomb. We wish to take steps to prevent having our bridges wantonly destroyed by saboteurs like Smith. In examining the causal ancestry of the collapse of the bridge, we find no factor such that we could reasonably expect to be able to manipulate its analogue in similar situations to prevent destruction of our bridges until we get to Smith's action. Since it turns out to be the case more often than not that the first manipulable factor we encounter in tracing the causal antecedents of states of affairs of types we desire to promote or prevent happens to be a human agent, it is rational for us to have the practice of typically pushing the assignment of responsibility all the way back to the first agent we encounter. For practical reasons and not reasons of principle, things we encounter before the first agent along the causal track are treated as transparent to ascriptions of responsibility.

Though plausible enough as an account of the transparency of the bomb, a story of this kind seems to leave the opacity of Smith mysterious, if not utterly inexplicable. According to determinism with respect to human actions, there is a causally sufficient condition for Smith's action of placing the bomb on the bridge and setting the timing mechanism. If Smith's action is not causally overdetermined, manipulation of some necessary part of a sufficient causal condition for Smith's action would have served our purposes as well as, if not better than, manipulating Smith himself in the present case. In at least some similar cases, it is likely to turn out to be cost-effective or otherwise more efficient to manipulate some such factors if we can identify them. Though we cannot now reliably identify such factors due to the fact that our human sciences are such paltry things, the progress of science will provide a remedy. Thus, as the human sciences progress, we may expect the cases in which human agents too are treated as transparent to ascriptions of responsibility to become more numerous. It will become rational for us to push the assignment of responsibility all the way through the agent onto antecedent conditions that are more easily manipulable to serve our purposes. The very same practical reasons that impel us to push assignments of responsibility back to the agent will lead us to shove them back beyond some agents. Agent opacity will yield to semi-transparency and, perhaps, ultimately to full transparency.

The trouble with this scenario, I believe, is that it is based on a misunderstanding of the function of agent opacity in our moral practices. We treat human agents as stubbornly opaque to transmission of responsibility further up the line because we insist upon acknowledging them as authors of actions, and accountable for them and some of their consequences, when the actions in question are fully voluntary, and not as mere conduits through which are channeled some of the forces that produce consequences of positive and negative value. To push assignments of responsibility through the agent and back to easily manipulable causal antecedents of his actions would be to treat him not as author of his deeds but merely as a channel through which the things that do produce his actions, whatever they may be, operate. To treat the voluntary actions of normal humans in this fashion would be to fail to treat humans with the respect they deserve as persons.[2] Human agents so conceived or so treated might well complain of being regarded or treated as lacking in dignity. In my opinion, in the absence of unusual defeating conditions, such complaints would be justified.

Can the determinist give a better account of the features of ascriptions of responsibility I have lumped together under the concept of agent opacity? I do not know, but I think it is incumbent on him to try to do so. For what is left unexplained and, perhaps, inexplicable on the determinist's view that all human actions are nothing more than intermediate links in causal chains (or nodes in nets or trees, to vary the image) is the special respect human agents deserve in virtue of being authors of some of their deeds, the respect we pay them by holding them and not other things accountable for their voluntary actions and some of the consequences of these actions.

PENALIZING PEOPLE JUST TO DO THEM GOOD

Holding people responsible for their misdeeds is connected with the practice of inflicting penalties upon them. Indeed, for the determinist, as Grünbaum sees it, in the case of misdeeds, "being held responsible is tantamount to *liability to reformative or educative punishment*" (Feigl *et al.*, 1972, p. 615). Punishment is educative just in cases "when properly administered, it institutes countercauses to the repetition of injurious conduct" (Feigl *et al.*, 1972, p. 615). Grünbaum asserts:

> For the humane determinist, the decision whether pain is to be inflicted on the culprit, and, if so, to what extent, is governed solely by the conduciveness of such punishment to the reform and re-education of the culprit and to repairing his damage, where possible, or to the deterrence of other potential criminals (Feigl *et al.*, 1972, p. 615).

When a choice is to be made between equally effective penalties of differing severity, an upper limit on severity is established by "the use of the moral requirement that *gratuitous* suffering be avoided" (Feigl *et al.*, 1972, p. 615). With respect to criminals suffering from certain sorts of brain damage who will not be reformed or educated by punishments of the usual sorts, Grünbaum supposes the humane determinist would enjoin that they "receive appropriate neurological therapy, if at all possible" (Feigl *et al.*, 1972, p. 615). In the case of injurious behavior which occurs under compulsion in the technical sense described above, the person's volitional dispositions require no reforming, and so "the determinist does not administer punishment" (Feigl *et al.*, 1972, p. 616). And in the case of misdeeds done under the influence of posthypnotic suggestion, the only appropriate punishment "would be deterrence of the individual against re-submitting to that kind of risky hypnosis" (Feigl *et al.*, 1972, p. 616).

If it is possible to draw together into a single set of principles the rather heterogeneous considerations Grünbaum introduces into his discussion of the determinist's rationale for punishment, the result would seem to be that Grünbaum supposes the determinist will hold and, perhaps, must hold two principles. First, a necessary condition for it being justified to inflict some penalty *P* on a person for an injurious bit of behavior of kind *K* is that the infliction of *P* will be, or is likely to be, causally efficacious either in deterring that person from future behavior of kind *K* or in deterring others, who would or might otherwise engage in it, from behavior of kind *K*. And, second, a necessary condition for it being justified to inflict some penalty *P* on a person for an injurious bit of behavior of kind *K* is that *P* is, among the available alternatives, one whose infliction would be minimally sufficient in the circumstances to have the deterrent effect in question. As far as I can tell, Grünbaum says nothing that commits him to supposing either that the determinist will or that he will not hold these two conditions to be jointly sufficient for the justification of punishment. If the determinist were to hold the two conditions to be jointly sufficient, his view would be open to an obvious objection. Consider again the case of the man who forces my finger to press the button and thereby literally compels me to blow up the bridge. Suppose that man immediately takes off for parts unknown, leaving behind in my neighborhood a nest of unknown co-conspirators ready and willing to force my finger to press other buttons and thereby to compel me to blow up various banks, post offices and department stores. According to Grünbaum, I should not be penalized for this bit of behavior under compulsion, and I certainly agree. If anyone should be penalized for this bit of behavior, it is the man who compelled me to blow up the bridge. But, to continue the story, suppose further that chopping off my hand is, in fact, likely to be causally efficacious in deterring the co-conspirators simply because it is likely to be causally efficacious in getting me to avoid all situations in which there is the slightest chance that anyone can grab my other hand and force one of its fingers to press a button and in getting everyone else in my neighborhood to do likewise. And, finally, suppose that chopping off my hand is, among the available alternatives, the only one minimally sufficient in the circumstances to have the deterrent effect in question because, by hypothesis, the man who forced my finger to press the button has already vanished into parts unknown and his co-conspirators are also unknown. Then, just as it would, on this view, be justified to penalize a man to deter him from submitting himself again to risky hypnosis so it would also be justified to penalize me to deter me from submitting myself again to the risk of forced button-pressing, unless there is some further relevant difference between the two kinds of case. Hence, perhaps the determinist would be prudent to refrain from asserting that the two conditions are jointly sufficient to justify punishment.

Unfortunately, such commendable prudence would not help the determinist much, for there are also serious objections to the claim that the two conditions are individually necessary for the justification of punishment. Suppose Robinson and Friday find themselves thrown together on the proverbial desert island. By day, Friday builds himself huts of palm fronds; by night, Robinson maliciously tears them down. After a while, Friday, understandably annoyed, wonders whether it would be appropriate for him to inflict some penalty on Robinson. Since there is no one else on the island, Friday need not consider the deterrent effects of various possible penalties on anyone other than Robinson. As it happens, Robinson is obdurate in his malice. If Friday thrashes him soundly, Robinson will continue tearing down the huts and, in addition, poison Friday's well out of spite. If Friday pens him inside a stockade, Robinson will contrive to escape, continue tearing down huts and burn Friday's breadfruit trees. If Friday finally tries to kill him, Robinson will flee to the jungle, only to return by night to continue his self-appointed task of destroying huts and, in addition, to steal Robinson's goat. As Friday eventually learns to his cost, there is no penalty he can inflict on Robinson that is likely to have the least causal efficacy in deterring him from tearing down huts and, hence, none whose infliction would be minimally sufficient in the circumstances to have that deterrent effect. So neither of the two conditions is satisfied, yet it surely seems that Friday is justified in thrashing Robinson soundly, at the very least. It is not that inflicting such a penalty would be inappropriate. It would be inefficacious, but that is quite another matter. If Friday were one of Grünbaum's humane determinists, whose decisions about inflicting penalties are governed solely by the conduciveness of the penalties to reform or deter, he would give up the thought of penalizing Robinson and, presumably, also stop building those annoying huts. And, no doubt, Friday would eventually stop putting up new huts, but out of frustration and not for humane reasons.

From examples like these, I conclude that Grünbaum's attempt to provide the determinist with a plausible justification for punishing the guilty is unsuccessful. But perhaps the defects I have so far pointed out could be remedied by the exercise of ingenuity. After all, we would surely expect the determinist to be able to justify some penalties for the guilty. Human actions contribute causally to conditions of negative value: on this the determinist and his opponent agree. It is *prima facie* desirable to discourage, inhibit or prevent actions of types that regularly contribute to bad consequences. For the determinist, each action that occurs has an antecedently obtaining causally sufficient condition; an action can be prevented from occurring only if all its causally sufficient conditions can be prevented from obtaining. If there is enough regularity in human affairs to insure that a person is likely to do similar things in similar circumstances, then it is feasible to act upon a person after he has performed one action that has contributed to bad consequences to change something about him enough so that subsequently no causally sufficient condition for an action of the same type is likely to obtain. Such intervention is *prima facie* desirable, and in some cases it will turn out to be actually justified. Moreover, in some of the cases in which such intervention is actually justified, it will involve, as a matter of empirical fact, inflicting penalties on the person, because the person will not desire such intervention, but such penalties, though bad, will be justified because in the absence of such intervention further actions of the original type would recur, which would be worse. What the determinist needs to do is specify the conditions under which such intervention is actually justified in the wake of an action with bad consequences. But it is hardly to be doubted that there are such conditions, and so there is no reason of principle why the determinist could not make progress toward a specification as detailed as one might wish. The determinist, then, ought sooner or later to be able to tell a plausible story explaining why some of the guilty may be justifiably punished.

A more recalcitrant problem for the determinist will be explaining, if he can, why only the guilty may be punished. One of Grünbaum's examples may be elaborated to indicate the seriousness of the problem. Grünbaum mentions, with apparent approval, the father of a murdered child who "nobly disavowed all cave-man revenge" but asked "for greater preventive efforts in diagnosing potentially homicidal but seemingly exemplary adolescents" (Feigl *et al.*, 1972, p. 615). Presumably it would be contrary to the desires of many seemingly exemplary adolescents to be subject to such diagnostic efforts. To submit them to such testing would be to inflict penalties on them. And if there were both reliable diagnostic means of identifying potentially homicidal adolescents and reliable therapies for treating such people, presumably it would still be contrary to the desires of some seemingly exemplary adolescents so identified as potentially homicidal, and to the desires of their parents, that they should be subjected to such therapy. To impose such therapy upon them would be to inflict penalties on them. Instituting such a system of preventive diagnosis and treatment of potentially homicidal adolescents would result in inflicting penalties upon some adolescents unless involvement in the system were completely voluntary at every stage. But if such a system were completely voluntary, it is extremely likely that it would be ineffectual in preventing murders by seemingly exemplary adolescents. If a determinist has justified imposing educative or deterrent penalties on previous offenders on the grounds that such penalties, though bad, will serve to prevent future offenses which would be worse, he should by the same line of argument be able to justify a mandatory system of testing and treating adolescents for homicidal tendencies provided the penalties the system inflicts, or is likely to inflict, though bad, serve, or are likely to serve, to prevent worse future murders. After all, the only difference between cases of the two kinds is that in the former the penalty is inflicted on an offender and in the latter it is not. But, from the point of view of the deterministic line of thought we are now exploring, this difference is morally insignificant. It has no bearing on whether the penalties will perform their educative, deterrent or preventive functions, and performance of such functions is all that is relevant to justifying the infliction of such penalties. Grünbaum says:

The humane determinist rejects as barbarous the primitive vengeful idea of retaliatory, retributive or vindictive punishment (Feigl *et al.*, 1972, p. 615).

If this is a correct account of the determinist's position, then the determinist can justify an efficacious and cost-effective system of mandatory diagnosis and treatment for homicidal tendencies in exactly the same way he can justify inflicting effective educative or deterrent penalties on criminals.

It is sometimes said that imposing penalties on the innocent would not be punishment even if it were justified. It is alleged that this is the case because it is a conceptual truth about punishment, or a definitional truth about the word 'punishment', that penalties count as punishment only when inflicted upon the guilty. A retributivist might add that, in order to qualify as punishment, such penalties must be inflicted because the guilty party deserves or merits them in virtue of having done wrong in the past. Having rejected retribution as barbarous and primitive, Grünbaum's humane determinist will have no such rationale for distinguishing punishment from other cases in which penalties are imposed for educative, deterrent or preventive reasons. From his point of view, the fact that only penalties imposed upon the guilty are to be called punishment, if this is a fact, reflects nothing deeper than a linguistic accident, perhaps merely that current usage was shaped by and contains fossil remains of retributivist theory. And someone who complains that the humane determinist is willing to punish the innocent under certain circumstances may be charitably interpreted, not as having fallen into linguistic or conceptual error, but as objecting to the determinist's willingness to inflict penalties on guilty and innocent alike, for exactly the same reasons, and irrespective of differences between guilt and innocence.

Once the determinist has admitted that inflicting penalties on the innocent is justified when such a course of action will serve to prevent still greater evils and is the only available course of action that will so serve, he seems to be on a slippery slope. In the absence of some non-arbitrary grounds for a distinction between preventing greater evils and producing greater goods which would forbid subsuming both under the rubric of maximizing the good, must he not also admit that inflicting penalties on the innocent is justified when such a course of action will serve to produce greater goods than would otherwise occur and is the only available course of action that will so serve? What awaits him at the bottom of the slope is the position that we may, and perhaps would be remiss if we did not, penalize people just to do them, or their society, good.[3] Having rejected the elements of mere retaliation and revenge he takes to be embedded in retributivism, the humane determinist seems in danger of being forced to accept the benevolent totalitarianism that would result from coupling effective social technologies with the educative justification of the infliction of penalties.

Is a determinist here forced to choose between the frying pan and the fire? Maybe not. My discussion so far has followed Grünbaum's in supposing that the determinist's justification for inflicting penalties on people will rest solely on broadly utilitarian considerations involving such things as the educative and deterrent tendencies of such penalties in promoting good and preventing evil. Must the determinist appeal to such considerations and to no others in defining conditions and circumstances under which the infliction of penalties on people is justified? Though determinism and utilitarianism may have been, historically speaking, natural allies, I can see no logical connection between them. So perhaps the determinist could appeal to a human right to be free from interference on the part of others to limit the extent to which educative penalties may be inflicted on people contrary to their desires, and maybe he could make out a case that it is because criminals have forfeited the immunity protected by this right that there is relatively greater scope for inflicting penalties justified by social utility on them contrary to their desires. In some such fashion, a line might be drawn between the kinds and degrees of penalty it is appropriate to inflict upon the guilty by way of punishment and those it is appropriate to impose upon the innocent. This suggestion, of course, provides only the barest hint of how a determinist might proceed. Much more needs to be said. In particular, a determinist following up this line of thought would owe us an explanation of why humans, whose actions are, one and all, every bit as subject to causal determination as the behavior of machines and other animals, have such a right to freedom from interference.

After all, we consider it no violation of the rights of automobiles to tinker with their engines even when they are running well and no infringement on the rights of rats to use them in some psychological experiments.

The determinist's opponent can point to a significant difference between the behavior of humans, on the one hand, and the behavior of machines and other animals, on the other. On his view, some human actions occur without being determined by antecedently obtaining causally sufficient conditions logically independent of the agent's action, but none of the behavior of machines and other animals is so generated. So he might argue for the ascription to humans, but not to machines and other animals, of a *prima facie* right to freedom from interference on the ground that the function of such a right is to protect the sphere of human life in which actions of this special kind occur. Such a right need not be so strong as to guarantee absolute immunity from interference in this sphere, but it would limit the extent to which prospective goods can serve to justify the imposition of present hardships on normal human agents. Such a strategy is not available to the determinist, but perhaps he can find some difference between the actions of humans and the behavior of machines and other animals, other than this distinction in the manner of their causal genesis, to serve as the ground for ascribing such a right to humans, preeminently if not exclusively.

One thing, however, seems fairly clear to me. If a determinist is able to pursue this line of thought to a successful conclusion, the story he comes up with will be different from and considerably more complicated than the one Grünbaum puts in the mouth of his 'humane determinist'. That story, as we have seen, contains some rather ominous suggestions and may not be so humane after all when elaborated in full detail. From this I conclude that Grünbaum's attempt to provide the determinist with a plausible rationale for inflicting penalties on people contrary to their desires is, at best, seriously incomplete and, at worst, radically defective. It is incumbent on the determinist to do something more if his view is to avoid being correctly charged with leaving itself open to or, perhaps, even leading to morally repugnant consequences.

INTRICACIES OF MORAL ATTITUDE

There are philosophers who have held that some of our moral attitudes would never be appropriate if determinism were true or that we would consider it irrational to have such attitudes if we believed determinism to be true. In discussing this view, it is important to distinguish between attitudes and emotions. Moral attitudes are a not very well-defined group of intentional attitudes directed upon such things as states of affairs, events, actions and agents on account of the moral value thought to be associated with them. Moral emotions are an equally fuzzy group of feelings or passions connected at best contingently with moral attitudes. Thus, for example, a person may, in a cool hour, regret having performed a certain discreditable action without undergoing an intense episode of felt sorrow. Or a person may feel intense sorrow at having disappointed a friend and yet not regret the action at all if he believes that it would have been seriously wrong to comply with the friend's wishes. It is possible for a person to suffer from either an excess or a deficiency of a moral emotion. Neurotic guilt feelings sometimes paralyze a person or, at least, present serious obstacles to efforts at reform, but a person who feels no sorrow at all over misdeeds he regrets having done is a cold fish. In ordinary English, a single word often does double duty in referring both to an attitude and to an emotion. A person may regret that he has given offense and yet have no special feelings of regret if he perversely continues to have a predominant feeling of pleasure at having scored off a pompous rival. If in the midst of a cruel joke at the expense of a third party that person enters the room, the teller may regret having started the joke but continue to the end driven by a nasty sort of glee. There are situations in which both a moral attitude and a moral emotion are present, and yet the one is appropriate and the other is not. A person may justifiably resent the good fortune of a great malefactor on the grounds that it is undeserved and yet also feel inappropriately vindictive if his feelings move him to take the law into his own hands. In this discussion, I will be concerned with the appropriateness of both moral attitudes and moral emotions or feelings.

Grünbaum restricts his discussion to a brief consideration of remorse and regret. About remorse he says only that:

> We sometimes experience remorse over past conduct when we reconsider that conduct in the light of different motives or of a new awareness of its consequences (Feigl *et al.*, 1972, p. 614).

About regret he says a bit more. The claim is:

> Regret is an expression of our emotion toward the disvalue and injustice which issued from our past conduct, as seen in the light of new motives. The regret we experience can then act as a *deterrent* against the repetition of past behavior which issued in disvalue. If the determinist expresses regret concerning past misconduct, he is applying motives of self-improvement to himself but not indulging in retroactive self-blame (Feigl *et al.*, 1972, p. 615).

From these passages, I gather that Grünbaum has in mind primarily and, perhaps, exclusively the emotions of remorse and regret. On his view, the determinist will hold that they are themselves caused by changes in motivational factors such as desires for things taken to be valuable and beliefs about the consequences of actions and that they are appropriate only to the extent to which they contribute causally to deterring actions of kinds that have or contribute to disvalue. And though I find it difficult to make sense of the notion of a person applying motives of self-improvement to himself, as opposed to coming to have motives of self-improvement, partly because the notion of applying motives to oneself suggests a power to decide which motives one will have that the determinist ought not to attribute to people, it does seem to me reasonable to hold that feelings of remorse or regret sometimes function as deterrents and appropriately so.

Grünbaum's account of regret is, however, open to two other objections. The first is that the feeling of regret need not be an emotion directed toward injustice. When a person acts justly but contrary to his own interests, he may feel regret that he has acted justly. But this objection could easily be met by redefining the feeling of regret as an emotion focused on something the agent at the time of the feeling takes to be an aspect of his past conduct that has disvalue of some sort. The second objection is more serious. If one person offends another and endeavors to earn forgiveness by expressing regret, rather than by offering an excuse, pleading mitigating circumstances or something of that ilk, then it seems the endeavor will fail if the offender does not hold himself accountable and blameworthy for the offense. If the determinist, when he expresses regret, merely means to indicate the presence in him of certain feelings tending to deter him from future bad conduct, and does not intend to acknowledge that his past conduct was blameworthy, then forgiveness is out of place. In the absence of excuses, mitigating circumstances and the like, forgiveness is appropriate, like other mercies, only after the offender has admitted that, strictly speaking, he deserves to be blamed for his offense. So if the determinist does not indulge in retroactive self-blame, he cannot expect others to find it appropriate to forgive his offenses. Might he think, though in my opinion this would bespeak a terrible pride, that he will never need or want forgiveness?

Be that as it may, it is noteworthy that Grünbaum discusses no moral feelings or attitudes other than the emotions of remorse and regret. Either he simply has no account of the others to propose on behalf of the determinist or he takes it for granted that the view that they are, when appropriate, potential deterrents can be in some straightforward way generalized to cover them. A brief reflection on some of the intricacies of moral attitudes will serve to show how implausible the second alternative is.

Consider first the contrast between regret and shame. A person can properly regret the occurrence of events that are not human actions and that have no human actions or failures to act among their causal conditions. Many of us, reasonably enough, regret the occurrence of earthquakes that kill hundreds of innocent people. But a person cannot properly be ashamed of the occurrence of an event unless it is one of his actions or unless something he did or failed to do contributed causally to its occurrence. No one is reasonably ashamed of the occurrence of earthquakes. And if someone is properly ashamed of some event such as the collapse of a skyscraper, it is because he either did something that contributed to its collapse or failed to do something that would have contributed to preventing its collapse. A person can properly regret his failure to perform some action he knows it was not within his power to perform. It makes perfectly good sense to say "I regret I was unable to attend your party last night, but I was very ill." But it is dubious that a person can properly be ashamed of his failure to perform some action he knows it was not within his power to perform unless, for instance, his inability was itself partially caused by some other action it was within his power not to perform or by his failure to perform some action it was within his power to perform. It is difficult to understand what could be meant by saying "I am ashamed I was unable to attend your party last night, but I was very ill", unless it is presupposed that I am somehow responsible for being or getting ill last night. So regret and shame are diverse attitudes. And if feelings of regret and feelings of shame are identified in terms of the attitudes they typically accompany, then feelings of regret differ from feelings of shame.

It is worth noting that nothing I have said about the conditions under which shame is appropriate rules out the possibility that one person be properly ashamed of the actions of another. When parents are appropriately ashamed of the behavior of their children, presumably it is because their actions have made some causal contribution to that behavior. If a client is reasonably ashamed of his lawyer's low tricks in conducting a courtroom defense, I suppose the client has contributed causally to the lawyer's action in virtue of having done something which made a causal contribution to generating by convention an authorization for the lawyer to act in that manner as his agent. Indeed, it seems that people can even be ashamed of the actions of corporate persons. Thus, for example, many Americans were once ashamed of their country's waging a brutal and sordid war in Southeast Asia.

I suggest that shame in such instances is appropriate only if it is appropriate to think of one's country as in some sense one's agent to the authorization of whose actions one has made some causal contribution by action or omission.

It is because the attitude of regret can be directed toward things other than outcomes within the agent's power to affect with complete propriety that the agent need not blame himself for everything he regrets or even everything about himself that he regrets. This feature of the attitude of regret lends some intuitive plausibility to Grünbaum's claim that feelings of regret would be proper even in the absence on the part of the agent of any tendency to blame himself for the object of regret. But it is otherwise with shame. An agent can properly direct an attitude of shame only toward actions within his power or events to which actions or failures to act within his power contributed causally. It is this feature of the attitude of shame which lends plausibility to the claim that feelings of shame would be inappropriate unless the agent properly held himself to blame for the object of shame or for some action or failure to act within his power and causally efficacious in producing that object. If the agent did not hold himself to blame for something he did at least causally connected to the production of the object of his attitude, then this attitude would not be shame and the associated feeling would not be feelings of shame. On account of these salient differences, Grünbaum's account of what the determinist could make of feelings of regret will not generalize easily, if at all, to cover feelings of shame and the attitude of shame. So the determinist owes us an explanation of how, on his view, the attitude of shame can ever be appropriate, unless he is prepared to bite the bullet and maintain the counterintuitive thesis that shame is always an improper attitude.

Consider next the attitudes of resentment and indignation. Suppose a black person resents his being treated unjustly on racial grounds alone. I take it there would be widespread agreement that such resentment of injustice is a justified moral attitude on the part of the victim. If the attitude of resentment is accompanied by feelings of resentment of a suitable degree of intensity, these feelings too will be considered appropriate to the situation. Yet such feelings are hardly likely to function as motives of self-improvement. For one thing, it is the oppressor's behavior, not the victim's, that stands in need of improvement. For another, feelings of resentment are dangerous even when appropriate; they have been known to harden the heart. Nor are such feelings, if they find expression, likely to deter the oppressor from future acts of injustice. They seem more likely to confirm the prejudice that blacks are 'uppity' and provoke more severe forms of oppression. Of course, resentment on a large scale might serve as a powerful stimulus to revolt, but revolt might merely lead to repression. In short it seems that resenting an injustice done to oneself is a moral attitude we consider justified quite independent of any belief that its occurrence is likely to ameliorate the situation of the victim of oppression.

Or suppose a spectator is moved to indignation at the sight of a terrorist killing innocent people in an airport massacre. Here too I take it there would be widespread agreement that such indignation at the slaughter of the innocent is a justified moral attitude. If the attitude of indignation is coupled with feelings of indignation or anger of a suitable degree of intensity, these feelings will be considered proper in the circumstances. But such feelings are unlikely to improve the terrorist's character or behavior; if he learns of them, it may only serve to reinforce his conviction that terror is an effective political weapon. Nor are such feelings, if they find expression, guaranteed to do the spectator any good; if he acts on them, he may simply throw away his life in a futile effort to stop the terrorist's killing spree. To be sure, widespread indignation might function to motivate a serious campaign to suppress terrorism, but such a campaign might merely provoke further terrorist outrages. Again, the moral is that indignation at the wanton slaughter of the innocent is a moral attitude we regard as justified quite apart from any belief that its occurrence is likely to deter terrorists.

The determinist's opponent explains why moral attitudes, such as resentment and indignation, which are directed at other people are sometimes justified quite apart from any tendency to deter vicious actions by saying that the wrong-doer could in the circumstances have done otherwise and is, therefore, accountable and blameworthy for his actions in those cases where such attitudes are proper. Can the determinist offer an explanation of equal or greater plausibility? Perhaps not. Maybe he must regard all racists and terrorists as sick people in need of therapy, and maybe he must regard pity or sorrow, rather than such things as resentment or indignation, as the only attitudes it is proper to direct toward them or their actions.[4] If so, his view is seriously at odds with some of our most deeply entrenched convictions about the moral life. For though some racists and terrorists doubtless are sick people, the supposition that all are boggles the mind. Can we even begin to imagine the changes in our lives and practices that would be necessary to bring them into line with such a supposition? Can we seriously entertain the hypothesis that we might consider such massive changes required of us by reason at any time in the foreseeable future? I doubt it. But if the determinist does not propose to condemn as irrational every instance of a moral attitude that seems to imply that a person is accountable and blameworthy for an action, then he surely owes us an explanation of how it can be that such moral attitudes are sometimes justified which is much more complicated than anything Grünbaum has provided for him.

DETERMINISM IN THE DOCK

Grünbaum has attempted to provide determinism with a defense against charges that it is a morally repugnant doctrine. His procedure is to rebut moral objections to determinism by showing either that the moral data are not as they are alleged to be or that the moral data are not inconsistent with determinism but are instead plausibly explained by it. In this paper I have argued that there are moral objections to determinism not successfully rebutted by Grünbaum's arguments. Specifically, I have charged that the determinist has not yet met three challenges: (1) to show that it is irrational for us to pay agents the respect of sometimes treating them as opaque to the transmission of responsibility for their actions even in the absence of manipulability arguments for opacity, or to provide a rationale for this practice from within his theoretical position; (2) to show that it is irrational for us to treat the guilty and the innocent in significantly different ways with respect to the infliction of penalties, or to provide a rationale for this practice from within his theoretical position; and (3) to show that it is irrational for us to have, apart from ameliorative tendencies, self-directed moral attitudes such as shame and other-directed moral attitudes such as resentment and indignation which involve the propriety of blaming agents for their actions, or to provide a rationale for having such attitudes from within his theoretical position. If this charge has merit, determinism is once again in the dock. There it awaits further defense.[5]

Brown University and
University of Notre Dame

NOTES

[1] Grünbaum's main publications on determinism are the following: (1) Grünbaum (1952). This paper was reprinted in the Bobbs-Merrill Reprint Series in the Social Sciences. It has also been reprinted in Feigl and Brodbeck (1953), pp. 766–778; in Mandelbaum *et al.* (1957), pp. 328–338; in Daniel (1959), pp. 328–336; in Ulrich *et al.* (1966), pp. 3–10; and in Kamil and Simonson (1973), pp. 18–25. (2) Grünbaum (1962). This

paper has been reprinted in Hammer (1966), pp. 55–66 and in Mandelbaum *et al.* (1967), pp. 448–462. The version in the latter anthology contains omissions and revisions. (3) Grünbaum (1969). Grünbaum (1971) is a revised and expanded version of this paper. A further revised and expanded version appears in Feigl *et al.* (1972), pp. 605–627. Abridgements of one or another version are reprinted in Struhl and Struhl (1975), pp. 211–228 and in Lipman (1977), pp. 421–431. Translations of one or another version into Italian, Russian and Spanish have also been published. Since the version of this paper included in Feigl *et al.* (1972) contains the most recent and most fully developed statement of Grünbaum's position, I shall cite it exclusively in the text.

[2] A similar view has recently been expressed by Robert Nozick. He says: "Without free will, we seem diminished, merely the playthings of external forces. How, then, can we maintain an exalted view of ourselves? Determinism seems to undercut human dignity, it seems to undermine our value" (Nozick, 1981, p. 291). A complication, deliberately neglected in the text in the interest of simplicity of exposition, deserves mention here. Sometimes we push part but not all of the responsibility back beyond the first agent we encounter to other agents whose voluntary actions are necessary parts of causally necessary conditions for the action of the first agent, particularly if we have good reason to believe that the actions of those other agents significantly affected the range of options open to the first agent. For example, our story could be supplemented with details about Smith's childhood in such a way that Smith's father would be assigned some responsibility for the destruction of the bridge. So it would be more precise to say that our practice involves treating single human agents as at least partially opaque to the transmission of responsibility but only partially opaque when responsibility diffuses over several agents in this manner. As far as I can see, though, this complication does not affect the main point of my argument, which is the sharp distinction we take ourselves to be justified in drawing, when it comes to assignments of responsibility, between agents and other individual things involved in the causal genesis of events that engage our moral interests. Only agents are to be held accountable.

[3] That abandoning retributivist views of punishment puts one on such a slippery slope is, of course, a common enough opinion. It is succinctly stated in the following piece of fictional dialogue:

" . . . In real life punishment may produce any result, it's wild guesswork. And retribution is only important as a check, it's necessary for the sort of rough justice we hand out here below. I mean, the chap's got to have *done* something, and we must have a shot at saying how large or small it is – " "Otherwise we might penalise people just to do them good." "Or to deter other people, yes."

The quote is from Murdoch (1982), p. 68.

[4] At least in private, Darwin was willing to draw this conclusion from the rejection of free will. In a manuscript, he has this to say about the practical consequences of the determinism he himself was inclined to accept:

"Effects. – One must view a wicked man like a sickly one – We cannot help loathing a diseased offensive object, so we view wickedness. – It would however be more proper to pity than to hate and be disgusted with them. Yet it is right to punish criminals; but solely to *deter* others. – It is not more strange that there should be necessary wickedness than disease."

The quoted passage is to be found in Manier (1978), p. 219.

[5] I read an earlier version of this paper to the Philosophy Colloquium at the University of Notre Dame; Ernan McMullin was my commentator on that occasion. I thank him and members of the audience for helpful comments.

REFERENCES

Daniel, R. S. (ed.). 1959. *Readings in General Psychology*. Boston: Houghton Mifflin.

Feigl, H. and M. Brodbeck (eds.). 1953. *Readings in the Philosophy of Science*. New York: Appleton-Century-Crofts.

Feigl, H., K. Lehrer, and W. Sellars (eds.). 1972. *New Readings in Philosophical Analysis*. New York: Appleton-Century-Crofts.

Grünbaum, A. 1952. 'Causality and the Science of Human Behavior,' *American Scientist* 40, 665–676.

Grünbaum, A. 1962. 'Science and Man,' *Perspectives in Biology and Medicine* 5, 483–502.

Grünbaum, A. 1969. 'Free Will and the Laws of Human Behavior,' *L'Age de la Science* 2, 105–127.

Grünbaum, A. 1971. 'Free Will and the Laws of Human Behavior,' *American Philosophical Quarterly* 8, 299–313.

Hammer, L. Z. (ed.). 1966. *Value and Man*. New York: McGraw-Hill.

Kamil, A. C. and N. Simonson (eds.). 1973. *Patterns of Psychology: Issues and Prospects*. Boston: Little, Brown and Co.

Lipman, M. (ed.). 1977. *Discovering Philosophy*. 2nd ed. Englewood Cliffs, N.J.: Prentice-Hall.

Mandelbaum, M., F. W. Gramlich, and A. R. Anderson (eds.). 1957. *Philosophic Problems*. 1st ed. New York: Macmillan.

Mandelbaum, M., F. W. Gramlich, A. R. Anderson, and J. B. Schneewind (eds.). 1967. *Philosophic Problems*. 2nd ed. New York: Macmillan.

Manier, E. 1978. *The Young Darwin and His Cultural Circle. Studies in the History of Modern Science*, vol. 2. Dordrecht and Boston: D. Reidel.

Murdoch, I. 1982. *Nuns and Soldiers*. New York: Penguin Books.

Nozick, R. 1981. *Philosophical Explanations*. Cambridge, Mass.: Harvard University Press.

Struhl, P. R. and K. J. Struhl (eds.). 1975. *Philosophy Now: An Introductory Reader*. 2nd ed. New York: Random House.

Ulrich, R., T. Stachnik, and J. Mabry (eds.). 1966. *Control of Human Behavior*. Glenview, Ill.: Scott, Foresman and Co.

NICHOLAS RESCHER

THE UNPREDICTABILITY OF FUTURE SCIENCE

ABSTRACT. (1) It is difficult – indeed impossible – to predict the future of science. For we cannot forecast in detail even what the questions on the agenda of future science will be. (2) Viewed not in terms of its *aims* but in terms of its *results*, science is inescapably *plastic*: it is not something fixed, frozen and unchanging, but something variable and protean – given to changing not only its opinions but its very form as well. (3) We thus cannot discern the substance of future science with sufficient clarity to say just what it can and cannot accomplish. (4) Setting domain limitations to science – putting entire ranges of phenomena outside its explanatory grasp – is a risky and unprofitable business.

1. DIFFICULTIES IN PREDICTING THE FUTURE OF SCIENCE

The landscape of natural science is ever-changing. Innovation is the name of the game. Not only do the theses and themes of science change, but so do the very questions with which it grapples. To be sure, once a body of science is seen as something settled and in hand, many issues become routine. Various problems become mere reference questions – a matter of locating an answer within the body of pre-established, already available information that somewhere contains it. (The 'mere' here is, of course, misleading in its downplaying of the formidable challenges that can arise in looking for needles in very large haystacks.) However, in pioneering science we face a very different situation. People may well wonder "what is the cause of *X*" – what causes cancer, say, or what produces the attraction of the loadstone for iron, to take an historical example – in circumstances where the concepts needed to develop a workable answer still lie beyond their grasp. Scientific inquiry is a creative process of theoretical and conceptual innovation – not a matter of pinpointing the most attractive alternative within the presently specifiable range, but of enhancing and enlarging this range of envisageable alternatives. Such issues pose genuinely open-ended questions of original research: they do not call for the resolution of problems within a pre-existing framework, but for a rebuilding and enhancement of the framework itself. Most of the questions with which present-day science grapples could not even have been raised in the state of the art that prevailed a generation ago.

Questions about the course of questioning itself are among the most

R. S. Cohen and L. Laudan (eds.), Physics, Philosophy and Psychoanalysis, 153–168.

interesting issues in the entire sphere of metascience. One would certainly like to be in a position to have prior insight into and give some advance guidance to the development of scientific progress. But grave difficulties arise at this point, where questions about the future, and, in particular, the *cognitive* future are involved.

Quite in general, when the prediction of *a specific predefined occurrence* is at issue, this forecast can go wrong in only one way, by proving to be incorrect – the particular development at issue simply does not happen as predicted. But the forecasting of *a general course of developments* can go wrong in two ways, either by way of positive errors of commission (that is, forecasting something quite different from what actually happens – say, a rainy season instead of a drought), or by way of negative errors of omission (that is, failing to foresee some significant development within the overall course – predicting an outbreak of an epidemic, say, without recognizing that one's own locality will be involved). In the first case, one forecasts the wrong thing, in the second, there is a lack of completeness, a failure of prevision, a certain blindness.

In cognitive forecasting it is these errors of omission – our blind spots, as it were – that present the most serious threat, yet it is exactly this sort of error that is most prominent in this domain. For the fact is that we cannot substantially anticipate the evolution of knowledge.

To begin with, we cannot say in advance what the specific answers to our scientific questions will be. It would, after all, be quite unreasonable to expect detailed prognostications about the particular *content* of scientific discoveries. It may be possible in some cases to speculate *that* science will solve a certain problem, but *how* it will do so – in the sense of specifying what the particular character of the solution is – generally lies beyond the ken of those who antedate that discovery itself. If we could *predict* discoveries in detail in advance, then we could *make* them in advance.[1] In matters of scientific importance, then, we must be prepared for surprises. Commenting shortly after the publication of Frederick Soddy's speculations about atomic bombs in his 1930 book on *Science and Life*, Robert A. Millikan, a Nobel laureate in physics, wrote that:

> The new evidence born of further scientific study is to the effect that it is highly improbable that there is any appreciable amount of available subatomic energy to tap.[2]

In science-forecasting, the record of even the most qualified practitioners is poor. We may well not even be able to conceive the explanatory mechanisms of which future science will make routine use. As one sagacious observer noted over a century ago:

There is no necessity for supposing that the true explanation must be one which, with only our present experience, we *could* imagine. Among the natural agents with which we are acquainted, the vibrations of an elastic fluid may be the only one whose laws bear a close resemblance to those of light; but we cannot tell that there does not exist an unknown cause, other than an elastic ether diffused through space, yet producing effects identical in some respects with those which would result from the undulations of such an ether. To assume that no such cause can exist, appears to me an extreme case of assumption without evidence.[3]

Science-prediction is inherently problematic.

The circumstance that we cannot predict the answers to the presently open questions of natural science means that we cannot predict its future questions either. For these questions will of course hinge upon these yet unrealizable answers. Indeed, we cannot even say what *concepts* the inquirers of the future will use within the questions they will raise. Maxwell's work was directed towards answering questions that grew out of Faraday's. Hertz devised his apparatus to answer questions about the implications of Maxwell's questions. Marconi's devices were designed to resolve questions about the application of Hertz's work. It is impossible to discern at earlier stages the substance of the questions that will only emerge at later ones – and thus the issues and conceptions to which these will give rise.

It is a key fact of life in this domain that ongoing progress in scientific inquiry is a process of *conceptual* innovation that always places certain developments outside the cognitive horizons of earlier workers because the very concepts operative in their characterization become available only in the course of scientific discovery itself. Short of learning our science from the ground up, Aristotle could have made nothing of modern genetics, nor Newton of quantum physics. The key discoveries of later stages are those of which the workers of a substantially earlier period (however clever) not only have failed to make, but which they could not even have *understood* because the concepts required for such understanding were simply not available to them. It is thus effectively impossible to predict not only the answers but even the questions that lie on the agenda of future science, because these questions themselves will grow out of the answers we obtain at yet unattained stages of the game. Detailed prediction is outside the realm of reasonable aspiration in those domains where innovation is pre-eminently conceptual.

A body of knowledge may well fail to answer a question at all, or else, even if an answer is forthcoming, this can be given in so tentative or indecisive a tone of voice as to suggest that further information is needed to settle the matter with any degree of confidence. But even our an existing 'knowledge' does confidently and unqualifiedly support a certain particular resolution, this circumstance can never be viewed as absolutely final. What is seen as the correct anwer to a question at one stage of the cognitive venture, may, of course, cease to be so regarded at another.

In inquiry as in other areas of human affairs, major upheavals can come about in a manner that is sudden, unanticipated, and often unwelcome. Major scientific breakthroughs often result from research projects that have very different ends in view. Louis Pasteur's discovery of the protective efficacy of inoculation with weakened disease strains affords a striking example. While studying chicken cholera, Pasteur accidentally inoculated a group of chickens with a weak culture. The chickens became ill but, instead of dying, recovered. Not wanting to waste chickens, Pasteur later reinoculated these chickens with fresh culture – one strong enough to kill an ordinary chicken. The chickens remained healthy. Pasteur's attention then shifted to this interesting phenomenon, and a productive new line of investigation was opened up.

It is an ironic but critically important feature of scientific inquiry that the unforeseeable tends to be of special significance just because of its unpredictability, its coming upon us unexpectedly. We face the circumstance that: *the more important the innovation, the less predictable it is, because its very unpredictability is a key component of its importance*. Thomas Kuhn has interestingly distinguished between the "normal science" being developed when things go along the tracks of a well-defined tradition of thought and investigation (proceeding within an established *paradigm* as he calls it), and "scientific revolutions" in which there is massive intellectual upheaval and one paradigm is overthrown and replaced by another.[4] Science forecasting is beset by a pervasive *normality bias*, because the really novel often seems so bizarre prior to the event itself, when the wisdom of hindsight is as yet unavailable. (Before the event, scientifically important innovations will, if imaginable at all, generally be deemed outlandishly wild speculation – mere science fiction, or perhaps just plain crazy.) Then too, there is the circumstance that our future scientific questions are engendered by our endeavors to answer those we have on hand. Our scientific questions exfoliate from the answers we give to previous ones, so that, in consequence, the issues of future science simply lie beyond our present horizons. Accordingly, we cannot predict science's solutions to its problems because we cannot even predict in detail just what these problems will be – let alone what solution the science of the future will provide for them.

Moreover, forecasts of scientific innovation conform to the vexatious general principle that, other things being equal, *the more informative a forecast is, the less secure, and conversely, the less informative, the more secure.*[5]

A degradation relationship circumscribes the domain of feasible prediction. The reliability of the claims we can responsibly make declines sharply with the exactness to which we aspire. We are led to the situation depicted in Figure 1, that if one insists on resolving a predictive question with informativeness (exactness, detail, precision), one's confidence in the response will have to be drastically diminished. And any exercise in science forecasting is bound to conform to this general relationship of futurological indeterminacy.

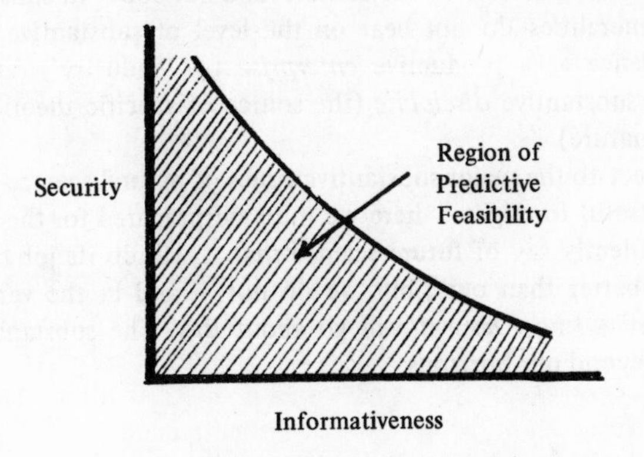

Fig. 1. The degradation of security with increasing informativeness.

The course of safety accordingly lies in degrading the informativeness of our forecasts. But this obviously undermines their utility as well and, in the case of science, renders them ineffectual and useless. Science scorns the security of inexactness. If we are content to say that the solution to a problem will be of 'roughly' this general sort or that it will be 'somewhat analogous' to some familiar situation, we will be on safe ground. But science is not like that. It does not deal in analogies and generalities and rough approximations. It strives for exactness and explicitness – for precision, detail, and, above all, for generality and universality of application. It is exactly this feature of its claims that makes science forecasting a very risky business. For the more we enrich our forecast with the details of just exactly what and just exactly how (etc.) the more vulnerable it becomes.

In maintaining that future science is inherently unpredictable, one must be careful to keep in view the distinction between substantive and structural issues, between particular individual scientific questions, theses, and theories, and generic features of the entire system of such individual items. At the level of generality, various inductions regarding the future of science, can, no doubt, be safely made. We can, for example, safely predict that future science will have greater taxonomic diversification, greater theoretical unification, greater substantive complexity, further high-level unification, low-level proliferation, increased taxonomic speciation of subject-matter specialties, etc. And we can certainly predict that it will be incomplete, that its agenda of availably open questions will be extensive, etc. But of course this sort of information tells us only about the *structure* of future science, and not about its *substance*. These structural generalities do not bear on the level of substantive detail: they relate to science as a productive *enterprise* (or 'industry') rather than to science as a substantive *discipline* (the source of specific theories about the workings of nature).

With respect to the major substantive issues of natural science there can be little or no useful foresight – here we must be prepared for the unexpected. We can confidently say of future science *that* it will do its job of prediction and control better than ours; but we do not – and in the very nature of things cannot – know *how* it will go about this. The substance of future science lies beyond our horizons.

2. PRESENT SCIENCE CANNOT SPEAK FOR FUTURE SCIENCE

Analysis of the theoretical general principles of the case and induction from the history of science both indicate that the theories accepted at one state-of-the-art stage of scientific inquiry need not carry over to another, but can be replaced by something radically different. The reality of past scientific revolutions cannot be questioned and the prospect of future scientific revolutions cannot be precluded.

If there was one thing of which the science of the first half of the seventeenth century was absolutely certain, it was that natural processes are based on contact-interaction and that there can be no such thing as action at a distance. Newtonian gravitation burst upon this science like a bombshell.

Newton's supporters simply stonewalled. Roger Coates explicitly denied there was a problem, arguing (in his Preface to the second edition of Newton's *Principia*) that nature was generally unintelligible and that the unintelligibility of forces acting without contact was thus nothing worrisome. Now, however unpalatable Coates' position may seem as a precept for science (given that making nature's workings understandable is, after all, one of the aims of the enterprise), there is something to be said for it – not, to be sure, as science, but as metascience. For we cannot hold the science of tomorrow bound to the standards of intelligibility espoused by the science of today. The cognitive future is inaccessible to even the ablest of present-day workers. After Pasteur had shown that bacteria could come only from pre-existing bacteria, Darwin wrote that, "It is mere rubbish thinking of the origin of life; one might as well think of the origin of matter."[6] One might indeed!

The inherent unpredictability of future scientific developments – the fact that no secure inference can be drawn from one state of science to another – has important implications for the issue of the limits of science. It means that *present-day science cannot speak for future science*: it is in principle impossible to make any secure inferences from the substance of science at one time to its substance at a significantly different time. We cannot say with unblinking confidence what sorts of resources and conceptions the science of the future will or will not use. And thus any attempts to set 'limits' to science – any advance specification of what science can and cannot do by way of handling problems and solving questions – is virtually destined to come to grief. Given that it is effectively impossible to predict the details of what future science will accomplish, it is no less impossible to predict in detail what future science will *not* accomplish. We can never confidently put this or that range of issues outside "the limits of science," because we cannot discern the shape and substance of future science with sufficient clarity to be able to say with any assurance what it can and cannot do.

A seeming violation of the rule that present science cannot bind future science is afforded by John von Neumann's 1932 attempt to demonstrate that all future theories of subatomic phenomena will have to contain an analogue of Heisenberg's uncertainty principle if they are to account for the data explained by present theory, so that complete predictability at the subatomic level was forever exiled from science. But von Neumann's 'demonstration' is at bottom circular: involving not merely presently accepted fact but also placing a heavy burden on presently accepted theory.[7] The salient circumstance remains that we cannot preclude fundamental innovation in science: present theory cannot delimit the potential of future discovery. In natural science we cannot erect our structures with a solidity that defies demolition and reconstruction.

3. THE PLASTICITY OF SCIENCE

The crucial fact is that science is not an atemporal phenomenon. We can certainly speak of 'scientific fact,' 'scientific questions,' 'scientific methods' at the level of abstract classificatory generality. But we can never speak concretely of 'the facts of science,' 'the questions of science,' 'the methods of science' flatly and categorically – without adding a temporal qualification such as 'as construed in the seventeenth century,' or 'as we see them today.' And the reason for this disability lies exactly in this fact that present-day science cannot speak for future science – that we are not, and never will be, in a position to delineate the issues and materials of *future* science.

It is only in *functional* terms that we can give a satisfying definition of natural science and provide a viable characterization of its nature. The one and only thing that is fixed about science is its mission or mandate – viz. the tasks of description, explanation, prediction and control of natural phenomena, and the commitment to proceed in these matters by the sort of rigorous, controlled methods of testing and substantiation for our assertions that has become known as 'the scientific method.' Everything else is potentially changeable. Entire subject-matter areas can come (quantum electrodynamics) or go (astrology). Questions can arise (the composition of quarks) or vanish (the structure of the luminiferous ether). Science is inherently changeable.

At the frontiers of scientific innovation, the half-life of theories and explanatory models is proverbially brief. Even particular devices of methodology can change drastically – witness the rise of probabilistic argumentation and use of statistics in the 'design of experiments.' Larry Laudan has illustrated this point nicely:

> In many cases, *it is the methodology of science itself which is altered*. Consider, as but one example, the development of Newtonian theory in the eighteenth century. By the 1720's, the dominant methodology accepted alike by scientists and philosophers was an *inductivist* one. Following the claims of Bacon, Locke, and Newton himself, researchers were convinced that the only legitimate theories were those which could be inductively inferred by simple generalization from observable data. Unfortunately, however, the

direction of physical theory by the 1740's and 1750's scarcely seemed to square with this explicit inductivist methodology. Within electricity, heat theory, pneumatics, chemistry, and physiology, Newtonian theories were emerging which postulated the existence of imperceptible particles and fluids – entities which could not conceivably be "inductively inferred" from observed data ... [and so, various] Newtonians (e.g., LeSage, Hartley, and Lambert) insisted *the norms themselves should be changed* so as to bring them into line with the best available physical theories. This latter group took it on themselves to hammer out a new methodology for science which would provide a license for theorizing about unseen entities. (In its essentials, the methodology they produced was the hypothetico-deductive methodology, which even now remains the dominant one.) This new methodology, by providing a rationale for "micro-theorizing," eliminated what had been a major conceptual stumbling block to the acceptance of a wide range of Newtonian theories in the mid and late eighteenth century.[8]

The quintessential requisites of scientific intelligibility can themselves suffer drastic transformation – witness the fate of the hallowed proscription of action at a distance in the age of Newtonian physics. Science can and does change "the rules of the game" on us. It can and does abandon whatever is unserviceable and co-opt whatever can serve its purposes to good effect.

Any attempt to specify the 'essential presuppositions' or 'unchangeable commitments' of science is thus virtually destined to come to grief. The principle that every event has a specific cause whose operation explains why it eventuated thus rather than otherwise has been an article of scientific faith since classical antiquity. *Nihil turpius physico quam fieri quidquam sine causa dicere* said Cicero,[9] and for two millennia this doctrine that "Nothing occurs by chance" was a received dogma in science, until quantum theory came along to sweep this so-called *principle of causality* away with a single stroke. Again, consider R. G. Collingwood's valiant attempt to specify certain absolute (or ultimate, or primal) presuppositions of the scientific enterprise.[10] He held, in particular, that *the uniformity of nature* must be presupposed if the scientific enterprise is to arrive anywhere. But who is to say that cosmology may not decide that the universe is partitioned into distinct compartments (and/or eras) within which different ground-rules apply – i.e., whose 'laws of nature' are nowise uniform. Any such position is undermined by the changeability of science. What seems an absolute presupposition of science at one point may be explicitly denied at another. In fact, this view involves an insubstantiatable absolutism – quite uncharacteristic of Collingwood – by assuming the changeless universality of these 'absolute presuppositions.'

Natural science is altogether ruthless and opportunistic – very much a fair-weather friend. If an old and heretofore useful theory can no longer do serviceable work, science has no hesitancy about dropping it. If a heretofore rejected theory can do serviceable work in altered circumstances, science has no hesitancy about taking it up. On the side of its procedures and theses science has no fixed nature, no stable commitments; it is prepared to turn whichever way the wind blows. Science is fickle: it is given to flirtations rather than lasting relationships. It is prepared on any and every day to wipe the slate clean if circumstances indicate this to be advantageous.

For science is not something fixed, frozen, and unchanging – it is variable and protean, capable of changing not only its opinions but its very form as well. Matters squarely placed outside the boundaries of science in one era (action at a distance dismissed as unintelligible nonsense, or hypnotism rejected as fakery) can enter within the orthodoxy of a later generation. Even more than popular American culture, science worships the power of success. It is incurably pragmatic, and settles for whatever works, abandoning its 'established general principles' with a shameless disloyalty when changed circumstances recommend this course. When something new comes along that proves more effective than existing 'knowledge,' science immediately changes course. Success is the guiding star: science immediately *co-opts* whatever succeeds in the sphere of its mission. If something other than existing science were to work out better, science itself would change to accomodate this. We cannot permanently exclude something from the boundaries of science because these boundaries are changeable, which is to say that they are effectively nonexistent.

The *supremacy* of natural science is closely bound up with its *plasticity*. If we could set limits to the shape and substance of science, then it could also be possible to set limits to what science can and cannot accomplish. If, for example, we could say (with Emil du Bois Reymond[11]) that the explanatory program of science is an atomistic Newtonianism – a mechanistic world in which we could, at the most and best, achieve the 'astronomical knowledge' of a 'Laplacean Spirit' who knows the total history of the motions of all particles throughout space for all of time – then we could indeed say (as Reymond does) that science cannot account for the source of motion, the nature of matter, the operations of consciousness, etc. If we could circumscribe the substantive nature of science, we could also circumscribe its potential achievements. But just this is impossible. We cannot defensibly project the present lineaments of science into the future as a whole.

Requirements for the 'scientific acceptability' of explanatory resources are always problematic. Once upon a time we were told that there can be no genuine science of human behavior because human acts can be unpredictable ('impulsive', fortuitous) while scientifically proper phenomena must in principle be predictable. But even before the rise of stochastic phenomena in quantum physics one might have asked: who has issued a guarantee that scientifically tractable phenomena must be predictable – that science cannot tread where predictability is absent?

While we can say with confidence what the state of science *as we now have it* does and does not allow, we cannot say what science as such will or will not allow. The boundary between the tenable and the untenable in science is never easy to discern. A. N. Whitehead has wisely remarked that:

> If you have had your attention directed to the novelties in thought in your own lifetime, you will have observed that almost all really new ideas have a certain aspect of foolishness when they are first produced.[12]

Future science can turn in unexpected directions. The realm of scientific possibility is unchartable. "There are more things in heaven and earth, Horatio . . . ."

There is indeed a line between science and pseudoscience – and also between competent science and poor science. But such boundaries cannot, however, be drawn on *substantive* grounds regarding the assertion-content of the theories at issue, but only on *methodological* grounds with respect to the process by which the assertions have been tested and substantiated. That some theoretically available scientific position fails to accord with the science of the day is readily established, but that it is inherently unscientific is something that it lies beyond our powers to show. The contention that this or that explanatory resource is inherently unscientific should always be met with instant scorn. For the unscientific can only lie on the side of process and not that of product – on the side of *modes* of explanation and not its *mechanisms*; of arguments rather than phenomena.

Science – as already noted – is simply too opportunistic to be fastidious about its mechanisms. Eighteenth century psychologists ruled out hypnotism. Nineteenth century biologists excluded historical catastrophes. Twentieth century geologists long rejected continental drift. Many contemporary scientists give parapsychology short shrift – yet who can say what the future bodes? The pivotal issue is not what is substantively claimed by an assertion but rather whether this assertion (whatever its content!) has been substantiated through the usual canons of scientific method. The line between science and pseudoscience cannot be defined in terms of content – in terms of what sorts of theses or theories are maintained – but only in terms of method, in terms of how these theories are argued for. Its very fixity and unyieldingness is a mark of pseudoscience in contrast to science proper.

There is, to be sure, rough wisdom in scientific caution and conservatism: it is perfectly appropriate to feel sceptical towards unusual phenomena-constructions and to view them with right-minded scepticism. Before admitting 'strange' phenomena as appropriate exploratory issues we certainly want to check their credentials, make sure they are well attested and appropriately characterized. The very fact they go against our understanding of nature's ways as best we can develop them renders abnormalities suspect – a proper focus of worriment and distrust. But of course to hew this line dogmatically and rigidly in season and out of season, is to risk throwing out the baby with the bath water. The untenable in science does not conveniently wear its untenability on its sleeve. We have to realize that throughout the history of science, stumbling on anomalies – on 'new phenomena', occurrences that just don't fit into the existing framework – has been a strong force of scientific progress and that there is no way in which we can write *finis* on the page. We have – and can have – no basis for claiming that we have reached the end of the road and that the day of new phenomena is over. And new theories can be ruled out even less than new phenomena – for to do so would be to claim that science has reached the end of its tether, something we certainly cannot and doubtless should not even wish to do.

One does well to distinguish two modes of 'strangeness' – the exotic and the counterindicated. The *exotic* is simply something additional, foreign – something that does not fall into the range of what is known, but that does not clash with it either. It lies in what would otherwise be an informational vacuum, *outside* our current understanding of the natural order. (Hypnotism and precognition are perhaps cases in point.) The *counterindicated*, however, stands in actual *conflict* with our current understanding of the natural order (action-at-a-distance for seventeenth century physics; telekinesis today). One would certainly want to apply rigid standards of recognition and admission to those phenomena that are counterindicated. But if and when they measure up, we have to take them in our stride. In science we have no alternative but to follow nature where it leads us. Throughout natural science we are poised in a delicate balance between reasonable assurance that what we believe is worth holding to and a heart-of-hearts recognition that we do not yet have 'the last word' – that the course of events may at any time shatter our best laid plans for understanding the world's ways. We can set no *a priori* restrictions. In natural science we have to be flexible – and prepared to be open-minded.

The epistemological situation of science points towards a theology-reminiscent view of the human condition that sees man as a creature poised between comfortable assurance and the abyss. Our hold on the things of the world – our science included! – is always tenuous. We must face it with the dual realization that it is at once the best we can do, and that nevertheless it is by no means good enough. Our scientific picture of the world is vulnerable to being overthrown at any instant by new, unlooked for and perhaps even unwelcome developments. The history of man's science, like the history of all his works, invites ruminations on the transiency of things human and the infeasibility of attaining perfection. (To be sure, in our cognitive as in our moral life, the unattainability of perfection affords no reason why we should not do the very best we can; the point is simply that we can never afford to rest justified in the smug confidence that what we have attained is quite good enough.)

The inherent unpredictability of scientific change is the very hallmark of science. It sets real science apart from the closed structures of pseudoscience. And it means that no sort of idea, mechanism, or issue can be placed with reasonable assurance outside the realm of science as such. Nobody can say what science will and will not be able to do.

4. AGAINST DOMAIN LIMITATIONS

The idea of cognitive limits has a paradoxical air. It suggests that we claim knowledge about something outside knowledge. Yet as Hegel was wont to insist, with the realm of knowledge we are not in a position to draw a line between what lies inside and what lies outside – seeing that, *ex hypothesi*, we have no cognitive access to the latter. And there is no doubt that the preceding deliberations strongly suggest that the project of stipulating boundaries for natural science – of placing certain classes of nature's phenomena outside its explanatory reach – has its problems.

Domain limitations purport to put entire sectors of fact wholly outside the effective range of scientific explanation, maintaining that an entire range of phenomena in nature defies scientific rationalization. And this is always a problematic contention.

The scientific study of human affairs affords a prime historical example. Various nineteenth century German theorists maintained that a genuine science of man is in principle impossible – one that affords explanation of the full range of human phenomena, including man's thoughts and creative activities. One of the key arguments for this position ran as follows: it is a presupposition or postulate of scientific inquiry that the *object* of investigation is essentially independent of the *process* of inquiry, being itself nowise altered or affected by the inquiry. But this is clearly not the case when our own thoughts, beliefs, and actions are at issue, since they are all affected as we learn more about them. Accordingly, it was argued, there can be no such thing as a science of the standard sort regarding characteristically human phenomena. One cannot understand or explain human thought and action on the usual procedures of natural science, but must introduce some special *ad hoc* mechanisms for coming to explanatory terms with the human sphere. (Thus Wilhelm Dilthey, for example, maintained the need for recourse to an internalized *Verstehen* as a peculiar, science-transcending instrumentality that sets the human 'sciences' apart from the standard, i.e., natural *sciences*.) And (so it was argued) seeing that such a process cannot, strictly speaking, qualify as *scientific*, there can be no such thing as a 'science of man'.

But if construed as a *delineation* of scientific capacity, this sort of critique is badly misguided. Science is characterized as such by a certain method of inquiry, a certain cognitive approach to problems. It cannot prejudge results nor does it predelimit ways and means. One cannot in advance rule in or rule out particular explanatory processes or mechanisms ('observation-independence,' etc.). If in dealing with certain phenomena they emerge as observer-indifferent, so be it. If we indeed need explanatory recourse to some sort of observer-correlative resource ('sympathy' or the like), then that's that. In science we cannot afford to indulge our *a priori* preconceptions.

To be sure, if *predictability* is seen as the hallmark of the scientific then there cannot be a science that encompasses all human phenomena (and, in particular, not a science of science). But of course there is no reason why, in human affairs any more than in quantum theory, the boundaries of science should be so drawn as to exclude the unpredictable.

When we encounter strange 'intractable' or 'inexplainable' phenomena it is folly to wring our hands and say that science has come to the end of its tether. For it is exactly here that science must roll up its sleeves and get to work. Long ago, Baden Powell got the main point right:

> When we arrive at any such seeming boundary of present investigation, still this brings us to no *new world* in which a different order of things prevails; it merely points to what will assuredly be a fresh starting point for future research. It is an unwarrantable presumption to assert, that at a mere point of difficulty or obscurity we have reached the boundary of the dominion of physical law, and must suppose all beyond to be arbitrary and inscrutable to our faculties. It is the mere refuge and confession of ignorance and indolence to imagine special interruptions, and to abandon reason for mysticism.[13]

It is poor judgment to jump from a recognition that the science of the day cannot handle something to the conclusion that science as such cannot handle it.

But the changeability and plasticity of science is also a source of its power. Its very instability is at once a *limitation* of science and a part of what frees it from having actual *limits*. We can never securely place any sector of phenomena outside its explanatory range.

Not only can we never claim with confidence that the science of tomorrow will not resolve the issues that the science of today sees as intractable, but one can never be sure that the science of tomorrow will not endorse what the science of today rejects. This is why it is infinitely risky to speak of this or that explanatory resource (action at a distance, stochastic processes, mesmerism, etc.) as inherently unscientific. Even if *X* lies outside the range of science as we nowadays construe it, it by no means follows that *X* lies outside science as such. We cannot but recognize the commonplace phenomenon that the science of one day manages to do what the science of an earlier day deemed infeasible to the point of absurdity ('split the atom', abolish parity, etc.).

The upshot of these deliberations is clear. To set domain limitations to natural science is always risky and generally ill-advised. The course of wisdom is to refrain from putting issues outside the explanatory range of science. To repeat: present science cannot speak for future science; it cannot establish what science as such can and cannot do. In the domain of scientific theory, the cognitive present can never extend absolute assurances for the future. The prospect of a change of mind in scientific matters can never be totally precluded. It makes no sense to set limits to natural science itself. Charles Sanders Peirce's dictum holds good: one must not bar the path of scientific inquiry.

University of Pittsburgh

NOTES

[1] As one commentator has sagely written:

But prediction in the field of pure science is another matter. The scientist sets forth over an uncharted sea and the scribe, left behind on the dock, is asked what he may find at the other side of the waters. If the scribe knew, the scientist would not have to make his voyage. (Anonymous, 'The Future as Suggested by Developments of the Past Seventy-Five Years,' *Scientific American* **123** (1920), pp. 320–321 (see p. 321).)

[2] Quoted in *Daedalus* 107 (1978), p. 24.
[3] Baden Powell, *Essays on the Spirit of the Inductive Philosophy, the Unity of Worlds and the Philosophy of Creation* (London: Longman, Brown, Green and Longmans, 1855), p. 23.
[4] See Thomas Kuhn, *The Structure of Scientific Revolutions* (Chicago: University of Chicago Press, 1962; 2nd ed., 1970).
[5] Consider some applications of this principle. (1) It is easier and safer to forecast general trends than specific developments. (2) It is easier and safer to forecast over the near future than over the longer term; long-range forecasts are inherently more problematic. (3) The fewer and cruder the parameters of a prediction, the safer it becomes: it is easier and safer to forecast aggregated phenomena than particular eventuations (e.g., how many persons will live in a certain city 10 years hence as compared with how many will belong to a particular family. (4) The more extensively it is laden with a protective shield of qualifications and limitations, the safer the prediction. (5) The more vaguely and ambiguously a prediction is formulated, the safer it becomes; particularly equivocal predictions have an inherent advantage. (6) The prediction of possibilities and prospects is more safe and secure than that of real and concrete developments. (It is one thing to predict what will be *feasible* at a given 'state-of-the-art' and another to predict what will be *actual*.)
[6] Quoted in Philip Handler, ed., *Biology and the Future of Man* (Oxford: Oxford University Press, 1970), p. 165.
[7] See the criticisms of his argument also in David Bohm, *Causality and Chance in Modern Physics* (New York: Van Nostrand, 1957), pp. 95–96.
[8] Larry Laudan, *Progress and its Problems* (Berkeley: University of California Press, 1977), pp. 59–60.
[9] Cicero, *De finibus*, I, vi, 19.
[10] See R. G. Collingwood, *An Essay in Metaphysics* (Oxford: Oxford University Press, 1940) and *The Idea of Nature* (Oxford: Oxford University Press, 1945).
[11] *Ueber die Grenzen des Naturerkennens* (11th ed., Leipzig: Veit Verlag, 1916).
[12] A. N. Whitehead as cited in John Ziman, *Reliable Knowledge* (Cambridge: Cambridge University Press, 1969), pp. 142–143.
[13] Baden Powell, *Essays on the Spirit of the Inductive Philosophy, the Unity of Worlds, and the Philosophy of Creation* (London: Longman, Brown, Green and Longmans, 1855), p. 111.

BENJAMIN B. RUBINSTEIN

FREUD'S EARLY THEORIES OF HYSTERIA

Freud's theories of hysteria are of interest for both the history and the philosophy of psychoanalysis. A critical study of the development of these theories is apt to throw light on Freud's way of thinking. Since the theories are predominantly clinical, a study of their development can be expected to be relevant also for psychoanalytic practice.

1. THE TRAUMATIC CAUSATION OF HYSTERIA

The beginnings of psychoanalysis can be traced back to the early 1980's when Freud for a few years collaborated with Breuer. The latter had more than ten years earlier discovered the so-called cathartic method of treating hysteria. According to the editor's introduction to the *Studies on Hysteria* (Breuer and Freud, 1893–1895/1955, p. xxii), Breuer may have been the one who originally introduced the concept of catharsis; but the first published use of the term was in the 'Preliminary Communication' by Breuer and Freud (1893, p. 8) that opens the *Studies*. The concept, of course, goes back to Aristotle's view of tragedy which, according to this author, includes, among other things, the presentation of "incidents arousing pity and fear, wherewith to accomplish its catharsis of such emotions" (*Poetics*, p. 348).

Breuer's findings are readily summarized. He found (1) that hysterical symptoms often are consequent on a psychologically traumatic event, i.e., an emotionally charged, in some way highly unpleasant experience that because of its unpleasant character had been split off from consciousness and (2) that the symptoms disappeared when the unpleasant experience and its associated emotions were again rendered conscious through hypnosis. We may note that the symptoms were related, not to one, but to several such experiences.

This was a momentous discovery. The use of hypnosis in the treatment of hysteria was by no means unknown but only as a vehicle for the removal of hysterical symptoms by suggestion. Although he for some years had been acquainted with Breuer's method,[1] until 1889 Freud himself used hypnosis largely that way.[2] Breuer's procedure, on the other hand, seemed to go to the root of the condition by enabling the patient to recall the particular

R. S. Cohen and L. Laudan (eds.), Physics, Philosophy and Psychoanalysis 169–190.

experiences that because of their traumatic character had been split off from consciousness and, as it appeared, had thus given rise to the hysterical symptoms. It is of interest that Breuer was led to discover this treatment method by the patient herself, a highly intelligent young woman, who, during the last phase of her treatment, when in hypnotic trance, spontaneously started to talk about certain traumatic experiences which led to the disappearance one by one of the related symptoms (Breuer and Freud, 1895, pp. 34–41).

For years Breuer did not publish this finding; but he talked a great deal about it with Freud who obviously was impressed. In his *Autobiography* Freud thus wrote:

> The state of things he [i.e., Breuer] had discovered seemed to me to be of so fundamental a nature that I could not believe it could fail to be present in any case if it had proved to occur in a single one (1925b, p. 21).

This is the *first* of a number of generalizations Freud formulated in the course of developing his theory of hysteria. On the face of it, there is nothing wrong with this one – considering in particular the statement he added to the passage just cited that "the question could only be decided by experience" (1925b, p. 21). Accordingly, Freud proceeded to apply the cathartic method to a number of his own patients.

2. HYPNOID STATES VERSUS SEDUCTION IN CHILDHOOD

Freud's results were similar to those of Breuer's. Soon, however, the views of the two authors began to diverge. Breuer believed that hysterical symptoms could be produced not only by psychological but also by physical traumata (Breuer and Freud, 1895, pp. 186–192). He referred to his first patient who had developed an "accidental paralysis of her right arm, due to pressure" (*ibid*., p. 42). The paralysis had occurred while she was sitting at the bedside of her sick father with the arm over the back of her chair (*ibid*., p. 38). Breuer concedes that in this case psychological factors were also involved; but he did not regard these as decisive.

As we shall see shortly, in contrast to Breuer, Freud believed that only psychological traumata could cause hysterical symptoms. A second disagreement between the two authors developed later. It centered around the concept of hypnoid states. In their *Preliminary Communication* they had both claimed that such states form "the basis and *sine qua non* of hysteria" (1893, p. 12; italics in text), that, in other words, the presence of a hypnoid state is a necessary condition for any hysterical symptom to develop. The significant point is that by adducing the hypothesis of hypnoid states it seemed possible to account for cases of hysteria in which neither a psychological nor a physical causative factor could be demonstrated.

Freud soon abandoned this hypothesis. As he indicated later, it allows us to conclude that in the presence of a hypnoid state "even an innocuous experience can be heightened into a trauma" (1896b, pp. 194–195). Freud thus took Breuer's hypothesis to mean that a hypnoid state is not only a necessary but at times also a sufficient condition for hysteria. That is a difficult position to defend, particularly since, in Freud's words, "there are often no grounds whatever for presupposing the presence of such hypnoid states" (*ibid*., p. 195). If we take this statement of Freud's at face value, we might conclude that Breuer generalized from a single case, the patient with the paralyzed arm mentioned earlier who, when she suffered the paralysis, was in one of her frequent states of auto-hypnosis (Breuer and Freud, 1895, pp. 42, 217).

Breuer's hypothesis is more complex than I have indicated. In the present connection, however, that is not relevant. What is relevant is that Freud somehow had misunderstood Breuer's position. Even though, as I mentioned, Breuer at first had regarded hypnoid states as necessary conditions for hysteria, he not long thereafter modified this view. In the theoretical part of the *Studies*, he stated, explicitly referring to Freud, that the amnesia Freud described as involved in defense had the same effect as a hypnoid state and quite independently of such a state (1895, p. 216). The notion that Breuer had based his theory of hypnoid states on an over-generalization is thus without foundation in fact.

Freud, on the other hand, did generalize, but in a radically different way than he thought Breuer had done. Freud held on to the belief I referred to as his first generalization, namely, that hysteria is always caused by traumatic *experiences*, i.e., by psychological causative factors. In the course of his work he did, however, come up against cases of hysteria in which a sufficient psychological cause could not be demonstrated.[3] But rather than bow to the data he had gathered and conclude that his first generalization after all was valid only for a limited range of cases he extended it in a particular way. His reasoning was that if a sufficient traumatic experience could not be shown to have occurred in the comparatively recent past, such an experience may nevertheless have occurred, but in the distant past of the patient's childhood.

Freud soon decided that that was indeed the case and, further, that even if a recent, apparently sufficient trauma had been discovered, behind it were in some way related early traumata which were the truly effective ones. In his view, if the analysis was pushed far enough such early traumata would be revealed. This was clearly a new hypothesis based on a *second* generalization which went far beyond the first. Freud thus wrote:

We have learned that *no hysterical symptom can arise from a real experience alone, but that in every case the memory of earlier experiences awakened in association with it plays a part in causing the symptom* (1896b, p. 197; italics in text).

And in concluding this passage he expressed his belief that "this proposition holds good *without exception*" (*ibid*., italics in text). Thus, instead of Breuer's hypnoid states, Freud implicated childhood traumata in the causation of hysterical symptoms.

Although he did not exclude heredity and constitution as auxiliary factors (1896b, p. 210), Freud made it quite clear that he was aiming for "a *psychological* theory of hysteria" (1896b, p. 197; italics added). In his view, the proposition just cited could well serve as the basis for such a theory (*ibid*.).[4] He thus took a position that he knew Breuer could not adopt. In the work they had jointly published a year or so earlier, the latter had criticized Möbius's theory that all hysterical symptoms are ideogenic as an overgeneralization (1895, pp. 186–191); and he could be expected to regard Freud's new concept the same way.

That did not shake Freud's belief in his theory. Accordingly, he proceeded to specify the nature of the pathogenic childhood experiences. In so doing he formulated a *third* generalization. Having described certain characteristics of the associations of hysterical patients, he put forward the claim that

[whatever] case and whatever symptom we take as our point of departure, *in the end we infallibly come to the field of sexual experience* (1896b, p. 199; italics in text).

And he added:

So here for the first time we seem to have discovered an aetiological precondition for hysterical symptoms (*ibid*.).

Freud based this third generalization on eighteen cases he had treated (1896b, p. 199). In these the traumatic sexual experience was seduction in one way or another of the child, mostly by an adult – often enough a close relative, a governess, or a tutor (1896b, pp. 207–208).[5] The fact that memories of such experiences were not easy to retrieve Freud explained by positing that they had been rendered unconscious and that it was precisely because they were unconscious that they could contribute to the development of hysterical symptoms later in life. He believed further that the infantile sexual experiences had no effect to begin with and only became pathogenic if unconscious memories of them were aroused after puberty (1896b, pp. 211–212).

But why had these memories been rendered unconscious? Freud averred that that had happened to avoid the painful affects they would give rise to if they were conscious. The avoidance operation he referred to as *defense*. He had observed that it often was very difficult for his patients to recall the pathogenic experiences and concluded that a *resistance* against remembering was operating and, further, that this resistance and defense were two sides of the same coin (Breuer and Freud, 1895, pp. 154, 268–270).

Freud did not arrive at his view of hysteria by exclusively using Breuer's cathartic method. Already while he was collaborating with Breuer he employed a modification of this method which in part represented an adaptation for his purpose of Bernheim's experiments with post-hypnotic suggestion (*ibid*., pp. 109–110, 145, 268, 270). When a patient turned out not to be hypnotizable, Freud insistently exhorted him/her to remember things related to the symptoms; and he added the suggestion that the patient would remember such things when he, Freud, pressed his hand firmly against his/her forehead. This procedure was repeated over and over so Freud eventually assembled a collection of variously interrelated memories and other ideas which he could put together "exactly like . . . a child's picture-puzzle." He could thus identify the scenes of seduction in childhood that, so to speak, fitted into the empty spaces of the puzzle (1896b, p. 205).

One does not need to read between the lines to realize that on occasion Freud actually *suggested* to his patients sexual scenes he believed they had experienced as children. He thus wrote that patients "are indignant as a rule if we warn them that such scenes are going to emerge" (1896b, p. 204). He wrote further:

> I have never yet succeeded in forcing on a patient a scene I was expecting to find in such a way that he seemed to be living through it with all the appropriate feelings. Perhaps others may be more successful in this (1896b, p. 205).

And years later, in the second of the five lectures he delivered at Clark University, he said in reference to the resistance to remember:

> The existence of this force could be assumed with certainty, since one became aware of an effort corresponding to it, *if, in opposition to it, one tried to introduce the unconscious memories into the patient's consciousness* (1910, p. 23; italics added).

Sulloway (1979, pp. 95–97) cites a number of additional examples of how Freud actively directed the associations of his patients to sexual themes.

That does not mean, of course, that seduction can never be demonstrated to have occurred. Most analysts have come across cases where such occurrences have been confirmed by independent observers. The point is merely that, in his eagerness to show that *every* case of hysteria is contingent on seduction in childhood, Freud seems to have resorted to methods that could not yield convincing evidence. We may note that in the last of the above quotes he claimed that the resistance to remember could be assumed *with certainty*, when the resistance hypothesis might be said to be at best quite probable. The three statements I have quoted were apparently meant to show that, when a patient actually did remember, his memories could not be ascribed to suggestion. It is questionable, however, that they show that. It is entirely possible that suggestion may have been involved in some cases but not in others.

Soon after the joint publication of the *Studies on Hysteria* a chill descended on the relationship between Breuer and Freud. The latter wrote in his *Autobiographical Study* that, although Breuer had publicly defended Freud's view of "the part played by sexuality in the aetiology of the neuroses, . . . it was easy to see that he too shrank from recognizing the sexual aetiology of the neuroses" (1925b, p. 26). Freud, it seems, had in part at least misunderstood Breuer's position. I noted above that Breuer had criticized Möbius for over-generalizing and that he would be likely to react in a similar manner to Freud's generalizations. That is exactly what he did. As quoted by Sulloway (1979), in a letter written in 1907 to August Forel, Breuer referred to Freud's work as "magnificent" (Sulloway, 1979, p. 98) but observed in the same letter that

> Freud was a man given to absolute and exclusive formulations: this is a psychical need which, in my opinion, leads to excessive generalization (quoted from Sulloway, 1979, p. 85).

In reference to the rest of this letter, Sulloway commented that Breuer in part accepted Freud's ideas of the role of sexuality in the etiology of the neuroses but "refused to accept only Freud's claim that *every* neurosis has a specific sexual cause" (1979, p. 98; italics in text).

In his biography of Freud, Ernest Jones (1953) expressed a quite similar view:

> That is the way Freud's mind worked. When he got hold of a simple but significant fact he would feel, and *know*, that it was an expression of something *general or universal,*

and the idea of collecting statistics on the matter was quite alien to him (p. 97; italics added).

In view if these characterizations of Freud as a theorist, it is of interest to note that in a letter to Fliess, dated December 8, 1895, Freud wrote in reference to the latter's work that "[we] cannot do without men with courage to think new things before they can prove them" (1950/1954, p. 137, letter 38). I agree with Sulloway's implication that in this statement Freud had himself in mind as much as he had Fliess (Sulloway, 1979, p. 86).[6] Freud thus clearly felt that he could not sufficiently substantiate at least some of his bolder ideas. It is of interest that years later he wrote in reference to his rejection of Breuer's concept of hypnoid states that he (i.e., Freud) "had taken the matter less scientifically" (1914a, p. 11). It is entirely possible that Freud therefore was particularly vulnerable to, and hence intolerant of, Breuer's after all quite reasonable criticism. I may mention that Sulloway (1979, p. 99) and Jones (1953, p. 255) have both commented on Freud's completely unreasonable reaction to this criticism. For example, as quoted by Jones (1953, p. 255), in an unpublished letter to Fliess, dated March 29, 1897, he thus wrote that the mere sight of Breuer would make him want to emigrate; and once he pretended not to see Breuer when they walked past one another on the street (Sulloway, 1979, p. 99).

3. FROM UNCONSCIOUS MEMORIES TO UNCONSCIOUS WISHES

The blow came sooner than one might have expected. In a letter dated September 21, 1897, i.e., less than two years after the letter just referred to, Freud informed Fliess about his discovery that his seduction theory of hysteria was mistaken (1950b/1966, pp. 259–260, letter 69). In this letter Freud claimed that he did not feel depressed about this discovery: "It is remarkable, too," he wrote, "that there has been an absence of any feeling of shame, for which, after all, there might be occasion . . . I have more of the feeling of a victory than of a defeat – and, after all, that is not right" (p. 260).

And that was not right. In his *History of the Psychoanalytic Movement* Freud wrote that when his seduction theory "broke down . . . the result at first was helpless bewilderment" (1914a, p. 17); and in his *Autobiographical Study* he admitted that

[when] . . . I was at last obliged to recognize that these scenes of seduction had never taken place, and that they were only phantasies which my patients had made up or

which I myself had perhaps forced on them I was for some time completely at a loss. My confidence alike in my technique and its results suffered a severe blow (1925b, p. 34; italics added).[7]

It is noteworthy also that it took Freud nearly eight years before he made his discovery public, at first half-heartedly in a brief remark in the *Three Essays on the Theory of Sexuality* (1905b, pp. 190–191) and at greater length and more explicitly in a paper devoted largely to this question (1906, pp. 274–275). Freud's silence on the disturbing discovery, which clearly refuted his earlier theory of hysteria, may be explained in part by the fact that during the intervening years he had been engrossed, among other things, in his work on dreams (published 1900), on slips of the tongue and other similar phenomena (published 1901), and on the construction of a general theory of sexuality (published 1905).

As I just indicated, eventually Freud returned to the discovery that in 1897 had upset the theory of hysteria he had formulated in 1896. He had found a way out of the dilemma. He realized that he did not have to discard the findings on which this theory was based. But he could no longer take them at face value. He in effect asked himself: when is a memory truly a memory, and when is it something else? Sifting his evidence he came to the conclusion that the memories of seduction in childhood that his patients had presented to him (and that he in part may have suggested to them himself) were not in fact memories of actual events but *fantasies* about such events appearing in the guise of memories.

Obviously, this was not an observation but a hypothesis. It was, however, a momentous hypothesis, as momentous in its way as Breuer's original discovery of the cathartic method. It did not come from nowhere. Freud's ongoing work on dreams and on the various seemingly meaningless lapses of everyday life must have made him increasingly sensitive to the idea that things were not always what they seemed to be. The hypothesis that what appears as memory often enough is sheer fantasy gave rise to a number of other hypotheses, not only, as is usually claimed, to the hypothesis – or discovery – of infantile sexuality[8] but also to a further generalization of previously formulated hypotheses about wishes and their fulfilments, particularly their imaginary and disguised fulfilments. Thus the theory of hysteria became closely related to dream theory and the theory of parapraxes, in both of which wishes play a central role. According to Freud, the fantasies on which hysteria is based represent wish fulfilments in much the way that dreams do.

I shall briefly sketch the basic ideas Freud proposed for a theory of hysteria after he had publicly announced the collapse of his seduction theory. As I just indicated, he had come to realize that in hysteria the pathogenic factors are fantasies, not memories, of seduction. He thought that the fantasies develop around puberty as a defense against memories of infantile masturbation. As such they are unconscious but achieve indirect expression in the form of hysterical symptoms (1906, p. 274). Since the unconscious fantasies represent wish fulfilments, the hysterical symptoms represent wish fulfilments also – although usually too distorted to be recognizable as such. It is in this sense, as Freud remarked, that the symptoms "constitute the patient's sexual activity" (1906, p. 278). I should mention that, as we shall see shortly he did not, however, believe that only fantasies of seduction may be pathogenic.

Freud thus successively formulated three theories of hysteria, the traumatic theory, the seduction theory, and the wish fulfilment theory. It is readily seen that *structurally* the two theories last mentioned are quite similar. We can discern the following sequence of events in the account just given of the wish fulfilment theory:

infantile masturbation
↓
unconscious puberal fantasy
↓
hysterical symptom

The seduction theory, on the other hand, is characterized by the following sequence:

childhood seduction
↓
unconscious puberal memory of the seduction
↓
hysterical symptom

Both theories identify two periods that are crucial for the emergence of hysterical symptoms, early childhood and puberty. It is hardly a coincidence that the stress on these periods is consistent with the most general feature of Freud's view of human sexual development, as explicated in the 'Three Essays' (1905b, p. 234).

Let us proceed. In a number of papers published after 1906 Freud elaborated on each of the events of the sequence characterizing the wish fulfilment theory and on the relationships between these events. He also introduced some additional points. I shall comment on these events – quite summarily, to be sure – roughly in the order in which they appear in the sequence. We will note that, as I have mentioned, the unconscious fantasies serve to repress the memory of infantile masturbation. Freud once remarked that the older child "spares himself shame about masturbation by retrospectively phantasying a desired object into these earliest times" (1917, p. 370). He thus clearly pointed to a putative motive for the formation of a seduction fantasy. But besides he may also have meant to explain in part how such a fantasy, if it should become conscious, may acquire the appearance of a memory. As for the repression, we might say that it is by thus being made to appear as a memory that the fantasy effects its repressive function. It is as if the puberal child had said to himself: "I didn't masturbate. He seduced me."[9]

This is of course highly speculative. For the moment, however, I will leave this part of the theory as it stands and focus instead on the fantasies. These, as Freud stated, have their "common source and normal prototype . . . in what are called the day-dreams of youth" (1908b, p. 159). He now emphasized that the fulfilment of a wish, and thus the fantasy representing such fulfilment, involves "a correction of unsatisfying reality" (1908a, p. 146). And he added:

> Mental work is linked to some current impression, some *provoking occasion* in the present which has been able to arouse one of the subject's major wishes (*ibid*., p. 147; italics added).

Freud pointed out further that a thus aroused wish tends to awaken memories of early fulfilments and that traces of these can be found in the fulfilment fantasies the aroused wish gives rise to (*ibid*., p. 147).

The consequences of frustrations of the indicated type may, however, be more radical. In a later connection Freud, expressing himself in terms of the libido theory, declared that

> the unsatisfied libido which has been repulsed by reality . . . will finally be compelled to take the path of regression and strive to find satisfaction either in one of the organizations which it has already outgrown or from one of the objects it has earlier abandoned. The libido is lured into the path of regression by the fixation it has left behind it at these points in its development (1917, p. 359).

Expressed in plain language this seems to mean that early fulfilments may lead to a *disposition*, triggered by later frustrations of age-adequate wishes, for a more or less full *reactivation* of the wishes corresponding to these fulfilments, in regard to the way in which the latter were manifested and/or to the objects they involved. Such reactivation thus meets what Freud saw as a necessary condition for symptom formation.

Another characteristic of the relevant wishes is that they are either *overtly* sexual or the *actual* fulfilment, say, of an ambitious wish in a man, which although in itself nonsexual, serves as a means to attain the fulfilment of a sexual wish (1908b, p. 159). In Freud's view, therefore, at least at bottom the *significant* fantasies related to wishes of the two kinds are *always* sexual.[10]

That, however, is a very broad characterization. After several years of further observation, Freud concluded that for the most part, but not exclusively, these fantasies are of seduction, of observing parental intercourse, and of castration. Not that the corresponding events did not sometimes occur in reality also. In a number of cases children had in fact been seduced by their elders, had observed parental intercourse, and had been threatened with castration. These events, however, were not frequent enough to explain the prevalence of the corresponding fantasies (Freud, 1917, pp. 369–370). In the 'Three Essays', in a footnote added in 1920, Freud emphasized that these fantasies "are distinguished . . . by being to a great extent independent of individual experience" (1905b, p. 226, fn. 1). In his Lamarckian way he, accordingly, proposed that these "*primal phantasies*," as he called them, are part of the "phylogenetic endowment" of man (1917, p. 371).

So much about the fantasies themselves. We will recall that, as was the case with the pathogenic memories in the seduction theory, in Freud's view the fantasies eventuate in hysterical symptoms only if they are unconscious (1908b, p. 160). Although Freud believed that some fantasies may be primarily unconscious (*ibid*., p. 161), he thought that for the most part an unconscious fantasy has been rendered unconscious by repression. Repression in this case obviously differs from the repression by an unconscious fantasy of memories of infantile masturbation. At this stage in the development of his theory Freud described the repression of wishes and their associated fulfilment fantasies in language reminiscent of his later descriptions of the repressing function of the ego-ideal (1914b, pp. 193–194) and partly of the superego (1923), but without using these terms.[11]

A particularly important point is that, according to Freud, hysterical symptoms represent a "compromise" between the repressing forces and the unconscious fantasies they are supposed to prevent from becoming conscious (1908b, p. 164). That is one reason why the meanings of hysterical symptoms are often difficult to unravel. Another reason is that a particular symptom may represent not one but several unconscious fantasies (*ibid*., p. 163) of which some may be homosexual (*ibid*., p. 164). Occasionally nonsexual fantasies also find expression in a symptom. But, as Freud saw it, such a symptom "can *never* be without sexual significance" (1908b, p. 164; italics added).

4. PRELIMINARY REMARKS ON THE EPISTEMIC STATUS OF FREUD'S WISH FULFILMENT THEORY OF HYSTERIA

The foregoing account has brought out some characteristic features of Freud's thinking. He undoubtedly was a bold and inventive thinker. But he also had a striking penchant for quick generalizations which more than once turned out to be ill founded. The preceding pages are replete with quotes illustrating this point. I have indicated that this characteristic of Freud's may well have been a primary factor in the cooling of his relationship with Breuer.

Freud apparently had an overblown belief in the power of inductive inference as he practised it. It would be of interest to know how much he was influenced by the philosophy of J. S. Mill. Sulloway (1979) is silent on this point. We know that he was acquainted with some of Mill's writings.[12] But we do not know to what extent. In any event, the statement of Freud's quoted in the beginning of this paper that he could not believe that what Breuer had discovered "could fail to be present in any case of hysteria if it had proved to occur in a single one" and many similar statements sound like echoes of Mill's general statement that "[the] universe, as far as known to us, is so constituted, that whatever is true in any one case, is true in all cases of a certain description" (1965, p. 133).[13]

If this statement indeed provided Freud with one of his basic premises, he failed to take note of the restrictions, expressed by Mill himself and others, against the application of this general statement to particular instances. Thus Day, in his critical discussion of Mill's concept of empirical laws, wrote that "generalizations like 'Lead-poisoning is a cause of death' . . . are all normally established by enumerative induction, the degree of probability depending on the variety and number of the observed instances" (1964, p. 351).

In part Freud seems to have been aware of the problem. Referring to the fact that he had based his theory of the sexual etiology of hysteria on a total of eighteen cases in which seduction was claimed to have occurred in childhood, he indicated that maybe the number of cases was not sufficient to establish the generalization (1896b, p. 200). And it was not. However, it turned out to be a question not only of numbers but also of the *reliability* of the observations on which the generalization was based. As we have seen, these were highly unreliable. In the end, in the majority of instances childhood seduction proved to be a fiction. Freud's use of inductive inference, thus, was badly flawed.[14]

We must turn now to the wish fulfilment theory of hysteria. We will recall that, according to this theory, hysterical symptoms are traceable in the first place to unconscious puberal fantasies and eventually to infantile masturbation. Freud originally believed that masturbation in early childhood is a "consequence of abuse or seduction" (1896a, p. 165). But with the abandonment of the seduction theory he had to give up this view. He gradually realized that masturbation is a regularly occurring spontaneous activity, starting in the early days of infancy and going through the well-known oral, anal, and genital (or phallic) phases of sexual development (1905b, pp. 185–189). Freud besides recognized two varieties of infantile masturbation, the one being associated with fantasy, the other not (1908b, p. 161).

Direct observation of children has fully confirmed Freud's views of infantile masturbation. Considering oral and anal sexual practices in adults, one can hardly quarrel with the idea that similar practices in children can in a broad sense be classified as sexual. It is, however, one thing to recognize infantile sexuality, including infantile masturbation, and quite another to identify the masturbation as a causative factor in the later occurrence of hysteria.

This is a tricky point. Let me note to begin with that originally Freud had adduced the notion of infantile sexuality to explain the occurrence of hysterical symptoms in the absence in the comparatively recent past of a sufficient psychological cause. When he, however, posited unconscious puberal sexual fantasies as causes of hysteria, it would seem that the notion of infantile sexuality and specifically of infantile masturbation as a causative factor had become redundant. Apart from whatever theoretical considerations that may have moved Freud to retain this notion, he presumably also had an empirical reason for doing it. *Conscious* sexual fantasies and fantasy-related actions occurring at puberty and later often show features characteristic of infantile sexual experience and may even repeat motifs of infantile masturbation fantasies. Freud may therefore have concluded that, *by analogy*, the puberal *unconscious* fantasies he regarded as giving rise to hysterical symptoms are in a similar way related to infantile sexuality including infantile masturbation.

It is not difficult to understand why Freud thought that the puberal sexual fantasies he had implicated in the causation of hysteria had to be unconscious. If they were conscious, they would simply serve as actual masturbation fantasies or they might be acted out in real life (1908b, p. 162). But they would not give rise to hysterical symptoms. If such symptoms are indeed derived from puberal fantasies, then these fantasies thus cannot be conscious.

The above observations and conjectures lead to the conclusion that both infantile masturbation and related unconscious puberal fantasies may be causal factors in the development of hysteria. But are they in fact? The answer clearly depends on whether or not the indicated conjectures have any merit. They cannot be tested directly. Our only option, therefore, is a frontal approach. Let us take the unconscious fantasies first. One way to find out if they are present and have the function the theory requires is to (1) make it possible for the patient to produce, or at any rate to recognize, the fantasies we believe are operating unconsciously, and (2) determine whether or not as a consequence of this maneuver the related hysterical symptoms will disappear. The basic logic here is essentially the same as that underlying Breuer's cathartic method. It is spelled out in great detail by Grünbaum in his discussion of what he refers to as Freud's *Tally Argument* (Grünbaum, 1980, pp. 321–322).

It follows from what I said above about infantile masturbation that its relevance can only be demonstrated by an examination of the character of the produced, or recognized, fantasies. From whichever point we start our query we thus come up against the problem how to track down the unconscious fantasies we presume underlie hysterical symptoms.

5. FURTHER REMARKS ON THE EPISTEMIC STATUS OF FREUD'S WISH FULFILMENT THEORY OF HYSTERIA

The key methods in this pursuit are those of free association and interpretation. The method of free association is too well known to require specific presentation. We should note, however, that, according to a basic assumption of Freud's, "[the] idea occurring to the patient must be in the nature of an *allusion* to the repressed element, like a representation of it in indirect speech" (1910, p. 30; italics in text). He had in fact made this assumption much earlier, at the time he, under the influence of the traumatic theory, was still using hypnosis. He noted then that when he pressed his hand against the patient's forehead, a "forgotten" recollection would not always emerge but often enough "an idea . . . which is an intermediate link in the chain of associations between the idea from which we start and the pathogenic idea which we are in search of" (Breuer and Freud, 1895, p. 271). Some years later, Freud said in a different connection that in formulating interpretations he follows "a number of rules, reached empirically, of how the unconscious material may be reconstructed from the associations" (1904, p. 252). He stated that these rules had not yet been published (*ibid.*). To the best of my knowledge they never were.

The striking point that emerges from these quotations is that the patient's associations were not expected to lead to the pathogenic material itself but to pointers, as it were, in accordance with which Freud could reconstruct, i.e., interpret, this material. The patient was thus brought to a point where he/she could recognize, or fail to recognize, Freud's interpretation as describing a fantasy he/she may well entertain, or have entertained. This is analogous to Breuer's cathartic method which, as I mentioned in the beginning of this paper, enabled the patient to recall the traumatic experience that, by being repressed, had caused the hysterical symptom. With the new method, however, the patient occasionally produced directly what seemed to be the pathogenic fantasy. Sometimes Freud did it himself without the benefit of the patient's associations.

To give an idea of the kinds of interpretation Freud worked with, I will choose a few examples that do not involve long associative chains.

A patient reported in the *Studies*, Frau Cäcilie M., suffered from a number of hysterical symptoms and also from occasional bouts of severe facial neuralgia. Suspecting that the neuralgia was likewise hysterical, Freud induced hypnosis during an attack. He gave the following dramatic account of what happened:

> When I began to call up the traumatic scene the patient saw herself back in a period of great mental irritability towards her husband. She described a conversation which she had had with him and *a remark of his which she had felt as a bitter insult. Suddenly she put her hand to her cheek, gave a loud cry of pain and said: 'It was like a slap in the face.'* With this her pain and her attack were both at an end (Breuer and Freud, 1895, p. 178; italics added).

The attacks, however, had not subsided forever. And even though both Freud and Breuer repeatedly referred to this case, as far as I can make out, they did not indicate whether or not the attacks eventually disappeared.

Although in this case Freud used hypnosis, I have cited it here because the symbolic nature of the interpretation the patient arrived at spontaneously has features that are typical of a great many interpretations formulated by Freud in other cases.

The analysis of Dora, Freud's eighteen-year-old patient, provides us with another example. Dora suffered from a number of hysterical symptoms, including what Freud termed a nervous cough. A wealth of associations, interspersed with interpretations offered by Freud, suggested that Dora was in love with both her father and his mistress, Mrs. K. A play on words led Freud to believe that Dora may have thought that her father was impotent and hence could only engage in oral sex with Mrs. K. Dora admitted that this was possible but could not see a connection, suggested by Freud, between the thought of oral sex and her nervous cough. After Freud had presented her with the interpretation her cough nevertheless disappeared which, however, it had often done spontaneously before (1905a, pp. 47–48). Because of the complexities involved I will not consider this symptom further.

A further element in Dora's story is quite interesting. Apart from her father and Mrs. K. she was also in love with Mr. K. who had unsuccessfully tried to seduce her when she was fourteen. Aphonia was another one of Dora's symptoms. It came in attacks which usually lasted from four to six weeks. Questioning by Freud brought to light that Mr. K. traveled a great deal in business, that he usually was away for from four to six weeks, and that this absence coincided with Dora's attacks of aphonia. The conclusion seemed obvious. As Freud put it: "Dora's aphonia, then, allowed of the following symbolic interpretation. When the man she loved was away she gave up speaking; speech had lost its value since she could not speak to *him*" (1905a, p. 40; italics in text).

That could not be all there was to it. Dora, who had prematurely interrupted her analysis, came for a consultation with Freud a little over a year later. Her attacks persisted but had become less frequent and she felt much better. She had broken off her relationship with both Mr. and Mrs. K. She also mentioned that about half a year before this consultation she had had an attack of aphonia after seeing a man knocked down on the street by a carriage. This man, it turned out later in the conversation, was Mr. K. (*ibid.*, pp. 120–121). Did the aphonia occur because Dora at that point could not, or would not, speak to him? Or was it because the accident frightened her? We do not know. And Freud did not speculate. He simply left the questions raised by the new data unanswered.

Apart from attacks of an ostensibly nonsexual character, like Frau Cäcilie's facial neuralgia and Dora's aphonia, Freud identified attacks that are frankly sexual, although not readily recognizable as such. He held that they represent sexual fantasies in a sort of "pantomime" that because of the "censorship" are distorted, very much in the manner of dreams (1909, p. 229). Among attacks of the indicated type Freud listed the so-called "*arc-de-cercle*" which he took to represent "an energetic repudiation . . . , through antagonistic innervation, of a posture of the body that is suitable for sexual intercourse" (*ibid.*, p. 230). Hysterical convulsions may represent coitus (*ibid.*, p. 231). He claimed further that involuntary passing of urine and biting of the tongue are not incompatible with the diagnosis of hysteria. He saw the former as "merely repeating the infantile form of a violent pollution " (*ibid.*, p. 233).

A number of the interpretations just indicated involve inferences of an unconscious fantasy from a hysterical symptom which are based on an analogy of one kind or another. In the case of Frau Cäcilie M. the analogy was a common one between a mental and a physical pain. The uncommon feature here was that the physical pain was not merely referred to but actually experienced as such. In the case of hysterical convulsions and involuntary passing of urine, Freud interpreted a nonsexual physical act, the symptom, in terms of a physically similar, and thus in this sense analogous, sexual act. The *arc-de-cercle*, on the other hand, he interpreted, not in terms of a hidden sexual fulfilment fantasy, but as the means to an end, a dramatization in bodily terms of a wish to avoid sexual intercourse. As interpreted by Freud, Dora's aphonia was also a dramatization in bodily terms, but of a wish not to speak.

It seems clear that the data are sufficient to justify the presentation of the listed interpretations as *hypotheses*. But are they sufficient without further evidence to render these hypotheses probable to any appreciable degree? I do not think so. For one thing, some of the analogies Freud relied on seem quite arbitrary. But that did not daunt him. If he felt intuitively that an interpretation was right, he felt little need for further validation. This attitude is revealed in a letter to Fliess, dated December 9 and 14, 1899, in which he wrote in reference to a philosopher friend's criticisms: "Intelligence is always weak, and it is easy for a philosopher to transform resistance into discovering logical refutations" (1950a/1954, p. 305, letter 125). To invoke the psychoanalytic concept of resistance in response to a logical argument is a prime example of a resistance, on Freud's part, to nonpartisan critical thinking.

We should note, however, that Freud was not completely insensitive to the problem of confirmation. Thus, in the Dora analysis, when, referring to her father, he told the young woman that early in life she must have been "completely in love with him," she said at first that she didn't remember that, but then told a story about a cousin who at age seven had said that when her mother dies she will marry her father. Freud commented:

> I am in the habit of regarding associations such as this, which brings forward something that agrees with the content of an assertion of mine, as a confirmation from the unconscious of what I have said (1905a, p. 57).

Taken by itself, Dora's response carries some, but only some, weight as evidence. But the principle that is involved is important. It is the principle of listing as evidence patient productions that can be seen as *indirect* expressions of unconscious fantasies. Such indirect expressions are comparable to the recall achieved with the cathartic method. Obviously, however, they do not have the same evidential force. But several patient productions that can be seen as different indirect expressions of the same unconscious fantasy lend the corresponding interpretation at least some degree of probability.

Freud did not directly touch on this problem. In one of his last papers (1937) he did, however, consider the question of indirect evidence, but mostly in regard to historical constructions. In that paper (1937, p. 263) he also referred to the significance as confirmation of an exclamation by the patient like 'I didn't think of that' in response to an interpretation. In reporting the Dora case, Freud in fact had made the same observation in a footnote added in 1923 (1905a, p. 57, fn. 2). Such turns of phrase *may* indicate a recognition on the part of the patient, say, of a fantasy Freud had interpreted. He also called attention to what can be seen as admission by negation of a thought or an impulse, as when a patient says: "'Now you'll think I mean to say something insulting, but really I have no such intention'" (1925a, p. 235).

These are interesting contributions to a theory of confirmation but they are not more than that. With the exception of the beginning Freud made in relation to Dora's response to his interpretation of her love for her father they are not of the kind that would help us validate interpretations of unconscious fantasies, such as the coital fantasies taken to be involved in hysterical convulsions. Freud referred quite scantily to the probative value of the disappearance of hysterical symptoms as a consequence of interpretations presented to the patient. As I have mentioned, his one major attempt to demonstrate the correctness of an interpretation that way (1896b, p. 200) miscarried because of the unreliability of his observations.

I cannot leave the problem of confirmation without noting Freud's claim that his interpretations reflect the particular mode of 'thinking' that leads to hysterical symptoms and that this mode of 'thinking' is quite similar to schizophrenic thinking. Suppose, for the sake of argument, that we accept this claim. Do we then have to conclude that the actual occurrence of schizophrenic thinking is good evidence for the actual occurrence of the hysterical mode of 'thinking'? Freud (implicitly) answered this question in the affirmative and concluded further that, since suggestion plays no part in schizophrenic thinking, it plays no part in the hysterical mode of 'thinking' either – nor in the interpretations that reflect it (1917, p. 453).

A last point before I proceed. As we will remember, Freud also claimed that infantile masturbation affects the character of the unconscious puberal fantasies and thus also of the related hysterical symptoms. He presented only scant evidence for this claim. I cannot, therefore, consider it here.

6. CONCLUDING REMARKS

The theory of hysteria is not all of psychoanalysis. But it epitomizes a number of the problems that beset this discipline. In a series of penetrating studies, Grünbaum (1980; 1981; 1983) has considered these problems in great detail. Critical psychoanalysts are in the habit of blaming Freud's metapsychology for all the difficulties inherent in the theory.[15] Metapsychology, however, presents a comparatively minor problem. Although we may be far from a solution, it need not disturb us once we recognize its proper function which is to remind us of the fact that mind cannot exist without a body and that, accordingly, the mental operations that interest psychoanalysis can also be viewed as physiological processes and sometimes, for proper understanding, have to be viewed that way.

It is the clinical part of psychoanalysis that is really disturbing. It is top-heavy with theory but has only a slim evidential base. I have used the theory of hysteria to illustrate the arbitrariness, because of lack of adequate confirmation, of a great many clinical interpretations. This statement holds also beyond hysteria. That, of course, does not mean that all clinical interpretations fail in this way. But it is those that do that interest us here.

In this situation we can decide on one of two ways out of the dilemma. Each has its proponents. One is to decide that psychoanalysis, since it deals with interpretations, is a hermeneutic discipline. In its extreme form this view leads to a relativistic conception of psychoanalysis. Interpretations do not have to be confirmed. Any given interpretation, or set of interpretations, is as good as any other, provided it fits the data. It can be used as the basis for a story about the patient. The question of whether the story is true or not does not even arise. There are no true stories in an absolute sense. Truth is a function of perspective and just as there are many different perspectives so there may be many different true stories about the same person. Schafer (1980) is the best-known proponent of this way of thinking.

The other way out of the dilemma is simply to find better methods of clinically confirming clinical interpretations than those used by Freud and most of his followers. As I have indicated, Freud himself made a beginning in this direction, but he did not develop it further. This is the only way if we believe that psychoanalysis after all is a science, or at least can grow to become one.

New York City

NOTES

[1] See Freud (1925b).

[2] See Breuer and Freud (1895), p. 48; Freud (1950/1954), p. 52, letter 2; and the editor's introduction to Breuer and Freud (1893–1895/1955), p. xi.

In regard to the Fliess letters I shall use primarily *S.E. 1* and refer to it in the text by the publication years 1950/1966. Letters not in *S.E. 1* I shall quote from the *Origins of Psycho-Analysis* which I will refer to in the text by the publication years 1950/1954.

[3] A sufficient psychological cause, according to Freud (1896b), is an event that is (1) suitable as a determinant and (2) has necessary traumatic force. Freud did not define these expressions but tried to clarify them with examples. A railway accident, thus, is not suitable as a determinant of hysterical vomiting. Eating a partly spoiled fruit, on the other hand, if disgusting enough, may be a determinant of a transient attack of vomiting but lacks the necessary traumatic force to produce persistent hysterical vomiting (*ibid.*, pp. 193–194). These criteria are clearly derived from everyday psychological generalizations which, to the extent that they apply, account for the understandability of symptoms of the indicated type.

[4] In spite of the wording of the statement just cited, Freud did not regard the theory of hysteria as exclusively psychological but as both psychological and physiological. See below, Notes 10 and 14 and also Sulloway's comments on this issue (1979, pp. 90–91).

[5] Freud had earlier dismissed objections to this thesis claiming that *only psychoanalysis* could reveal the causative childhood events by "making conscious what has so far been unconscious" (1896a, p. 164).

[6] Sulloway prefers to retranslate this passage and substitute the word "demonstrate" for the word "prove."

[7] Later in the same paragraph Freud wrote: "When I had pulled myself together, I was able to draw the right conclusions from my discovery." These included a disbelief that he had "forced the seduction-phantasies" on his patients.

[8] See Freud (1925b), p. 35; and also the editor's note to 'Further Remarks on the Neuro-Psychoses of Defence,' (1896a), p. 160.

[9] We may note that on the unconscious level the memory and the corresponding fantasy may largely share the same content; they would thus differ from one another primarily by a *readiness* of this content to acquire, when it becomes conscious, the imprint of something that really has happened, in the one case, and the imprint of something that really has not happened but is imagined to be happening in a world of unreality, in the other.

[10] To Freud, sexuality was the link between psychology and biology (see, e.g., 1925b, p. 35). That explains his emphasis on sexuality whenever he could discern even the faintest hint of it. In this paragraph, something that in reality is an end in itself, the fulfilment of an ambitious wish, is in effect degraded to function merely as the means to another, namely, a sexual, end. See also Note 4 above and Note 14 below.
[11] See Freud (1905b), p. 177 and (1910), p. 24.
[12] Although Jones (1953, pp. 55, 176) refers to Freud's reading and even translating into German some of Mill's works, like Sulloway he has nothing to say about Mill's possible influence on Freud.
[13] This statement is equivalent to what Mill referred to as the "principle of the uniformity of the course of nature" (1965, pp. 134–135) or, for short, the principle of the uniformity of nature.
[14] Apart from the basic premise of inductive inference just considered, Freud adduced a few other general notions as basic premises. The enormous influence on his thinking of the so-called constancy principle, which he attributed to Fechner, is well known. Freud used this principle to explain the "driving force" of wishes. Equally well known is his emphasis on both the postulate of psychic determinism and the notion that psychological as well as physiological factors are involved in neuroses. See, e.g., Freud (1896b), p. 200, and (1905b), p. 113; and also Sulloway (1979). In essence the physiological factors were set forth in the libido theory which Freud first sketched in a number of letters to Fliess. A further discussion of these basic premises of psychoanalysis must be left for another occasion.
[15] See Gill and Holzman (1976).

REFERENCES

Aristotle. 'Poetics.' In J. D. Kaplan (ed.), *The Pocket Aristotle*, pp. 342–379. New York: Washington Square Press, 1958.

Breuer, J. and Freud, S. 1893. 'On the Psychical Mechanism of Hysterical Phenomena: Preliminary Communication.' *S.E. 2* (1955), 3–17.

Breuer, J. and Freud, S. 1895. *Studies on Hysteria. S.E. 2* (see below). London: Hogarth Press, 1955.

Day, J. P. 1964. 'John Stuart Mill.' In *A Critical History of Western Philosophy*. Ed. D. J. O'Connor, pp. 341–364. London: The Free Press of Glencoe.

Freud, S. 1953–74. [*S.E.*]. *The Standard Edition of the Complete Psychological Works*. Trans. under the General Editorship of James Strachey, in collaboration with Anna Freud. 24 vols. London: Hogarth Press and The Institute of Psycho-Analysis.

Freud, S. 1950a. *The Origins of Psycho-Analysis*. New York: Basic Books, 1954.

Freud, S. 1950b. 'Extracts from the Fliess Papers.' *S.E. 1* (1966), 177–280.

Freud, S. 1896a. 'Further Remarks on the Neuro-Psychoses of Defence.' *S.E. 3* (1962), 162–185.

Freud, S. 1896b. 'The Aetiology of Hysteria.' *S.E. 3* (1962), 191–221.

Freud, S. 1904. 'Freud's Psycho-Analytic Procedure.' *S.E. 7* (1953), 249–254.

Freud, S. 1905a. 'Fragment of an Analysis of a Case of Hysteria.' *S.E. 7* (1953), 7–122.

Freud, S. 1905b. 'Three Essays on the Theory of Sexuality.' *S.E. 7* (1953), 135–243.

Freud, S. 1906. 'My Views on the Part Played by Sexuality in the Aetiology of the Neuroses.' *S.E. 7* (1953), 271–283.

Freud, S. 1908a. 'Creative Writers and Day-Dreaming.' *S.E. 9* (1959), 143–153.

Freud, S. 1908b. 'Hysterical Phantasies and Their Relation to Bisexuality.' *S.E. 9* (1959), 159–166.

Freud, S. 1909. 'Some General Remarks on Hysterical Attacks.' *S.E. 9* (1959), 229–234.

Freud, S. 1910. 'Five Lectures on Psycho-Analysis.' *S.E. 11* (1957), 9–55.

Freud, S. 1914a. 'On the History of the Psycho-Analytic Movement.' *S.E. 14* (1957), 7–66.

Freud, S. 1914b. 'On Narcissism: An Introduction.' *S.E. 14* (1957), 73–102.

Freud, S. 1917. 'Introductory Lectures on Psycho-Analysis.' Part III. *S.E. 16* (1963).

Freud, S. 1923. 'The Ego and the Id.' *S.E. 19* (1961), 12–66.

Freud, S. 1925a. 'Negation.' *S.E. 19* (1961), 235–239.

Freud, S. 1925b. 'An Autobiographical Study.' *S.E. 20* (1959), 7–74.

Freud, S. 1937. 'Constructions in Analysis.' *S.E. 23* (1954), 257–269.

Gill, M. M. and Holzman P. S. (eds.). 1976. *Psychology versus Metapsychology: Psychoanalytic Essays in Honor of George S. Klein. Psychological Issues* 9. New York: International Universities Press.

Grünbaum, A. 1980. 'Epistemological Liabilities of the Clinical Appraisal of Psychoanalytic Theory,' *Noûs* **14**, 307–385.

Grünbaum, A. 1981. 'Logical Foundations of Psychoanalytic Theory.' Paper presented at the Annual Meeting of the Eastern Division of the American Philosophical Association in Philadelphia, Pa., December 1981.

Grünbaum, A. 1983. *The Foundations of Psychoanalysis: A Philosophical Critique*. Berkeley: University of California Press. Forthcoming.

Jones, E. 1953. *The Life and Work of Sigmund Freud*, vol. 1. New York: Basic Books.

Mill, J. S. 1965. 'The Ground of Induction.' (Selected from Mill's *A System of Logic*, 1843.) In *A Modern Introduction to Philosophy*. Ed. P. Edwards and A. Pap. Revised edition, pp. 133–141. New York: The Free Press.

Schafer, R. 1980. 'Narration in the Psychoanalytic Dialogue,' *Critical Inquiry* 7, 29–53.

Sulloway, F. J. 1979. *Freud, Biologist of the Mind*. New York: Basic Books.

KENNETH F. SCHAFFNER

CLINICAL TRIALS: THE VALIDATION OF THEORY AND THERAPY

1. INTRODUCTION

In recent years there has developed an increased general interest in the methods by which the efficacy of clinical therapies may be judged. The biomedical literature both in this country and abroad has contained many sophisticated essays on the methodology of the classical randomized controlled clinical trial and various alternatives to it.[1] Curiously, one specific clinical discipline, psychoanalysis with its extensive schools and variant forms of therapy, has until very recently been most reluctant to mount a clinical trial to determine the effectiveness of its modalities, though it has been criticized for not doing so by a number of philosophers, psychologists, and psychiatrists. It is not irrevelant on the more practical side to note that health insurance companies and the Federal government have become intensely interested in the question whether psychotherapy has any efficacy.[2]

Adolf Grünbaum has over the past five years produced a series of essays (and is completing an extensive monograph) which dissect the logical and epistemological foundations of a major branch of psychotherapy, namely Freudian psychoanalysis, with special attention to the discipline's clinical validity. In this paper I want to discuss the historical and philosophical foundations of various methods which have been proposed as providing probative force for clinical therapies. The main part of this article will be quite general, but some specific comments on and applications to psychoanalytic validation will be offered in the last section.

2. THE RATIONAL BASES FOR THERAPY

In his eminently clear *Rational Diagnosis and Treatment*, H. R. Wulff asks the (rhetorical) question why a clinician might expect a treatment to have a desired effect. Wulff offers three possible reasons:

1. It is the uncontrolled experience of the clinician himself or of other clinicians that the treatment is effective.
2. The treatment is considered rational from an assumed or a well-established knowledge of the pathogenesis or the aetiology of the disease.
3. Controlled clinical trials have shown the treatment to be effective (1976, p. 118).

R. S. Cohen and L. Laudan (eds.), Physics, Philosophy and Psychoanalysis, 191–208.

Wulff prefers the third alternative which he considers as "very firm ground," but since this is usually an "ideal requirement," the other sources of clinical knowledge require discussion.

It is quite clear that as *heuristic* generators of novel therapies, uncontrolled anecdotal experience is important. Though less obvious, I will offer evidence below that the same can be said of an application of our scientific pathogenetic or etiological knowledge. It is only the controlled clinical trial which provides sufficient justificatory evidence for the efficacy of clinical therapies. Tversky and Kahneman (1974) among others have analyzed the extent to which humans are inefficient and biased information processors. We employ heuristics to assist our reasoning which frequently leads us astray in our assessments. The remedy for these biased inferences involves starting from a systematic application of knowledge that is grounded in the basic sciences of biochemistry, histology, microbiology, and physiology.

In the clinical area, however, a rational extension of basic scientific knowledge to the domain of etiology and therapy often falls short of the mark. A good case in point was discussed in the prestigious British medical journal, *The Lancet*, in 1970. The editors noted that:

> Duodenal ulcer is associated with hypersecretion of acid; anticholinergic drugs reduce the output of gastric acid; therefore, anticholinergic drugs should be used in treating patients with duodenal ulcer. The logic is incontrovertible, but, as in much of clinical practice, the gap between logic and results is large.[3]

Though preliminary studies appeared to support the above quoted logical argument based on a pathogenesis and a hypothesized etiology of duodenal ulcer, a careful analysis of the evidence using double blind controlled clinical trials did *not* support a claim for the therapeutic effectiveness of anticholinergic drugs. The editors of *The Lancet* summarized several trials run in the late 1960's, writing: "The conclusion must be that anticholinergic drugs probably have little or no value either in short-term symptomatic treatment of duodenal ulcer or as long-term suppressors of acid output: in fact, they might be called logical placebos."[4]

Accordingly, logical derivations based on pathogenesis (and *a fortiori*, uncontrolled experience) are insufficient; only clinical trials can provide us with a secure basis for clinical knowledge. There are a number of reasons for this claim which have their roots in the nature of biological organisms and the special variability and interdependence encountered in living organisms. A partially historical account of methodologies for discovering and warranting causal generalizations in the biomedical sciences may make this clearer.

3. HISTORICAL BACKGROUND TO JUDGING THE SOUNDNESS OF BIOMEDICAL GENERALIZATIONS

We may profitably begin our inquiry into the appropriate methods for discovering and warranting biomedical generalizations, including therapies, with the work of John Stuart Mill and of Claude Bernard. This work constitutes an early methodologically sophisticated inquiry into the notion of a comparative instance or 'control' which is a central notion in contemporary analyses of biomedical causation.

A. *Mill's Method of Difference*

Ultimately I want to argue that the method of difference supplemented with a statistical interpretation, which bears certain weak analogies to Mill's method of concomitant variation, is often the most appropriate method of experimental inquiry to establish empirically and directly scientific claims. One way to approach this issue is to outline Mill's understanding of the method and in passing to touch on criticisms leveled by the distinguished French physiologist, Claude Bernard, against this method. The method of difference was stated by Mill in the following manner:

> If an instance in which the phenomenon under investigation occurs, and an instance in which it does not occur, have every circumstance in common save one, that one occurring only in the former; the circumstance in which alone the two instances differ is the effect or the cause, or an indispensable part of the cause of the phenomenon (1959, p. 256).

It should be noted that the application of the method of difference, like Mill's other methods of agreement and of residues, is not automatic in any experimental situation and, as has been observed by a number of thinkers including Whewell, *presumes* an analysis of the situation into all relevant factors which can be examined one at a time.

B. *Bernard's Methods of Comparative Experimentation and Its Relation to Mill's Views*

In his influential monograph, *Introduction to the Study of Experimental Medicine*, Claude Bernard noted that these presumptions of an antecedent analysis of the experimental situation and the ability to examine factors one at a time were not necessarily valid in experimental medicine, and he urged that the method of difference be further distinguished into (i) the method of "counterproof" and (ii) the method of "comparative experimentation". According to Bernard, in the method of counterproof one assumes that a complete analysis of an experimental situation has been made, i.e., that all complicating and interfering factors have been identified and controlled. Subsequently one eliminates the suspected cause and determines if the effect in which one is interested persists. Bernard believed that experimental medical investigators avoided counterproof as a method since they feared attempts to disprove their own favored hypotheses. Bernard defended a strong contrary position, and argued that counterproof was essential to avoid elevating coincidences into confirmed hypotheses. He argued, however, that those entities which fell into the province of the biomedical sciences were so complex that any attempt to specify *all* of the causal antecedents of an effect was completely unrealistic. As a remedy for this problem he urged the consideration of *comparative experimentation*. It is worth quoting him on his notion *in extenso*. Bernard wrote:

> Comparative experimentation bears . . . solely on notation of fact and on the art of disengaging it from circumstances or from other phenomena with which it may be entangled.
>
> Comparative experimentation, however, is not exactly what philosophers call the method of differences. When an experimenter is confronted with complex phenomena due to the combined properties of various bodies, he proceeds by differentiation, that is to say, he separates each of these bodies, one by one in succession, and sees by the difference what part of the total phenomenon belongs to each of them. But this method of exploration implies two things: it implies, first of all, that we know how many bodies are concerned in expressing the whole phenomenon, and then it admits that these bodies do not combine in any such way as to confuse their action in a final harmonious result. In physiology the method of differences is rarely applicable, because we can never flatter ourselves that we know all the bodies and all the conditions combining to express a collection of phenomena, and in numberless cases because various organs of the body may take each other's place in phenomena, that are partly common to them all, and may more or less obscure the results of ablation of a limited part . . .
>
> Physiological phenomena are so complex that we could never experiment at all rigorously on living animals if we necessarily had to define all the other changes we might cause in the organism on which we were operating. But fortunately it is enough for us completely to isolate the one phenomenon on which our studies are brought to bear, separating it by means of comparative experimentation from all surrounding complications. Comparative experimentation reaches this goal by adding to a similar organism, used for comparison, all our experimental changes save one, the very one which we intend to disengage.
>
> If, for instance, we wish to know the result of section or ablation of a deep-seated organ which cannot be reached without injuring many neighboring organs, we necessarily risk confusion in the total result between the effects of lesions caused by our operative procedure and the particular effects of section or ablation of the organ whose physiological

rôle we wish to decide. The only way to avoid this mistake is to perform the same operation on a similar animal, but without making the section or ablation of the organ on which we are experimenting. We thus have two animals in which all the experimental conditions are the same, save one, – ablation of an organ whose action is thus disengaged and expressed in the difference observed between the two animals. Comparative experimentation in experimental medicine is an absolute and general rule applicable to all kinds of investigation, whether we wish to learn the effects of various agents influencing the bodily economy or to verify the physiological rôle of various parts of the body by experiments in vivisection (Bernard, 1957, pp. 127–128).

This recommendation of Bernard's is a characteristic feature of most biomedical inquiries into the functions of organisms' parts and was, for example, utilized in the 1960's experimental investigations into the function of the thymus. Comparative experimentation, termed 'sham operation' in the current literature on ablation, is thus one of the basic methods to insure a rational determination of causes or functions in the biomedical sciences. Bernard referred to comparative experimentation as "the true foundation of experimental medicine" (Bernard, 1957, p. 129).

Bernard's account of comparative experimentation is useful but askew in important respects. It misrepresents the logic of causal inquiry in essentially two ways: (i) it overstates the distinction between the 'method of difference' as classically conceived and the method of comparative experimentation, and (ii) it ignores, as Mill also did, the crucial necessity for a *statistical* approach in biomedical inquiry.

Let me first argue for the essential identity of the method of differences and the method of comparative experimentation. To see clearly what the relations between the methods are I shall first reformulate the method in terms of an abstract case.

Let us suppose that an investigator wishes to apply the method of comparative experimentation to determine if an organ O plays a role in bringing about some physiological process I. Let us also suppose that ablation of O involves at this point of time in medical science, the need to severely injure or perhaps ablate an additional organ X, which is adjacent to or surrounds O. (This introduces one aspect of the complexity, with which Bernard was concerned in articulating his method of comparative experimentation.) The operative procedure also subjects the animal to general stress, S, which *may* also have an effect on process I. Implementation of Bernard's suggestions, then, will involve the investigator choosing (at least) two animals as similar as possible and creating the following experimental situation:

(1) $\bar{X}$ & S & $\bar{O}$ & $\bar{I}$

to the first animal, and:

(2) $\bar{X}$ & S & O & I

in the second. [Here the bar over the capital letter asserts ablation of that organ, or cessation or severe attenuation of the process, in the case of I. This notion of attenuation raises a point to which I shall have to return to below. Such ablation is normally surgical, but in some cases could be done by genetic (i.e., by selecting mutants without the organ) or by hormonal methods applied in the development of the organism (with appropriate modification of the stress term).] (1) accordingly reads: animal 1 has organs X and O ablated, has been subject to operational stress S, and does not manifest (the usual level of) I.

Now it is important to note that the *minutely specific* details of $\bar{X}$ (the ablation of X) such as detailed reports of the trauma to surrounding tissues and blood vessels, and the details of stress S, do *not* need to be known as long as there are reasonable grounds for assuming that $\bar{X}$ and S in the two animals are equivalent. This is emphasized by Bernard in his comments quoted above. It must be added, however, as a criticism of Bernard's overly strong distinction between the methods of comparative experimentation and of difference, that (i) he was not entirely fair to Mill, and (ii) when one examines the logic of the method represented by (1) and (2) above, it appears equivalent to the method of difference.

Bernard was not fair to Mill since Mill himself made the following comment concerning the application of the methods of experimental inquiry (including the method of difference). Mill wrote:

> The extent and minuteness of observation which may be requisite, and the degree of decomposition to which it may be necessary to carry the mental analysis, depend on the particular purpose in view. As to the degree of minuteness of the mental subdivision, if we were obliged to break down what we observe into its very simplest elements, that is, literally into single facts, it would be difficult to say where we should find them: we can hardly ever affirm that our divisions of any kind have reached the ultimate unit. But this, too, is fortunately unnecessary. The only object of the mental separation is to suggest the requisite physical separation, so that we may either accomplish it ourselves, or seek for it in nature; and we have done enough when we have carried the subdivision as far as the point at which we are able to see what observations or experiments we require. It is only essential, at whatever point our mental decomposition of facts may for the present have stopped, that we should hold ourselves ready and able to carry it farther as occasion requires, and should not allow the freedom of our discriminating faculty to be imprisoned by the swathes and bands of ordinary classification (Mill, 1959, p. 249).

Our second subpoint contending that, logically and epistemologically, the method of comparative experimentation is equivalent to the method of difference, can be substantiated by noting what the experimenter is attempting to do in (1) and (2) above is to fulfill Mill's conditions for application of the method of difference, namely the experimenter is searching for *one* difference (O versus $\bar{O}$) associated with the presence (I) or absence ($\bar{I}$) of the process I.

Let me now turn to my second general point, namely that Bernard (and Mill as well) ignores the crucial need for a statistical approach in biomedical inquiry.

Bernard's dislike of statistical analyses in scientific inquiry, and his scorn of statistics as illustrated by several colorful examples, is generally well known, almost to the point of notoriety. For example, I. Bernard Cohen in his forward to Bernard's *Introduction to the Study of Experimental Medicine* wrote:

> Anyone would agree with the absurdity of making a "balance sheet" of every substance taken in and excreted by a cat during eight days of nourishment and nineteen days of fasting, if on the seventeenth day kittens were born and counted as excreta . . . Another example given by Bernard is the physiologist who "took urine from a railroad station urinal where people of all nations passed, and who believed he could thus present an analysis of *average* European urine!" (Bernard, 1957, 'Foreword').

John Stuart Mill was similarly hostile to the use of probabilisitc and statistical reasoning in science. In his *A System of Logic*, Mill wrote:

> It is obvious too that even when the probabilities are derived from observation and experiment, a very slight improvement in the data, by better observations or by taking into fuller consideration on the special circumstances of the case, is of more use than the most elaborate application of the calculus to probabilities founded on the data in their previous state of inferiority. The neglect of this obvious reflection has given use to misapplications of the calculus of probabilities which have made it the real opprobrium of mathematics (Mill, 1959, p. 353).

Unfortunately for Bernard's and Mill's position, statistical analysis has become a cynosure in the methodological armamentarium of the biomedical sciences. For reasons having to do with the complexity and genetic and environmental variability of organisms, statistical assessments of the empirical evidence corraborating hypotheses have become mandatory even in those areas where it is believed the processes are ultimately non-stochastic. This is the case in inquiry into the functions of organs and other biological entities as well as into the causal efficacy of drugs and other more complex therapeutic interventions.

The incorporation of statistical methodology into the biomedical sciences took place after the demise of Mill and Bernard but appears to have been due in large measure to the earlier work of the French scientist, Quetelet, on social statistics. Quetelet himself was heavily influenced by Laplace's investigations, and was in turn brought to English attention through the work of John Herschel. In the later part of the nineteenth century Frances Galton combined the notions of Quetelet with the evolutionary theories of Charles Darwin. At the hands of Karl Pearson, a mathematical statistical approach to Darwinism evolved into the influential biometrical school which brought statistical reasoning into the mainstream of the biological sciences. Pearson's general theoretical contributions to statistics such as his χ^2 distribution and his work on correlation are well known. Gosset's discovery of the t distribution and its statistical importance occurred in 1908 and was further developed and embedded in a powerful general approach in R. A. Fisher's seminal book *Statistical Methods for Research Workers* in 1925. Jerzy Neyman and Egon Pearson's theories began to appear in the 1930's and, as I will note below, have now become classical.[5]

A brief simple example might help illustrate the way a statistical comparative methodology is employed. In his classical article on 'Immunological Function of the Thymus', J. F. A. P. Miller (1961) reported that there were significant differences between thymectomized and sham-thymectomized mice. [It should also be noted that similar conclusions were reported independently and almost simultaneously from R. A. Good's laboratory by Good, Dalmasso, Archer, Pierce, Martinez, and Papermaster; see Good and Gabrielsen (1964).] Miller analyzed both groups of mice from two aspects of cellular immunity. He provided evidence showing that skin grafts on genetically dissimilar mice survived significantly longer on thymectomized mice compared with their sham-thymectomized counterparts. No formal statistics were used *here* even though the 'comparative experiment' was clearly statistical, presumably because the comparison is not (at this point in time) distinctly quantifiable. The second aspect of Miller's experiment involved counts of the average lymphocyte: polymorph (or polymorphonuclear leukocyte) ratio of his two groups of mice. Here Miller reported specific statistical differences in the standard notation. His graph is reproduced as Figure 1.

It would take us beyond the scope of this essay to dwell extensively on the interpretation of the statistics in Miller's experiment, though these issues will be readdressed further below in the context of a discussion of clinical trials. Suffice it to say that in complex biological organisms, even of highly inbred strains, variability (whether the source be genetic, environmental, or

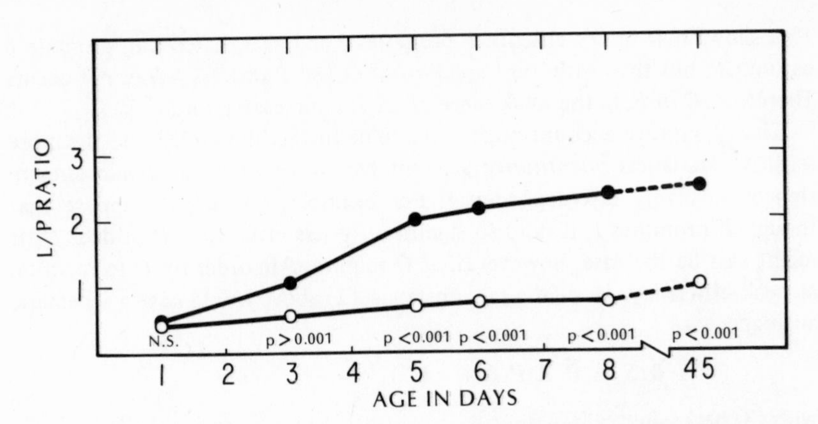

Fig. 1. Average lymphocyte: polymorph ratio of mice thymectomised in the neonatal period compared with sham-thymectomised controls. Statistical differences indicated. (o——o Thymectomised mice. •——• Sham thymectomised mice.) [From Miller, 1961.]

due to the measuring instruments) almost always requires a statistical analysis of the results of ablations in operated and sham-operated biota.

Let me now consider the question of complexity (and complex redundancy) from several other perspectives, from which it might be alleged one could call into question the methodology thus far developed.

First, let us imagine the existence in our above schema of some alternative organ P which assumes O's role only when O is ablated or damaged. P then will mask O's function, as ablation of O will still permit I to occur. This biological phenomenon whereby there exist alternative pathways to the same end has been termed *parastasis.*[6] I would argue in this case that the only way to determine experimentally that O's function is being masked by P is by applying a *reiterated method of comparative experimentation* to an augmented experimental situation. Accordingly, the way to determine causes experimentally (or to assess the effects of some therapeutic intervention) is not to eschew the method of comparative experimentation in such circumstances, but rather to apply it more vigorously. Schematically we can represent this reiterated application as follows: if P (or any other organ) is suspected of masking O's function then one creates experimentally the following situation:

$$\bar{X} \mathbin{\&} S \mathbin{\&} O \mathbin{\&} P \mathbin{\&} I \quad (1)$$
$$X \mathbin{\&} S \mathbin{\&} \bar{O} \mathbin{\&} P \mathbin{\&} I \quad (2)$$
$$\bar{X} \mathbin{\&} S \mathbin{\&} O \mathbin{\&} \bar{P} \mathbin{\&} I \quad (3)$$
$$\bar{X} \mathbin{\&} S \mathbin{\&} \bar{O} \mathbin{\&} \bar{P} \mathbin{\&} \bar{I} \quad (4)$$

This shows that in O's absence P promotes I and also that O can promote I without P, but that with joint ablation of O and P process I does not occur. Therefore, *O or P*, in the weak sense of 'or', are necessary for I.

This qualitative account might have to be further developed into the more sensitive statistical *quantitative* account hinted at above in certain circumstances involving interdependence. For example, one might suppose that though P promotes I, it does so significantly less efficiently than does O. It might also be the case, however, that O requires P in order for O to function at peak efficiency. In such a circumstance, (2) above would have a schematic analogue

$$\bar{X} \;\&\; S \;\&\; \bar{O} \;\&\; P \;\&\; \tilde{I} \qquad (2')$$

with (3) becoming:

$$\bar{X} \;\&\; S \;\&\; O \;\&\; \bar{P} \;\&\; \tilde{I} \qquad (3')$$

where $\tilde{I}$ reflects a significant attenuation of I. Such a situation would warrant asserting that I promotes or causes O, but would depend on a quantitative analysis and almost certainly involve statistical calculations.

Having defended the suitability of a generalized method of comparative experimentation in the case of parastasis, let us now turn to another type of criticism which might be directed against our thesis of the pre-eminence of a generalized method of comparative experimentation.

C. *The Deductive Method*

We have examined the method of difference above, and have mentioned in passing some of the other classical methods such as agreement, residues, and concomitant variation. It has been largely overlooked by students of methodology that John Stuart Mill believed that in certain complex sciences involving the intermixture of effects, for example, none of these four methods was sufficient.[7] For such inquiries he recommended the *deductive* method. This method he characterized as follows:

> ... in its most general expression the Deductive Method ... consists of three operations – the first one of direct induction; the second, of ratiocination; the third, of verification (Mill, 1959, p. 299).

Mill believed that one had to ascertain the simple laws and tendencies by direct induction and then, from a conjunction of these simple tendencies, deduce consequences which could in turn be verified. He noted, however, that there were considerable difficulties involved in applying this prescription to the biomedical sciences of physiology and pathology. He wrote:

> Accordingly, in the cases, unfortunately very numerous and important, in which the causes do not suffer themselves to be separated and observed apart, there is much difficulty in laying down with due certainty the inductive foundation necessary to support the deductive method. This difficulty is most of all conspicuous in the case of physiological phenomena: it being seldom possible to separate the different agencies which collectively compose an organized body, without destroying the very phenomena which it is our object to investigate:
>
> "Following life, in creatures, we dissect,
> We lose it in the moment we detect."
>
> And for this reason I am inclined to the opinion that physiology (greatly and rapidly progressive as it now is) is embarrassed by greater natural difficulties, and is probably susceptible of a less degree of ultimate perfection than even the social sciences, inasmuch as it is possible to study the laws and operations of one human mind apart from other minds much less imperfectly than we can study the laws of one organ or tissue of the human body apart from the other organs or tissues (1959, pp. 300–301).

Mill added that "pathological facts, or, to speak in common language, diseases in their different forms and degrees, afford in the case of physiological investigation the most valuable equivalent to experimentation properly so called, inasmuch as they often exhibit to us a definite disturbance in some one organ or organic function, the remaining organs and functions being, in the first instance at least, unaffected". Even these observations, however, were contingent on *early* observations, inasmuch as diseases quickly had complicating systemic effects.

Mill further noted that "Besides natural pathological facts, we can produce pathological facts artificially; we can try experiments . . . by subjecting [for example] the living being to . . . the section of a nerve to ascertain the functions of different parts of the nervous system" (Mill, 1959, p. 301). Such experiments, Mill suggested "are best tried . . . in the condition of health, comparatively a fixed state . . . [in which] the course of the accustomed physiological phenomena would, it may generally be presumed, remain undisturbed, were it not for the disturbing cause of which we introduce" (Mill, 1959, p. 301). Mill summed up the utility of these approaches, *which all appear to be the application of the method of difference or comparative experimentation* as we have characterized it above, as follows:

> Such, with the occasional aid of the Method of Concomitant Variations (the latter not less encumbered than the more elementary methods by the peculiar difficulties of the subject), are our inductive resources for ascertaining the laws of the causes considered separately, when we have it not in our power to make trial of them in a state of actual separation (1959, p. 301).

In the work of Mill and of Bernard we have an examination of the idea of *controlled* experimental inquiry developed with a considerable amount of philosophical sophistication. As noted, however, they were prisoners of their own time with respect to statistical analyses of biomedical experimentation.

4. THE ROLE OF STATISTICAL ANALYSES IN CLINICAL TRIALS

Thus far we have said relatively little about the role of statistics in biomedical inquiry, except to note in passing (1) that both Claude Bernard and John Stuart Mill tended to denigrate the subject, and (2) that current investigators value the subject highly. It remains to indicate the connection between the method of difference/comparative experimentation discussed above and the role of statistics vis à vis clinical trials.

My discussion will focus on the subdiscipline (within statistics) of 'hypothesis testing'. I intend here to present only summary highlights of the methodology and the interested reader may consult many fine texts and articles for details and examples.[8]

The central concepts that will be of concern to us involve two types of error, the probabilities associated with these types of error, and the concept of a *clinically impressive difference*.

We can begin by introducing the notion of a 'null hypothesis'. In the context of our inquiry a null hypothesis is simply the claim that there is *NO* real difference between a purported therapy T and an old 'control' therapy (or *no* therapy except perhaps that with *placebo* value). If there are two hypothetical groups, one of which receives T and the other C, we must examine the proportions on the groups, P_T and P_C, that achieve some predefined notion of improvement or cure, where presumably P_C is the currently accepted percentage of cures or improvement.

The null hypothesis in this notation is:

$$P_C = P_T \text{ or}$$
$$P_C - P_T = 0.$$

Our hypothetical groups are (potentially) infinitely large populations and we must make do with obtaining actual data from a sample of the populations. We assign new treatment T to one 'randomly' formed group and C to the other 'randomly' formed class, after controlling for well-established interfering variables. We examine the results after an 'appropriate' time and obtain *observed* proportions $\hat{P}_C$ and $\hat{P}_T$, where the circumflex indicates that these values are *estimates* of the true proportions P_C and P_T. We look for differences between P_C and P_T that would allow us to reject the null hypothesis and 'accept' the new therapy.

Now there is a possibility based on individual variability and irreducible measuring instrument error that (due to inadvertent sampling of a non-representative part of the population) an observable difference is recorded, i.e.,

$$\hat{P}_T - \hat{P}_C > 0.$$

This difference thus *could have arisen by chance, even though in truth* $P_T - P_C = 0$. We would like to keep this 'false positive' type of error low, and conventionally we can specify that it should not occur more than once in twenty such trials. This yields a probability α of this type of error, known as Type I error, $\leqslant 0.05$. (We could of course require that $\alpha \leqslant 0.01$, or $\leqslant 0.001$ which would be a more stringent requirement.)

Intuitively this kind of error represents the commission of an 'acceptance' mistake, since we accept a therapy that is in reality *not* better than the control, due to random sampling error. This probability α is generally known as a significance level or a P value.

There is, however, another kind of mistake we may make which intuitively can be called a *rejection* error, in that we may mistakenly *reject* as therapeutically worthless a substance, or technique, which is *in truth* better than that given to the control group. More specifically it may turn out that the observed difference:

$$\hat{P}_T - \hat{P}_C$$

is such that our P value is *greater* than α, therefore the difference is deemed *not statistically significant*. This would lead us to accept (or fail to reject) the null hypothesis, i.e., we would *reject* a hypothesis that claimed T *was more efficacious than* C. The probability of committing this type of 'rejection' error (or Type II error) is termed β.

It is here that the situation becomes quite complex, for the probability β, of committing a rejection mistake is not one single value like α. The α probability was computed on the basis of the assumed truth of the null hypothesis; in the case of calculating β we assume some *other* hypothesis is true, namely that

$$P_T = P_C + \Delta,$$

where we can interpret Δ as a true clinical difference between P_T and P_C. Since Δ can vary, β will also vary.

This true clinical difference Δ is a measure of the effect of the new T compared to old C. It can be shown that there are important interdependencies between α, β, Δ, and the number of patients (or sample size) who have to be examined to achieve preselected levels of α, β, and Δ. (Tables are generally available to facilitate such computations, though it is not too difficult to calculate these subject to appropriate statistical assumptions.) The Δ in therapeutic contexts will indicate the improvement (or lack thereof) of the therapy tested. A Δ value that is regarded as 'clinically impressive' will be a comparative function of other available therapies in the specific discipline or for the specific disorder being treated. The number of patients to be enrolled in a clinical trial should be large enough to detect clinically significant improvements, Δ, with low ($\alpha \leqslant 0.05$, $\beta \leqslant 0.1$) error probabilities.

What has been just sketched is an account of classical statistical hypothesis testing, generally following the pioneering work of Neyman and Pearson. A number of criticisms can and have been directed against this approach and alternative methods do exist. (One quite general and powerful approach known as the Bayesian statistical school has additional flexibility as regards choices of α and β, but in the absence of a well-defined prior probability, the results of the classical and Bayesian school converge as regards the probabilities of committing the types of mistakes reviewed above.) Nonetheless even though there is no currently *unanimously* accepted theory of statistical hypothesis testing, the methodology just sketched represents important advances on the notion of the method of difference/method of comparative experimentation as understood in the nineteenth century. Most important in my view is that the statistical tools allow a means rationally to detect clinically impressive differences due to a therapeutic intervention in populations where there is reasonably considerable variation. The (randomized) controlled clinical trial does not reject the insights of Mill and Bernard, but rather accepts them and embeds them in a statistical methodology which is of considerable power.[9]

5. CLINICAL VALIDATION IN PSYCHOANALYSIS

Lasagna (1962, Ch. 1) and Wulff (1976, Ch. 9) paint a depressing historical picture of clinical therapies which won the uncritical acceptance of physicians. Cauterization with boiling oil, enemata with an extraordinary variety of substances, and massive blood-letting are only a few of the therapies which were widely employed. More recently radiation for infected tonsils (extensively practiced in the 1940's in the Midwest) and synthetic estrogens (diethylsibesterol or DES) used in the 1950's have been shown to have no benefits and a considerable carcinogenic effect. As noted earlier, anticholinergic drugs for duodenal ulcers were also without significant therapeutic effect.

There are several difficulties encountered in rationally assessing putatively beneficial clinical therapies. As argued earlier, given the poor record of uncontrolled experience in clinical judgment, and the complexity and variability of the human organism, a controlled clinical trial is the only method likely to provide valid evidence for or against a potential substance or process intended to ameliorate a disease.

In his critical assessments of the clinical validity of psychoanalysis, Grünbaum has noted that:

> Clinical studies that purport to validate causal claims [of psychoanalysis] rely heavily on reasoning akin to Mill's method of agreement. But the latter has essentially heuristic rather than probative value. However great the heuristic value, it provides quite inadequate safeguards against the ravages of *post hoc ergo propter hoc*. To guard against inductive fallacies of causal inference, methods of *controlled* inquiry are needed. Such inquiry employs reasoning akin to that of Mill's joint methods of agreement and difference or of the method of concomitant variations not to speak of the more utopian method of difference. But *by and large*, the control required by these methods either has not or just cannot be instituted within the confines of the psychoanalytic interview (1980, p. 368).

Grünbaum (1980, 1981) and other investigators such as Frank (1973) have imputed only a *placebogenic* effectiveness to psychotherapy, and one of Grünbaum's important contributions to the placebo literature has been to propose a clarifying vocabulary for discussing "inadvertent" placebos. (In an inadvertent placebo, which Grünbaum relativizes to a target disorder D and to "those characteristic factors F that are singled out ... by one specified theory ψ ...," the truly causally efficacious factors are those "unknown incidental factors C of a therapeutic process" which result in a patient's improvement.) It is thus important to control for such factors, which can be done by varying the therapeutic process and the theory ψ which governs the process.

Recently the National Institute of Mental Health initiated a large multi-institutional project designed to determine the "efficacy, safety, and cost-effectiveness of particular treatment techniques and approaches applied to particular categories of mental disorders" (NIMH, 1980, p. 2). This project focuses on psychotherapy of depression and is intended to provide a carefully controlled prospective assessment of the clinical efficacy of several competing modes of therapy. One approach will be what is termed Cognitive/Behavioral Therapy (C/B) developed by Beck and his colleagues, and the other is the Interpersonal Psychotherapy (IPT) method of Klerman and his associates.[10] The latter approach is more psychodynamical in character. Both techniques will be compared not only with each other, but importantly with (1) an antidepressant drug (Imipramine) and 'clinical management' group, and (2) a pill-placebo and 'clinical management' cohort. 'Clinical management' involves supportive but not actively psychotherapeutic intervention, and provides an opportunity to monitor patients both for side effects and psychological emergencies.

Such a study utilizes a number of sophisticated scales for measuring depression in the various groups both prior to treatment and after the course of therapy. All data will be subject to statistical analyses following the general approach indicated above to show calculation of sample sizes based on preselected α and β values.

This prospective study illustrates the complex, costly, but necessary methods which are required to validate clinical efficacy. It demonstrates both the importance of controlled or comparative experimentation, and the extent to which statistical reasoning has become incorporated into the methodology of clinical assessment. The study also attempts to control for two types of placebo effects, one implicitly by employing two diverse therapies, the other explicitly by use of the pill-placebo. Though expensive and time-consuming, the study and the methodology it embodies leads one to hope that it will save patients from the burdens of inefficacious anecdotally founded expensive therapeutic regimens. Just as important, an awareness of the different types of statistical error inherent in such controlled clinical trials allows the study to be so designed as to detect a clinically significant improvement if it exists, by controlling for *both* the Type I *and* Type II kinds of errors, and thus hopefully will point the way toward safe, efficacious, and cost-effective treatment modalities.*

University of Pittsburgh

NOTES

* Grateful acknowledgement is made to the National Endowment for the Humanities for support of my research in the philosophy of medicine. I also want to thank Larry Laudan and Peter Machamer for reading an earlier draft of this material.

1 See, for example, the articles by Byar *et al.* (1976), Frieman *et al.* (1978), Kadane and Sedransk (1979), Weinstein (1974), and Zelen (1979). For an introductory discussion of 'case control' and other alternatives to the randomized clinical trial, see Fletcher, Fletcher and Wagner (1982), especially Chapter 10.

2 See, for example, Kolata's (1981) discussion of this issue and also *Science*, February 1, 1980, p. 506.

3 See *The Lancet* (1970), p. 1173.

4 *Ibid.*

5 Introductions to some of the history can be found in Merz (1904–12), Vol. II, Ch. XII, and in Fisher (1970), pp. 20–23.

6 See Murphy (1974), pp. 234–236, for a discussion of 'parastasis'.

7 For example, in his masterful discussion of Mill's methods in the Appendix of his (1974) Mackie does not discuss the 'deductive method' though he does mention in passing the 'hypothetico-deductive method', which Mill believed was different.

8 An elementary, autodidactically accessible source is P. Hoel's (1976). In the more clinical area Feinstein's (1977) is a fine collection of essays on this subject.

9 Disputes flourish among different schools of statistics concerning the value of some of the terminology such as Type I and Type II errors and the value of randomization. For an introduction to these issues see Giere (1977) and also Savage (1977). For a criticism of an uncritical commitment to P values $\leqslant 0.05$, see Schaffner (1981).

10 This study described by Kolata (1981) is in progress but no results have yet become available. I have been permitted to study the research design of this trial with the permission of several of its investigators, Dr. Stanley Imber and Dr. Paul Pilkonis, to whom grateful acknowledgement is made. The two trial psychotherapeutic approaches are discussed in detail in Beck *et al.* (1979), and in Klerman *et al.* (forthcoming).

REFERENCES

Beck, A. T., Rush, A. G., Shaw, B. F., and Emery, G. 1979. *Cognitive Theory of Depression*. New York: Guilford.

Bernard, C. 1957. *An Introduction to the Study of Experimental Medicine* (1865). Trans. H. C. Green. New York: Dover.

Byar, D. P., Simon, R. M., and Friedewald, W. T. *et al.* 1976. 'Randomized Clinical Trials,' *New England Journal of Medicine* **295**, 74–80.

Feinstein, A. 1977. *Clinical Biostatistics*. St. Louis: Mosby.

Fisher, R. A. 1970. *Statistical Methods for Research Workers* (1925). New York: Hafner.

Fletcher, R. H., Fletcher, S. W., and Wagner, E. H. 1982. *Clinical Epidemiology – The Essentials*. Baltimore: Williams and Wilkins.

Frank, J. 1973. *Persuasion and Healing*. Baltimore: Johns Hopkins Press.

Frieman, J. A., Chalmers, T. C., Smith, H. and Kuebler, R. R. 1978. 'The Importance of Beta, the Type II Error and Sample Size in the Design and Interpretation of the Randomized Control Trial,' *New England Journal of Medicine* **299**, 690–694.

Giere, R. 1977. 'Testing Versus Information Models of Statistical Inference.' In *Logic, Laws and Life*, R. Colodny (ed.). *University of Pittsburgh Series in the Philosophy of Science*, vol. 6. Pittsburgh: University of Pittsburgh Press, pp. 19–70.

Good, R. A., and Gabrielsen, A. (eds.). 1964. *The Thymus in Immunobiology*. New York: Harper and Row.

Grünbaum, A. 1980. 'Epistemological Liabilities of the Clinical Appraisal of Psychoanalytic Theory,' *Noûs* **14**, 307–385.

Grünbaum, A. 1981. 'The Placebo Concept,' *Behaviour Research and Theory* **19**, 157–167.

Hoel, P. 1976. *Elementary Statistics*. 4th ed. New York: Wiley.

Kadane, J. B. and Sedransk, N. 1979. 'Toward a More Ethical Clinical Trial,' *Proceedings of International Meeting on Bayesian Statistics*, Valencia, Spain 1979.

Klerman, G. L., Roundville, B., Chevron, E., Neu, C., and Weissman, M. *Manual for Short-Term Interpersonal Psychotherapy*. 4th draft. New Haven, forthcoming.

Kolata, G. B. 1981. 'Clinical Trial of Psychotherapies is Under Way,' *Science* **212**, 432–433.

Lancet. 1970. 'Anticholinergics and Duodenal Ulcer,' editorial in *The Lancet*, p. 1173.

Lasagna, L. 1962. *The Doctors' Dilemmas*. New York: Harper and Brothers.

Mackie, J. 1974. *The Cement of the Universe*. Oxford: Oxford University Press.

Merz, J. T. 1904–1912. *A History of European Scientific Thought in the Nineteenth Century*. Reprint. New York: Dover, 1965.

Mill, J. S. 1959. *A System of Logic* (1843). London: Longmans, Green and Co.

Miller, J. F. A. P. 1961. 'Immunological Function of the Thymus,' *Lancet* **2**, 748.

Murphy, E. 1974. *Logic of Medicine*. Baltimore: Johns Hopkins Press.

NIMH. 1980. 'Psychotherapy of Depression Collaborative Research Program (Pilot Phase).' Revised Research Plan, January 1980. Clinical Research Branch, National Institute of Mental Health.

Savage, L. 1977. 'The Shifting Foundations of Statistics.' In *Logic, Laws, and Life*, R. Colodny (ed.). Pittsburgh: University of Pittsburgh Press, 1977.

Schaffner, K. 1981. 'The Last Patient in a Drug Trial.' *Hastings Center Report* **2**, no. 6 (December 1981), 21–23.

Tversky, A. and Kahneman, D. 1974. 'Judgement Under Uncertainty,' *Science* **185**, 1124–1131.

Weinstein, M. C. 1974. 'Allocation of Subjects in Medical Experiments,' *New England Journal of Medicine* **291**, 1278–1285.

Wulff, H. R. 1976. *Rational Diagnosis and Treatment*. Oxford: Blackwell.

Zelen, M. 1979. 'A New Design for Randomized Clinical Trials,' *New England Journal of Medicine* **300**, 1242–1245.

ABNER SHIMONY

REFLECTIONS ON THE PHILOSOPHY OF BOHR, HEISENBERG, AND SCHRÖDINGER*

Many of the pioneers of quantum mechanics – notably Planck, Einstein, Bohr, de Broglie, Heisenberg, Schrödinger, Born, Jordan, Landé, Wigner, and London – were seriously concerned with philosophical questions. In each case one can ask a question of psychological and historical interest: was it a philosophical penchant which drew the investigator towards a kind of physics which is linked to philosophy, or was it rather that the conceptual difficulties of fundamental physics pulled him willy-nilly into the labyrinth of philosophy? I shall not undertake to discuss this question, but shall cite an opinion of Peter Bergmann, which I find congenial: he learned from Einstein that "the theoretical physicist is . . . a philosopher in workingman's clothes" ([1], q. v).

The questions with which I am preoccupied concern the philosophical implications of quantum mechanics – either epistemological, bearing on the extent, validity, and character of human knowledge; or metaphysical, bearing on the character of reality. Although quantum mechanics is not a system of philosophy, one can wonder whether it is susceptible to coherent incorporation in a philosophical system. I propose to examine the thought of three masters of quantum mechanics – Bohr, Heisenberg, and Schrödinger – not with a critical or historical intention, but in hope of finding some enlightenment concerning the problems posed by contemporary physics. I can say in advance that enlightenment will continue to elude us; nevertheless, the ideas of Bohr, Heisenberg and Schrödinger are rich and evocative for new studies.

Certain general principles of Bohr's philosophy can be sketched without any reference to quantum mechanics, even though it was his efforts to interpret the discoveries of the new physics which gave definitive form to his principles. Bohr always insists that scientific knowledge requires unequivocal description, a necessary condition for which is a distinction between the subject and the object ([2], p. 101). The success of our communication in everyday life concerning the positions and motions of macroscopic objects shows *a posteriori* that we can use these descriptions unequivocally, but one finds no assertion in the essays of Bohr that such concepts are *a priori*, like the categories of the understanding of Kant. In Bohr's opinion, the clarity of

R. S. Cohen and L. Laudan (eds.), Physics, Philosophy and Psychoanalysis, 209–221.

this macroscopic description does not at all imply that atomic objects are less existent than macroscopic objects ([2], p. 16). Rather, because of the indirectness of our knowledge of atomic objects and, even more, because of the quantum of action, an unequivocal description of an atomic phenomenon must "include a description of all the relevant elements of the experimental apparatus" ([2], p. 4).

From time to time Bohr indicates that his epistemological theses do not commit him to a metaphysics: for example, "the notion of an ultimate subject as well as conceptions like realism and idealism find no place in objective description as we have defined it" ([2], p. 79). In place of a metaphysics Bohr proposes a purely epistemological strategy – the mobility of the separation between the subject and the object ([2], pp. 91–92). I am very grateful to Aage Petersen, who was Bohr's assistant for seven years, for his testimony [3] – supplementary to the writings of Bohr but agreeing with them – concerning his renunciation of metaphysics. Bohr believed that even psychology must recognize this renunciation, because "in every communication containing a reference to ourselves we, so-to-speak, introduce a new subject which does not appear as part of the content of the communication" ([2], p. 101). The general project of elaborating an epistemology which rejects in principle the support of a metaphysics is reminiscent of the epistemological system of Kant. Although Bohr disagrees with some Kantian ideas concerning the structure of human knowledge, like the possibility of synthetic *a priori* judgments, he shares with Kant the renunciation of all knowledge of the 'thing-in-itself'.

The well-known proposals of Bohr concerning quantum mechanics follow, for the most part, from his epistemological theses in conjunction with the physical discovery of the quantum of action. The latter prevents the observation of all the properties of a physical object by a single experimental arrangement, or even the combination of all these properties in a single picture. But because of the mobility of the separation between the object and the subject, one can give complementary descriptions to a physical system. The range of possible descriptions is so rich that no experimental predictions can in principle exceed the means of the quantum formalism, and in this sense the formalism is complete. As for the analysis of Einstein, Podolsky, and Rosen, Bohr says:

Of course there is in a case like that just considered no question of a mechanical disturbance of the system under investigation during the last critical stage of the measuring procedure. But even at this stage there is essentially the question of an influence on the very conditions which define the possible types of predictions regarding the future behavior of the system. Since these conditions constitute an inherent element of the description of any phenomenon to which the term 'physical reality' can be properly attached, we see that the argumentation of the mentioned authors does not justify their conclusion that quantum-mechanical description is essentially incomplete ([2], pp. 60–61).

Bohr is saying essentially that the argument of Einstein, Podolsky, and Rosen is fallacious, because it is founded upon the supposition that we can speak intelligibly of the state of a physical system without reference to an experimental arrangement, which is equivalent to speaking of the 'thing-in-itself'.

There is some good sense in these proposals. In my opinion, Bohr is one of the great phenomenologists of science, showing a rare subtlety concerning the connections between theoretical concepts and experimental procedures. His 'thought experiments', which disentangle phenomena from inessential complications, clearly exhibit this subtlety. Nevertheless, something is missing in his overall interpretation of quantum mechanics. Perhaps he has renounced prematurely and without definitive reasons one of the great projects of Western thought, which is to establish the mutual support between epistemology and metaphysics. Bohr advises us to renounce the explanation of conscious activity, because introspection modifies the mental content which one wishes to examine ([4], pp. 13–14). But we can object that "the explanation of conscious activity" consists of a theory which sets forth principles governing the mind, rather than in a chronicle of mental content. One can see that there is a *threat* of paradox in the acquisition of knowledge of the principles governing the acquisition of knowledge; but all the reasoning that I have seen along this line lacks the force of the well-known set-theoretical and semantical paradoxes which are based upon self-reference.

There is another reason for my skepticism concerning Bohr's renunciation of metaphysics. If this renunciation is presented as a matter of principle, how does it differ from a kind of positivism, according to which the content of an assertion is completely exhausted by its implications for experience? To be sure, in disavowing idealism ([2], pp. 78–79) Bohr probably rejects all kinds of positivism; and moreover, Bohr shows a very strong attachment to the presence of ordinary things, which he does not wish to interpret as packets of sense impressions. But one arrives at a point where Bohr's renunciation of metaphysics begins to appear like an artifice: he wants to avoid the assault launched by positivists on our realistic preconceptions, and at the same time the obligation to examine questions of ontology. One can wonder whether such an artifice will not lead to more obscurity than illumination.

Heisenberg often identifies himself with the Copenhagen interpretation of quantum mechanics ([5], pp. 3 and 8), and he shares a large part of the philosophical theses of Bohr. There are, however, at least some differences of emphasis between them, and perhaps also some more profound differences, which deserve to be pointed out.

One sees an affinity to Bohr, and also to Kant, in the following passage: "what we observe is not nature in itself but nature exposed to our method of questioning" ([5], p. 58). Nevertheless, Heisenberg does not accept as completely as Bohr the Kantian idea of the renunciation of knowledge of the 'thing-in-itself'. Circumspectly and yet significantly he says, "The 'thing-in-itself' is for the atomic physicist, if he uses this concept at all, finally a mathematical structure; but this structure is – contrary to Kant – indirectly deduced from experience" ([5], p. 91). To the extent that he accepts the attribution of the quantum state to the atomic particle in itself ([5], p. 185), he weakens the renunciation of metaphysics, which is one of the hallmarks of Bohr's philosophy.

I shall risk expressing an even stronger opinion: that Heisenberg enunciates a metaphysical implication of quantum mechanics more explicitly than the other pioneers of this science. Quantum mechanics requires, according to Heisenberg, a modality which is situated between logical possibility and actuality, which he calls "potentia" ([5], p. 53). (It should be noted that Margenau ([6], p. 300) used the similar concept of "latency" to characterize the quantum state prior to Heisenberg.) This modality is relevant in considering the question of what happens between two observations, a question to which Heisenberg's answer is "the term 'happens' is restricted to the observation" ([5], p. 52). In spite of this response, Heisenberg does not wish to present himself as a positivist, because according to quantum mechanics the system is characterized between two observations by a quantum state, in other words by a wave function. This state evolves continuously in time in a manner determined by the initial conditions, and because of the independence of this state from the knowledge of any observer it deserves the characterization 'objective'. The quantum state describes nothing actual, but "It contains statements about possibilities or better tendencies ('potentia' in Aristotelian philosophy)" ([5], p. 53). The historical reference should perhaps be dismissed, since quantum mechanical potentiality is completely devoid of teleological significance, which is central to Aristotle's conception. What it has in common with Aristotle's conception is the indefinite character of certain properties of the system. One does not find Aristotle saying, however, that a property becomes definite because of observation and that the probabilities of all possible results are well determined, whereby the quantum mechanical potentialities acquire a mathematical structure. These probabilities, which Heisenberg characterizes as "objective" (*ibid.*), do not result from the ignorance of the observer, as is the case in classical statistical mechanics. The following is a remarkable passage, in which Heisenberg allows himself the use of the metaphysical term "ontology" and indicates the structural complexity of the set of potentialities:

> This concept of 'state' would then form a first definition concerning the ontology of quantum theory. One sees at once that this use of the word 'state', especially the term 'coexistent state', is so different from the usual materialistic ontology that one may doubt whether one is using a convenient terminology. On the other hand, if one considers the word 'state' as describing some potentiality rather than a reality – one may even replace the term 'state' by the term 'potentiality' – then the concept of 'coexistent potentialities' is quite plausible, since one potentiality may involve or overlap with other potentialities ([5], p. 185).

Heisenberg's interpretation of the wave function as a collection of potentialities is based in large part upon a consideration of the interference of amplitudes in the two-slit experiment. He acknowledges the formal success of hidden variables models of de Broglie and Bohm, but he objects to the reality of waves in a configuration space of more than three dimensions ([5], pp. 131–132). Heisenberg's objections are lacking in rigor, but his intuition was correct and was justified by the profound theorems of Gleason [7] and Bell [8, 9, 10]. As a result of their careful work one now knows that for a hidden variables theory to be both free from mathematical contradictions and in agreement with experiment it must have two properties: (1) it must be 'contextualist', that is, the values of quantities must be determined in part by the measuring apparatus, and (2) it must be 'non-local' in the sense of Bell. If one does not find these properties to one's taste (particularly the non-locality, which violates relativistic conceptions of space-time), one is obliged to admit that the wave function gives a complete description to a physical system. Then, if one does not want to renounce metaphysics, there is no other reasonable ontological conception of the wave function than that of Heisenberg.

Justifying the conception of the wave function as a collection of potentialities leads, however, to another metaphysical problem: how does the transition from potentiality to actuality take place? In other words, how does the reduction of the wave packet occur? It seems to me that Heisenberg offers two solutions, although he does not clearly distinguish them.

The first is essentially that of Bohr. Knowledge requires a separation between the subject and the object, even though the location of this separation is movable. Since the measuring apparatus is situated on the side of the subject, it is described in classical terms, in which one does not find quantum mechanical potentialities ([5], pp. 57–58). To the extent that this solution belongs to the general philosophy of Bohr, it has already been discussed above. It is difficult to see, however, how a strict adherence to the philosophy of Bohr would be compatible with Heisenberg's metaphysical doctrine concerning potentialities.

The second solution is in better agreement with this doctrine. It differs from the first solution in that it applies certain conceptions of quantum mechanics to the measuring process. Heisenberg suggests that the microscopic state of the measuring apparatus is indeterminate as a result of its interaction with the rest of the world ([5], p. 53). At this point he seems to say that the appropriate description of the apparatus ought to make use of a statistical operator (which is equivalent to the density matrix). He suggests further that the final statistical operator of the composite system consisting of atom plus apparatus is diagonal in a certain basis of vectors, each one of which is an eigenvector of a designated observable of the apparatus ([5], pp. 54–55). He hopes to capitalize upon the fact that the initial microscopic state of the apparatus is indefinite in order to arrive at the end of the measuring process at this diagonal statistical operator. If this were so, he could regard the designated apparatus observable as having a definite but unknown value, as in classical physics, and in that case the consciousness of the observer would not be the agent of the reduction of the wave packet: "we may say that the transition from the 'possible' to the 'actual' takes place as soon as the interaction of the object with the measuring device, and thereby with the rest of the world, has come into play; it is not connected with the registration of the result by the mind of the observer" ([5], pp. 54–55). This proposed explanation of the transition from potentiality to actuality is so clear that it is susceptible to being evaluated mathematically. Here is one of those rare cases in which a metaphysical question admits of a mathematical answer, as Leibniz hoped. Unfortunately from Heisenberg's point of view the result of the evaluation is negative. The dynamical law of quantum theory does not permit the statistical operator to evolve in the manner required by Heisenberg's proposed solution – a result first established by Wigner under special conditions [11], and then generalized by d'Espagnat [12], Shimony [13], and others.

To conclude, Heisenberg has drawn from quantum mechanics a profound and radical metaphysical thesis: that the state of a physical object is a collection of potentialities. But his discovery is incomplete, in that the transition from potentiality to actuality remains mysterious. T. S. Eliot has said (à propos of other things) that

> Between the potency
> And the existence
> . . . .
> Falls the Shadow. ('The Hollow Men', [14], p. 104.)

I turn now to Schrödinger, who I believe was the most remarkable philosopher among the physicists of our century. I propose to extract from his works three very different groups of remarks. Although there are no contradictions among them, there are some tensions which deserve close study.

(1) The first group of remarks concerns the implications of the quantum mechanical formalism, when it is considered to be an objective description of nature and not just a means for making predictions. The most celebrated remark concerns the cat which is prepared in a superposition of a state of being alive and a state of being dead [15]. Schrödinger accepts – at least provisionally – the interpretation of the wave function as a collection of potentialities, but he insists upon the fact that the dynamical law of quantum mechanics prohibits a transition from potentiality to actuality. It is clear from context that Schrödinger is giving a *reductio ad absurdum* argument. He wishes to signal that the quantum mechanical formalism needs to be changed in some way.

In his comment upon the experiment of Einstein, Podolsky, and Rosen, Schrödinger emphasizes the non-separability of the state of two particles: "they can no longer be described in the same way as before, viz., by endowing each of them with a representative of its own. I would not call that *one* but rather *the* characteristic trait of quantum mechanics" [16]. In this comment he is not attempting a *reductio ad absurdum*. He recognizes that one is concerned with a radical metaphysical thesis, the experimental evidence for which was incomplete at the time of his writing. He asks whether a non-separable state of particles spatially distant from one another is realizable in nature and leaves the answer to this question to experiment. So far as I know he never commented upon the positive answer which Bohm and Aharonov [17] derived from the experiment of Wu and Shaknov [18], while he was still alive.

(2) The second group of Schrödinger's philosophical remarks is his polemic defending realism against a positivist interpretation of science. He grants that

postulating material bodies, governed by the laws of physics, achieves order in our experience and produces an "economy of thought"; but he insists that the success of this postulate reveals something important which goes beyond our conventions: "The fact alone that economy and a successful mental supplementation of experience, in particular extrapolation to the future, are at all possible, presupposes a definite quality of experience: *it can be ordered*. This is a fact that itself demands an explanation" ([20], p. 183).

Schrödinger considers the Copenhagen interpretation of quantum mechanics to be a positivist exercise ([21], pp. 202–205). In his opinion, the principle of complementarity evades the ontological problems posed by quantum mechanics, by insisting primarily upon the mutual exclusiveness of the conditions of different types of observations. By contrast, Schrödinger himself speaks of the physical reality of the quantum mechanical waves: "Something that influences the physical behavior of something else must not in any respect be called less real than the something it influences – whatever meaning we may give to the dangerous epithet 'real'" ([21], p. 198). He grants that up till now no one has constructed a faithful picture of physical reality (*ibid*., p. 204). Nevertheless, he hopes that the renunciation of the conception of an individual particle endowed with individuality and the recognition of the primacy of waves will guide us towards the desired picture (*ibid*., pp. 205ff). In any case, Schrödinger is unwilling to abandon his grand vision in favor of the Copenhagen interpretation: an 'either-or' which seems to him too facile ([20], p. 160).

(3) The third group of philosophical remarks which I shall cite is chosen from Schrödinger's speculative writings, which are concerned with appearance and reality, the self, God, and above all the relation between matter and mind. It is noteworthy that he almost never makes use of physics or of his philosophical analysis of physics in dealing with these questions. An explanation for his abstention can be found in the thesis that science is founded upon 'objectivation' – that is, "simplification of the problem of nature by preliminary exclusion of the cognizing subject from the complex of what is to be understood" ([20], p. 183); but the most profound philosophical problems are precisely those concerning the subject which has been excluded from the scientific picture of the world.

Briefly, his principal philosophical theses are the following.

(i) The dichotomy between mind and matter is ultimately artificial, even though it is useful for the conduct of our lives. "The 'real world around us' and 'we ourselves', i.e., our minds, are made up of the same building material, the two consist of the same bricks, as it were, only arranged in a different order – sense perception, memory images, imaginations, thought" ([22], pp. pp. 91–92).

(ii) The difficulty of finding the place of mind in the scientific picture of the world is precisely that mind and matter are composed of the same elements: "To get from the mind-aspect to the matter-aspect or vice versa, we have, as it were, to take the elements asunder and to put them together again in an entirely different order" (*ibid.*, p. 92).

(iii) In spite of the illusion of a multiplicity of subjects in the world, each with its own feelings and thoughts, there is in fact only one Mind. In *What is Life*? ([19], [20]) – a book which argues powerfully for the reduction of biology to physics and which advocated the idea of the chemical character of the genetic code a decade before Watson and Crick – the epilogue contains only one equation: "ATHMAN = BRAHMAN (the personal self equals the omnipresent, all-comprehending eternal self)."

The tension among the elements of Schrödinger's philosophy which I mentioned earlier ought now to be evident. On the one hand he defends physical realism against a positivist interpretation of science; on the other hand he proposes an idealist metaphysics which recalls that of the *The Analysis of Sensations* [23] of the great positivist Ernst Mach, as well as Indian idealism. I have mixed reactions to the cornucopia of philosophical ideas which Schrödinger offers us.

First of all, his criticism of a positivist interpretation of science is excellent. It may not have been entirely fair to classify the Copenhagen interpretation as positivist, especially since both Bohr and Heisenberg reject this epithet for their views. However, the analysis made in the first part of this paper showed that the philosophical positions of Bohr and Heisenberg are, so to speak, 'metastable', and positivism is one of the stable states into which they could fall. Consequently, Schrödinger's criticisms are relevant, even if one has some doubts about his exegesis of their texts.

I share his hope for a completely physical solution to the problems of quantum mechanics, notably the problem of the reduction of the wave packet and the problem of non-locality. It is necessary to recognize a level of description on which physical discourse is appropriate, even if the fundamental ontology of the universe is idealist. In the language of Schrödinger, the world is susceptible to the operation of 'objectivation'. For the most part, the processes described by quantum mechanics do not go beyond this physical level. The apparatus by which the typically quantum phenomena are exhibited – interferometers, spectrometers, coincidence counters, etc. – belong as much to the realm of matter as any ordinary object. The essential question is whether the 'interphenomena' (Reichenbach's term) have a character as physical as that of the apparatus. Schrödinger agrees with Bohr and Heisenberg that no variant of classical physics can describe the 'interphenomena', but he insists, in strong opposition to them, that the purely physical means for achieving this description are far from being exhausted.

A little explored line of research is to replace the usual linear law governing the evolution of the wave function (that is, the time-dependent Schrödinger equation) by a non-linear law. This replacement would be consistent with the ontological primacy of waves, which Schrödinger never abandoned. There is even some textual indication ([24], p. 451) that Schrödinger considered the possibility of a non-linear modification of quantum dynamics, although he seems never to have made a specific suggestion. I find this line of research attractive, but must make two negative comments. The first is that this line of research at best promises to resolve the problem of the reduction of the wave packet and offers nothing concerning the problem of non-locality. The other is that the experimental results recently obtained [25] do not support the conjecture of non-linear dynamics.

Another line of research which does not go beyond the physical level is the study of relations between quantum mechanics and space-time structure. So far as I know, Schrödinger never made a suggestion along these lines, even though he was a profound investigator of space-time structure. There are, however, two reasons for taking seriously this line of research. One is the difficulty of applying the procedures of quantization to space-time itself. The other is the non-locality (in the sense of Bell) of certain experimental predictions corroborating quantum mechanics. I spoke elesewhere [26] of the possibility of "peaceful coexistence" of the non-locality of quantum mechanics and the relativistic structure of space-time, a possibility which is suggested by the fact that one cannot make use of quantum non-locality for the purpose of sending a message instantaneously. If, however, this peaceful coexistence does not succeed, then it would be necessary to take the radical step of postulating a modification of relativistic space-time structure.

It seems to me possible that all attempts to explain the reduction of the wave packet in a purely physical way will fail. There would then remain only one type of explanation of the transition from quantum mechanical potentiality to actuality: the intervention of the mind. I wish to emphasize that in my opinion it is very improbable that we shall be pushed to this extremity. Nevertheless, I think that Schrödinger was wrong in excluding this possibility *a priori*. Perhaps physical evidence will exhibit to us new restrictions upon the operation of objectivation (to use Schrödinger's own terminology) and will reveal some imperfections in the physical level – some fissures, so to speak, through which the essentially mental character of the world reveals itself. There are many people who embrace with enthusiasm the thesis of the indispensability of the mind for the reduction of the wave packet, a thesis to which I circumspectly allow only the possibility that it may be true. I allude particularly to the authors of a collection of parapsychological articles entitled *The Iceland Papers* [27]. Before the thesis in question could attain a status above pure speculation, it is essential to have careful experiments which are capable of repetition. I doubt that such experiments have already been carried out. One might say that my doubts indicate conservatism or even conformity, but to this accusation I offer the following response. With the help of three students I attempted to transmit a message by means of the reduction of the wave packet, an attempt which should have succeeded had the thesis in question together with an auxiliary hypothesis been true. Our result [28] was negative and it presents an obstacle to the thesis of the indispensability of the mind for the reduction of the wave packet, although it is far from being definitive.

Returning to Schrödinger, I evidently cannot do justice to his metaphysics in a few pages. I wish to indicate, however, the possibility of formulating an idealist metaphysics which remedies some of the imperfections of his own. I find his thesis that the mind cannot be included in a scientific picture of the world quite unconvincing. The science of psychology – by which I mean the study of thought, sensations, and feelings, and not merely the study of behavior – has made enough progress to cast doubt upon this thesis. And Schrödinger's doctrine of a single Mind is difficult to reconcile with the immense body of evidence concerning private sorrows, hidden hopes, and secret conspiracies. He discusses briefly ([29], pp. 94–95) only one example of a pluralistic idealism, namely the monadology of Leibniz, which he dismisses because the monads are windowless and therefore cannot account for language and other communication. So far as I know, Schrödinger never mentions the pluralistic idealism of Whitehead, according to which the monads are endowed with windows, so to speak, since one of them can contribute to the sensations of another. Whitehead supposes that the instantaneous state of an elementary particle must be characterized in mental terms, like 'feeling', even though the sense of these terms must be extrapolated far beyond their normal usage. His great design is to integrate physics into a generalized psychology, as Maxwell integrated optics into electromagnetic theory. Whitehead rejects the above-cited thesis of Schrödinger that it is necessary "to take the elements asunder and to put them together again in an entirely different order" in order to relate the mind-aspect of the world to the matter-aspect. Whitehead rather regards the matter-aspect as an abridged version of the mind-aspect. "The notion of physical energy, which is at the base of physics, must then be conceived as an abstraction from the complex energy, emotional and purposeful, inherent in the subjective form of the final synthesis in which each occasion completes itself" ([30], p. 188).

I do not wish to deny the obscurity of Whitehead's exposition. Despite its obscurity, however, it offers a possibility which Schrödinger has denied: the possibility of integrating the mind into a scientific picture of the world.

In conclusion, I would like to make one additional speculation. Perhaps the great metaphysical implications of quantum mechanics – namely, non-separability and the role of potentiality – have made the unification of physics and psychology somewhat less remote. Perhaps we are confronted with structural principles, which are applicable as much to psychological as to physical phenomena. If this should turn out to be the case, then the physical discoveries of Schrödinger would be more closely connected with his metaphysical preoccupations than he himself recognized.**

Boston University

NOTES

* This paper is dedicated to Adolf Grünbaum in honor of his lifetime of explorations of the interdependence of philosophy and the natural sciences.

** This paper is a translation of 'Réflexions sur la philosophie de Bohr, Heisenberg et Schrödinger,' which was part of a symposium entitled *Les Implications Conceptuelles de la Physique Quantique*, published in *Journal de Physique* **42**, Colloque C–2, supplément au no. 3 (1981), pp. 81–95. Permission for publishing this translation was kindly granted by Les Editions de Physique. The research on which the paper is based was supported in part by the National Science Foundation, Grant no. SOC–7908987. I wish to thank Dr. Andrew Frenkel for his helpful suggestions.

REFERENCES

[1] Bergmann, P. 1949. *Basic Theories of Physics* 1. New York: Prentice-Hall.

[2] Bohr, N. 1958. *Atomic Physics and Human Knowledge*. New York: Wiley.

[3] Petersen, A. 1968. *Quantum Physics and the Philosophical Tradition*. Cambridge, Mass.: M.I.T. Press.

[4] Bohr, N. 1966. *Essays 1958–1962 on Atomic Physics and Human Knowledge*. New York: Vintage.

[5] Heisenberg, W. 1962. *Physics and Philosophy*. New York: Harper.

[6] Margenau, H. 1949. *Philosophy of Science* **16**, 287.
[7] Gleason, A. 1957. *Journal of Mathematics and Mechanics* **6**, 885.
[8] Bell, J. S. 1965. *Physics* **1**, 195.
[9] Bell, J. S. 1966. *Reviews of Modern Physics* **38**, 447.
[10] Bell, J. S. 1971. In *Foundations of Quantum Mechanics*, ed. B. d'Espagnat. New York: Academic Press.
[11] Wigner, E. 1963. *American Journal of Physics* **31**, 6.
[12] d'Espagnat, B. 1966. *Supplemento al Nuovo Cimento* **4**, 828.
[13] Shimony, A. 1974. *Physical Review* **D9**, 2321.
[14] Eliot, T. S. 1936. *Collected Poems 1909–1935*. New York: Harcourt, Brace and Co.
[15] Schrödinger, E. 1935. *Naturwissenschaften* **23**, 807, 823, 844.
[16] Schrödinger, E. 1935. *Proceedings of the Cambridge Philosophical Society* **31**, 555.
[17] Bohm, D. and Aharonov, Y. 1957. *Physical Review* **108**, 1070.
[18] Wu, C. S. and Shaknov, I. 1950. *Physical Review* **77**, 136.
[19] Schrödinger, E. 1967. *What is Life?* and *Mind and Matter*. Cambridge: Cambridge University Press.
[20] Schrödinger, E. 1956. *What is Life? and Other Scientific Essays*. Garden City, N.Y.: Doubleday.
[21] Schrödinger, E. 1957. *Science, Theory, and Man*. New York: Dover.
[22] Schrödinger, E. 1954. *Nature and the Greeks*. Cambridge: Cambridge University Press.
[23] Mach, E. 1959. *The Analysis of Sensations*. New York: Dover.
[24] Schrödinger, E. 1936. *Proceedings of the Cambridge Philosophical Society* **32**, 446.
[25] Shull, C. G., Atwood, D. K., Arthur, J., and Horne, M. A. 1980. *Physical Review Letters* **44**, 765.
[26] Shimony, A. 1978. *International Philosophical Quarterly* **18**, 3.
[27] Puharich, A. (ed.). 1979. *The Iceland Papers*. Amherst, WI.: Essentia Research.
[28] Hall, J., Kim, C., McElroy, B. and Shimony, A. 1977. *Foundations of Physics* **7**, 759.
[29] Schrödinger, E. 1964. *My View of the World*. Cambridge: Cambridge University Press.
[30] Whitehead, A. N. 1965. *Adventures of Ideas*. New York: New American Library of World Literature.

BRIAN SKYRMS

ZENO'S PARADOX OF MEASURE

INTRODUCTION

Zeno of Elea is perhaps best known for his four paradoxes of motion: The Dichotomy, Achilles and the Tortoise, The Arrow, and the Stadium. These are, however, supporting pieces in a grand argument against plurality whose keystone is Zeno's paradox of measure.[1] Adolf Grünbaum has, on several occasions,[2] called attention to the importance of this paradox, and explained how the standard measure-theoretic resolution of it became possible only after Cantor's theory of infinite sets.

The argument in question has emerged from Zeno's fragments only after considerable scholarly reconstruction.[3] It is a *reductio ad absurdum* that an extended thing (for simplicity we can think of a unit line segment) can be thought of as made up of an infinite number of parts. The argument goes roughly as follows:

(I) If the parts had finite magnitude, the whole would have infinite magnitude.

(II) If the parts had no magnitude, then the whole would have no magnitude.

It is set up by a prior argument to the effect that if the thing in question is divisible *ad infinitum*, then it can be partitioned into an infinite number of parts.

In the first part of this paper, I will take a closer look at the logic of Zeno's argument, and of the ways in which various schools attempted to escape its embarrassments. In the second part, I will argue that despite the profound achievements of Cantor and Lebesgue, the fundamental spirit of Zeno's paradox is still capable of mischief.

I. ZENO TO EPICURUS

Zeno on Infinite Divisibility

Zeno argues that if the line segment is divisible *ad infinitum* it can be partitioned into an infinite number of parts. His construction is as follows:

R. S. Cohen and L. Laudan (eds.), Physics, Philosophy and Psychoanalysis, 223–254.

1: Partition the line into two segments by bisecting it.

n: Refine the partition gotten at stage $(n - 1)$ by bisecting each member of it.

ω: Take the common refinement of all the partitions gotten at finite stages of the process.

The number of parts at stage ω must be greater than any finite number.

This construction uses something slightly stronger than the stated assumption – it requires not only that each part can be divided but also that it can be divided into two equal parts. The bisection of a line segment with ruler and compass is an elementary construction which was familiar at the time. It was perhaps an important part of the motivation for holding that the line segment is divisible *ad infinitum*. For the stated conclusion bisection is not essential, but it can assume importance if one wants to argue by symmetry that each member of the resulting partition has equal length.

Zeno's construction makes the daring leap from potential to actual infinity. Given that one can, for any finite n move from the $(n - 1)$th stage in the construction to the nth stage, Zeno proposes an additional move from the totality of the finite stages to an infinite stage. Aristotle resists this move, charging Zeno with misunderstanding the nature of infinity, but from a modern point of view it is perfectly legitimate.

It is worth looking at this construction from an unabashedly modern view to see what it yields. Let us regard the unit line as a set of points, one for each real number in the closed interval $[0, 1]$. The bisection of this line partitions it into two point sets: $[0, 1/2)$; $[1/2, 1]$. (Likewise at each stage of refinement of the partition we throw the midpoints in the right hand set. This decision as to where the midpoints go is arbitrary, but some decision must be made if we are to have a genuine partition.) A sequence of sets, the first a member of the first partition . . . the nth a member of the nth partition, such that for each n the $(n + 1)$th set is a subset of the nth set, will be called a *chain*. Zeno's construction comes essentially to this: each intersection of such a chain counts as a part of the line on level ω. The collection of such 'ω-parts' is held to be an infinite partition of the unit line.

This construction gets us *something*. What does it get? Do we get all the points on the unit line? Do we get a partition? First, let us notice that *each point on the unit line is a member of some ω-part*. For any point consider the set containing for each finite n, the member of the nth level partition of which that point is a member. This set is a chain, and the given point is in its intersection. Next we see that *each ω-part contains no more than one point*. Between any two real points, there is some finite distance. Therefore, there is some finite stage *n* in the construction, such that at that stage the points fall into different elements of the partition. Thus they fall into different ω-parts since for them to fall into the same ω-part they would have to both be members of all elements of the chain of which that ω-part is the intersection. The foregoing two propositions show that the ω-parts do indeed form a *partition* of the unit line; i.e., they are a collection of disjoint sets whose union is the set of points on the line. Are the ω-parts gotten by this construction then exactly the sets containing one point? Not quite. All such sets are ω-parts, but some ω-parts are empty. For example, the intersection of the chain: [0, 1], [0, 1/2), [1/4, 1/2), [3/8, 1/2), [7/16, 1/2), . . . is empty. Any point distinct from the midpoint eventually gets squeezed out, and the point 1/2 is not itself in any member of the chain. So the ω-parts, at least in that version of the construction that I have pursued, consist of the empty set together with the unit set of each point on the real line.

I can't resist quoting here part of a passage from *De generatione et corruptione* where Aristotle is recounting a version of Zeno's argument as the argument which persuaded Democritus of the necessity of indivisible magnitudes:

> Suppose then, that it *is* divided; now, what will be left? Magnitude? No, that cannot be, since there will then be something left not divided, whereas it was *everywhere* divisible. But if there is to be no body or magnitude [left] and yet [this] division is to take place, *then either the whole will be made of points*, and then [parts] of which it is composed will have no size, *or* [that which is left] will be *nothing at all*. [A 2, 316^{a}24–30, quoted in Furley (1967), p. 84. Parenthetical insertions are Furley's, emphasis mine.]

It is tempting to ask – but only in the spirit of speculation – how much of the foregoing analysis of Zeno's construction would have been possible for a first-rate mathematician contemporary with Aristotle provided he accepted that the points on a unit line segment can be associated with the numbers, rational and irrational, from zero to one inclusive, and provided he accepted (perhaps only for the sake of argument) the conception of actual infinity implicit in the construction. I think that the answer is "All of it!" The facts used in the reasoning are all fairly elementary: e.g. there is a finite distance between any two distinct points. In particular, nothing special about the structure of the reals is used. If the construction were applied to the rational line, it would still generate an infinite partition; we would simply come up with the empty set more often. There is one slightly sticky issue regarding the foregoing construction, and that regards the seemingly trivial issue of where to throw the midpoint. One might argue that the line is not really divided into two equal parts if the midpoint is thrown on one side or the other. However, if the midpoint is put on both sides there is no partition. Such questions were, in fact, discussed. This one is raised for a slightly different purpose in the pseudo-Aristotelian treatise, 'Concerning Indivisible Lines.'[4] The objection could be met by modifying the construction. We could, at each stage, divide the line segment into three parts: the midpoint and two intervals. At the cost of some complication, Zeno's argument could still be carried through. Returning to the discussion in 'On Generation and Corruption':

But suppose that, as the body is being divided, a minute section – a piece of sawdust, as it were – is extracted, and that in this sense a body 'comes away' from the magnitude, evading the division. Even then the same argument applies (316^{a}34–316$_{b}$3).

Zeno's Paradox of Measure

Suppose that the line segment is composed of an infinite number of parts. Zeno claims that this leads to absurdity in the following way:

(I) Either the parts all have zero magnitude or they all have positive magnitude.

(II) If they have zero magnitude, the line segment will have zero magnitude, since the magnitude of the whole is the sum of the magnitudes of its parts.

(III) If they have positive magnitude, then the line segment will have infinite magnitude, for the same reason.

(I) is never explicitly stated, but it is certainly an implicit supposition of the argument. In the first place, it assumes that the parts in question are the sorts of things which *have* magnitude; that questions of magnitude meaningfully apply. Call this the assumption of *Measurability*. It is challenged by Aristotle. There is something more being assumed. There would be no paradox if an infinite number of the parts had zero magnitude and a finite number had appropriate positive magnitudes. This possibility is ruled out. On what basis? One might argue that if the infinite partition of the line segment is generated by bisection, as discussed in the last section, then each part should have *equal* magnitude because of the equality in magnitude of all members of the finite partitions at each stage of the process which generates the infinite partition.

Call this the assumption of *Invariance*. The assumptions of *Measurability* and *Invariance* legitimize (I) in a plausible way, and do a bit more as well. Invariance will enter again later in the reasoning.

(II) and (III) share the assumption that the principle that the magnitude of the whole is the sum of the magnitudes of its parts continues to hold good when we have a partition of the whole into an infinite number of parts. This requires for its intelligibility that some general sense be given to the notion of the sum of an infinite number of magnitudes. We have no such definition from Zeno. We could, of course, *supply* one adequate to his purposes. Let S be an infinite set of magnitudes, and let S^* be the set of *finite* sums of magnitudes in S. A real number is an *upper bound* for S^* if and only if it is greater than or equal to every member of S^*. Let the *sum* of S be defined as the *least upper bound* of S^* if a real least upper bound exists, and as infinity otherwise. I will call this the principle of *Ultra-Additivity*. Zeno probably simply thought that *any* reasonable principle relating the magnitude of the whole to that of an infinite partition of it would have to satisfy (II) and (III). Nevertheless, it may be useful to have an explicit formation of such a principle at hand while considering the argument.

On the principle of Ultra-Additivity, (II) is perfectly correct. From a modern viewpoint, we would say so much the worse for the principle. This does not appear to be a line that was taken in ancient times. Neither the school of Plato, nor that of Aristotle, nor the Atomists appear to have challenged (II).

At first glance, (III) may appear to be a mathematical blunder. Does Zeno really believe that any infinite sum of finite magnitudes is infinite? His own paradoxes of the Dichotomy, and of Achilles and the Tortoise provide a counterexample, a fact that did not escape Aristotle.[5] However, if Zeno is here presupposing that he has in hand an *invariant* partition by means of the previously discussed construction, there is no blunder. As we noted, such an argument may already be required for (I). If so, it costs no more to use it twice.

There is a more delicate question to be raised about (III). What if the magnitudes involved were infinitesimal? Then (III) could fail. Infinitesimals are ruled out by the *Axiom of Archimedes* (probably originated by Eudoxus). This axiom entails that for any quantity, e, no matter how small, and any integer m, no matter how large, there is an integer n such that n times e is greater than m. If the magnitudes under consideration are Archimedean, then the relevant S^* will have *no* upper bounds, and (III) will be justified.

The question of who, if anyone, at this time held a theory of infinitesimal magnitudes is a matter of controversy. It is of some interest that Cajori takes one of Zeno's fragments to be an ironic dismissal of the theory of infinitesimal magnitudes: "Simplicius reports Zeno as saying: 'That which when added to another does not make it greater, and being taken away from another does not make it less, is nothing'. According to this, the denial of the existence of the infinitesimal goes back to Zeno" (Cajori, 1919, p. 51). However, it should be noted that Cajori's interpretation of this problematic fragment is not one favored by most classicists.[6] However this may be, it remains that (III) requires the assumption of the Archimedean axiom.

Zeno has shown that the supposition of all the following leads to paradox: the line segment (or more generally any body with positive magnitude) can be partitioned into an infinite number of parts such that: (I) the concept of Magnitude applies to the parts (Measurability); (II) the parts have equal magnitude (Invariance); (III) there are no infinitesimal magnitudes (Archimedean Axiom); (IV) the magnitude of the whole is the sum of the magnitudes of the parts in the sense given (Ultra-Additivity). The argument as we have reconstructed it contains no fallacy. A consistent theory of non-trivial magnitudes must give up some part of the foregoing. That Zeno had raised a genuine problem seems to have been well enough understood by his contemporaries. Different solutions were explored by different schools.

Aristotle's Answer to Zeno

Aristotle blocks the paradox of measure at two places. As we have already noted, he denies that infinite divisibility allows the construction of an infinite partition. Aristotle will follow the construction to any finite level n, but not to level ω. Thus: "A thing is infinite only potentially, i.e. the dividing of it can continue indefinitely . . . " (Aristotle, 'On Generation and Corruption,' 318a21. See also Aristotle's discussion in 317a1–16 where he claims that the argument rests on a confused assumption that "point is 'immediately next' to point." It appears, however, that it is Aristotle himself who is confused here. As we have seen, the argument requires no such assumption, and it is nowhere to be found in Aristotle's lucid statement of the argument in 316a1–316b14.) Again, in the *Physics*, Bk. III Ch. 6, we have:

> Now, as we have seen, magnitude is not actually infinite. But by division it is infinite. (There is no difficulty in refuting the theory of indivisible lines.) The alternative then remains that the infinite has a potential existence.

But the phrase 'potential existence' is ambiguous. When we speak of the potential existence of a statue we mean that there will be an actual statue. It is not so with the infinite. There will not be an actual infinite ($206^a 16$–20).

Now, as Fränkel (1942) and Owen (1957–8) point out, Zeno has available a devastating ***argumentum ad hominem*** for anyone who, like Aristotle, will grant each finite n but will resist the move to level ω on the grounds that it presupposes the completed actual infinity of all the n stages. That is: "How does Achilles catch the Tortoise?", "How can I move from here to there?" Achilles does not catch the Tortoise at any finite stage of the process that Zeno describes; likewise with respect to the Dichotomy. Let us make the connection very explicit. Let me walk a unit distance. First, I walk half the distance, then half the remaining distance, etc. At each stage consider the set of points containing my (center of mass) location, the endpoint of the interval, and all the points in between these sets form the chain: [0, 1], [1/2, 1], [3/4, 1] .... I do not arrive at the endpoint at any finite stage of the process, but only at the ωth stage where the set in question contains only one member – the endpoint.

Could the possibility of this line of argument have escaped Aristotle? He comes perilously close to conceding its validity later in the same discussion in the *Physics* ($206_b 3$–10):

In a way the infinite by addition is the same as the infinite by way of division. In a finite magnitude, the infinite by addition comes about in a way inverse to that of the other. For in proportion as we see division going on, in the same proportion we see addition being made to what is already marked off. For if we take a determinate part of a finite magnitude and add another part *determined* by the same ratio (not taking the same amount of the original whole) and so on, we shall not traverse the given magnitude.

But we *do* traverse a given magnitude. Motion is real enough for Aristotle. Zeno appears to have demonstrated from Aristotelian premises, the existence of actual completed infinity. (Indeed the method of argument used in the dichotomy can be adapted to show how wiggling one's finger instantiates all sorts of countable ordinals.) Aristotle appears to grant the point in Bk. VI, Ch. 2 of the *Physics*: "Hence it is that Zeno's argument makes a false assumption when it asserts that you cannot traverse an infinite number of things one by one, in a finite time ... and the contact with the infinities is made *by means of moments not finite but infinite in number*" (*ibid.*, $233^a 13$–31, emphasis mine), only to retract it in a curious discussion in Bk. VIII, Ch. 8 $263^a 4$–$263_b 9$. After a discussion of what happens to the midpoint when a line segment is bisected, he returns to the question of potential vs. actual infinity:

Therefore to question whether it is possible to pass through an infinite number of units either of time or of distance we must reply that in a sense it is and in a sense it is not. If the units are actual, it is not possible: If they are potential it is possible. For in the course of a continuous motion the traveler has traversed an infinite number of units in an accidental sense but not in an unqualified sense: For though it is an accidental characteristic of the distance to be an infinite number of half-distances, this is not its real and essential character (*ibid*., 263_b4–9).

Aristotle has a second related reply to the paradox: he denies *Measurability*. Points are not to be thought of as *parts* of lines at all, and thus are not sorts of things that have magnitude. The things that qualify as genuine parts of a line segment are non-degenerate subsegments. This is argued in Book VI of the *Physics* on the basis of Aristotle's theory of continuity as a kind of contiguity: "nothing that is continuous can be composed of indivisibles: e.g. a line cannot be composed of points, the line being continuous and the point indivisible" (*ibid*., 231^a24–25). (Aristotle holds that "things are continuous if their extremities are one." Presumably an example would be the closed intervals [0, 1/2] and [1/2, 1] which are continuous and make up the interval [0, 1].[7] Aristotle does not seem to have the concept of an open interval.)

Aristotle has another reason for holding that points are not measurable parts of the line, and that is that aside from the objections stated here, he appears to hold that Zeno's argument is valid. In 'On Generation and Corruption,' (317^a1–9) where he explains what *the* fallacy of the argument is, he points only to the argument that the line can be thought of as composed of points.

Whatever Aristotle's reasons for holding that points are not legitimate parts of lines or bodies, the idea has been attractive for a number of thinkers, e.g. Whitehead. Points can, in a sense, be eliminated in favor of intervals, of course. We can take intervals as basic, and taking our cue from Zeno's construction, let chains of intervals stand for the points respectively that are the unit sets of their intersections. Points then re-emerge as logical reconstructions. (For a general account of this sort of approach see Tarski, 1956.) Such points would naturally be assigned as the limit of the lengths of the members of the chain, i.e. length zero. So while this approach retains a faint Aristotelian flavor, it partakes of strongly non-Aristotelian elements as well.

Aristotle's answer to Zeno's paradox of measure shows the temperament of the empirical scientist rather than that of the mathematician. His response involves a theory of some subtlety, but that theory is mathematically conservative if not reactionary. With regard to both infinity and measurability his instinct was to restrict the subject to a narrow and safe domain rather than to explore uncharted waters.

Indivisible Magnitudes

There are at least two schools that postulated some kind of indivisible magnitudes. Aristotle, in 'On Generation and Corruption,' cites what is essentially Zeno's argument as a motivation for the adoption of indivisible magnitudes by the Atomists. A doctrine of indivisible magnitudes was also held in the Platonic academy. Aristotle ascribes it to Plato himself:

> Further, from what principle will the presence of *points* in the line be derived? Plato even used to object to this class of things as being a geometrical fiction. He gave the name of the principle of the line – and this he often posited – to the indivisible lines (*Metaphysics*, 992^{a}19–21).

It is commonly ascribed to his student Xenocrates, and the pseudo-Aristotelian polemic 'Concerning Indivisible Lines' is thought to be directed at this doctrine. Details of either theory are hard to come by. The evidence available is sparse and mostly circumstantial.

One wonders, in each case, whether the indivisible magnitudes are meant to be infinitesimal or finite. Each alternative would block Zeno's argument, but they would block it in different ways. Finite indivisible magnitudes would block the construction of the partition of a thing of finite magnitude into an infinite number of equal parts. Infinitesimal indivisible magnitudes are consistent with the existence of an infinite partition, but allow for the possibility that an infinite number of positive magnitudes add up to a finite magnitude. Some commentators attribute a doctrine of infinitesimals to both schools (not to mention the Pythagoreans and Anaxagoras) but the evidence seems far from conclusive.

The Archimedean manuscript 'On Method,' discovered by Heiberg in 1906, shows that Archimedes used infinitesimals in a method of discovery, although he did not consider that infinitesimal methods provided a strict proof. Archimedes attributed the discovery of the volume of the pyramid and of the cone to Democritus, but says that he did not prove it rigorously. Boyer (1968, pp. 88–89) remarks: "This creates a puzzle, for if Democritus added anything to the Egyptian knowledge here, it must have been some sort of demonstration, albeit inadequate." Boyer suggests that this may point to a theory of infinitesimals. In the light of this speculation the following fragment of Democritus becomes even more tantalizing:

If a cone is cut along a plane parallel to its base, what must we think of the surfaces of the two segments – that they are all equal or unequal? If they are unequal, they will make the cone uneven, with many step-like indentations and roughnesses: If they are equal, then the segments will be equal and cone will turn out to have the same properties as the cylinder, being composed of equal and not of unequal circles – which is quite absurd [from Plutarch, *De communibus notitiis*, tr. Furley in Furley (1967), p. 100].

But whatever the truth of the matter is regarding Democritus, it appears clear that his follower Epicurus did not believe in infinitesimal magnitudes. His letter to Herodotus contains a passage which is inconsistent with that doctrine:

For when someone says that there are infinite parts in something, however small they may be, it is impossible to see how this can still be finite in size; for obviously the infinite parts must be of *some* size, and whatever size they may happen to be, the size (of the total) would be infinite [tr. Furley in Furley (1967), p. 14].

It is, of course possible that Epicurus modified the doctrine of Democritus, but in that case we might expect to find a more explicit discussion of the whole issue.

The main evidence regarding Xenocrates' theory is the treatise which was presumably written to refute it: 'Concerning Indivisible Lines.' Unfortunately the evidence is not univocal. Owen reads the treatise as directed against finite indivisible magnitudes:

It is not certain whether the proponents of this theory [of indivisible lines] thought that every measurable distance contained a finite or an infinite number of such distances. An argument for thinking the former is that this is assumed in the fourth-century polemic *On Indivisible Lines*. An argument for thinking the contrary is that the theory was held at a time when the difficulties of incommensurable lines were fully realized. It was a commonplace that the side and diagonal of a square cannot both be *finite* multiples of any unit of length whatsoever [Owen (1957–8), p. 150].

On the other hand, Boyer does not hesitate to interpret the treatise as directed against a theory of infinitesimals:

The thesis of the treatise ['On Indivisible Lines'] is that the doctrine of indivisibles espoused by Xenocrates . . . is untenable. The indivisible, or fixed infinitesimal of length or area or volume, has fascinated men of many ages; Xenocrates thought that this notion would resolve the paradoxes, such as those of Zeno, that plagued mathematical and philosophical thought [Boyer (1968), p. 108].

Indeed, the treatise is such a scattershot affair that it is hard to detect the target. Perhaps there was more than one target. Otherwise the author was hopelessly confused.

In the beginning of the treatise, the author lists the arguments which the proponents of indivisible lines use to support their doctrine. The fourth argument is only consistent with the indivisible magnitudes being finite rather than infinitesimal:

> Again, Zeno's argument proves that there must be simple magnitudes. For the body, which is moving along a line must reach the half-way point before it reaches the end. And since there is always a half-way point in any 'stretch' which is not simple, motion – unless there be simple magnitudes – involves that the moving body touches successively one-by-one an infinite number of points in a finite time which is impossible ('Concerning Indivisible Lines,' 968^{a}18–21).

But later in the treatise, we have already noted the passage which appears to be inconsistent with the indivisible lines being of finite magnitude:

> Further, the addition of the line will not (on the theory) make the whole line any longer than the original line to which the addition was made: for simples will not, when added together, produce an increased total magnitude (*Ibid.*, 970^{a}21–23. See also 970$_{b}$23–25, 972^{a}12–14; Aristotle, *Physics*, 220^{a}15–20; 263^{a}4–9; *On Generation and Corruption*, 316^{a}24–34).

Perhaps it is best just to say that both sorts of theory were in the air, without trying to be too positive about who held which theory.

Neither of these theories leaves Zeno bereft of argument. The theory of infinitesimals was not put on a firm foundation until the work of Abraham Robinson in this century. It would not be much of a problem to embarrass whatever preliminary ideas there were about them at the time. Indeed, if the theory in question *did* claim that the addition of a single infinitesimal magnitude would not make the line segment any longer, then Zeno had already made the reply: "If when it is taken away, the other thing is to be no smaller, and is to be no bigger when it is added, it is clear that what was added or taken away was nothing." Owen suggests that Zeno's paradox of the *Stadium* is also directed against infinitesimal magnitudes. His interpretation is only possible, however, if the theory of infinitesimals in question has the simple parts of the line discretely ordered. Any respectable theory of infinitesimals would have the simple parts of infinitesimal magnitude densely ordered. But since we are not in possession of the theory of infinitesimals in question (if there *was* one in question) we do not know how respectable it was.

However that may be, both the *Arrow* and the *Stadium* raise difficulties for an account of motion on a theory of *finite* indivisible magnitudes. The flying arrow moves but does not move at any instant. Motion becomes "first here, then there." Average velocity may make sense but instantaneous velocity does not. The Stadium shows us how considerations of relative motion almost *force* us into infinite divisibility. Even before considering relative motion there is this problem: If something is travelling a space unit for every two time units, where is it after one time unit? Conversely, if it is travelling two space units per time unit, how much time has elapsed after it travels one space unit? (Aristotle advances such arguments in the *Physics*, Bk. VI, Ch. 2, 232_b20 ff.) A theory of finite indivisible magnitudes might reject such questions, or failing that, might try to get by with a theory that only allowed rest and motion of one *speed*, i.e. that of one time unit per space unit. Zeno's paradox of the *Stadium* shows that this strategy does not escape the problem. By considering one series of bodies at rest, and two having unit speed but opposite direction, the embarrassing questions can be asked again in terms of relative motion.

Aristotle did not hesitate to use the Arrow against indivisible magnitudes. He argues that on the rival theory " . . . the motion will consist not of motions but of starts, and will take place by a thing's having completed a motion without being in motion . . . So it would be possible for a thing to have completed a walk without ever walking . . . " (*Physics*, Bk. VI, Ch. 1, 232^a9–11). Epicurus took the point. Furley reports the following comment of Themistius:

But our clever friend Epicurus is not ashamed to use a remedy more severe than the disease – and this inspite of Aristotle's demonstration of the visciousness of the argument. The moving object, he says, *moves* over the whole distance, but of each of the indivisible units of which the whole is composed it does not move but has *moved* [tr. Furley in Furley (1967), p. 113].

With regard to degree of motion, Epicurus again accedes to Aristotle. He actually maintains that all atoms *do* move through the void with the same speed [though not the same direction – see Furley (1967), p. 121 ff.]. That the *Stadium* shows that little or nothing is gained by this desperate move, appears to have escaped Epicurus as it did Aristotle.

Conclusion of Section I

We have seen that Zeno's paradox of measure rests on the following premises:

(I) *Partition*: the line segment can be partitioned into an infinite number of parts such that:
(II) *Measurability*: the concept of magnitude applies to the parts.
(III) *Invariance*: the parts all have equal positive magnitude, or zero magnitude.
(IV) *Archimedean Axiom*: there are no infinitesimal magnitudes.
(V) *Ultra-Additivity*: the magnitude of the whole is the sum of the magnitudes of the parts in the sense given.

Ancient attempts to answer Zeno focused largely on (I) and (II). Doctrines of finite indivisible magnitudes (certainly Epicurus and probably Democritus and Leucippus) rejected (I). Aristotle rejected (I) and (II). It is possible that a doctrine of infinitesimal indivisible magnitudes was also current (possibly held by Xenocrates, possibly by Democritus) which rejected (IV). (III) could have also been challenged by a holder of a doctrine of infinitesimal magnitudes. (V), Ultra-Additivity, appears to have been accepted without question by every party to the dispute. It is ironic that it is just here that the standard modern theory of measure finds the fallacy.

II. POST CANTOR

Measure According to Peano and Jordan, Borel, and Lebesgue

It is no accident that Zeno was first taken seriously in the modern era by mathematicians (see Tannery, 1885) at a time when problems in the theory of integration were leading to the development of measure theory. The concept of measurability was introduced (putting aside Aristotle's response to Zeno) by Peano in 1883 and generalized by Jordan in 1892 (for details see Hawkins, 1970). In discussing areas in the plane, Peano considers (I) the class of polygons which contain the region in question and (II) the class of polygons which are contained in the given region. The area of the region should be less than or equal to the areas of the polygons in the first class and greater than or equal to the areas of the polygons in the second class. If these conditions determine a unique number [i.e. if the greatest lower bound of the areas of polygons in class (I) equals the least upper bound of polygons in class (II)], then that is the area of the region. If not "then the concept of area would not apply in this case" (from Peano, 1883; quoted in Hawkins, 1970, p. 87).

Thus, on the line, an interval [a, b] is assigned its length, $b - a$, as measure.[8] This includes points which as degenerate intervals [a, a] are assigned measure zero. These measures are fundamental, and the concept of measure is extended to other point sets as follows. Consider finite sets of intervals which *cover* the set of points in question in that it is contained in their union. Associate with each such covering set the sum of the lengths of intervals in it. The greatest lower bound of these numbers is called the *outer content* of the set. Working from the other side, consider finite sets of non-overlapping (pairwise disjoint) intervals whose union is contained in the set in question. Associate with each such set the sum of the lengths of its members. The least upper bound of these numbers is called the *inner content* of the set. If the outer and inner content of a point set are equal, then the set is *measurable in the sense of Peano and Jordan* and that number is its measure. If not, then the set is not measurable – the concept of measure simply does not apply.

Jordan showed that measure, so defined, is *finitely additive*. That is, if each of a finite collection of mutually disjoint sets is measurable, then their union is also and its measure is the sum of theirs. The appropriate principle of additivity is not assumed in the definition, but rather proved from the definition; and it is a rather modest kind of additivity. The stronger principle of *countable* additivity fails for Peano–Jordan measure. The union of a denumerable collection of measurable sets may not itself be measurable. For instance, the set of rational points in [0, 1] is not Peano–Jordan measurable. Its outer content is 1, while its inner content is 0. Yet, as Cantor had shown, it is the union of a denumerable collection of unit sets.

The basic ideas of Peano–Jordan measure could have been introduced in Aristotle's time. They depend only on finite sums of intervals. The restriction to finite additivity allows a rich theory of measurable sets. Not every set of points becomes measurable, but the assignment of measure zero to unit sets of points causes no difficulties. Finite sets of points must have measure zero, but the Greek geometers knew well enough that the line was not exhausted by any finite set of points.

Borel took a rather different approach to measure and measurability in 1898. Borel *constructs* the Borel measurable sets out of the intervals by finite and denumerable set theoretic operations, and defines their measure by postulating a stronger form of additivity: i.e. countable additivity. A collection of sets is called a *sigma-algebra* if it is closed under countable union and intersection, and complementation. The Borel-measurable sets on the line segment can be defined as the smallest sigma-algebra of point sets containing the open intervals. Intervals have their length as their measure. Measure is taken to be countably additive (or sigma additive). That is, a countable union of mutually disjoint intervals has as its measure the infinite sum of the lengths of the intervals. Sigma additivity can be thought of as the restriction of the fancied principle of ultra-additivity of Part I to denumerable collections. Any denumerable set of points, e.g. the rationals in [0, 1], has Borel measure zero since it is a countable union of singletons each of which has measure zero. Since [0, 1] has measure 1, the set of irrational points in [0, 1] has measure 1. As Grünbaum has emphasized, this causes no problems because this set had been shown by Cantor to be uncountable. For this reason, Borel's theory of measure was only conceivable after Cantor's fundamental investigations of infinite cardinality.

This is not to suggest that only countable point sets have measure zero. Consider the famous Cantor ternary set. It can be constructed by starting with [0, 1] and then removing the middle third open interval (1/3, 2/3). Thus we have at stage 1 of the construction the points in [0, 1/3] and [2/3, 1]. To move from stage n of the construction to stage $n + 1$ we delete the middle open thirds of the closed intervals of stage n. The intersection of the sets at finite stages of the construction is Cantor's ternary set. It has Borel measure zero since we started with a set of measure 1 from which we have subtracted a set of points which by countable additivity has measure one. Alternatively, it has measure zero in the sense of Peano and Jordan, since each stage n in the process of construction provides a finite covering with measure $(2/3)n$, the outer content of the Cantor set is zero. Nevertheless the Cantor set is non-denumerable. The interval [0, 1] can be mapped 1-to-1 into the Cantor set. Remember Zeno's construction by infinite bisection of the line in Section I of this paper. For each point on the line, at each bisection it was either on the left or the right. The intersection of each chain, in Zeno's construction, contained at most one point. So each point on the line corresponds to a unique infinite sequence of 'left' and 'right'. Applying such a sequence to the stages in the construction of the Cantor set, we select the indicated left or right third which remains, so we have corresponding to each point in the original interval a unique chain of closed intervals. The intersection of such a chain must be non-empty by the Heine-Borel theorem. The chains are constructed in such a way that no point can be in the intersection of two such chains. So each point on the original line corresponds to a unique point in the Cantor set.

Borel's bold move to countable additivity was not received without some qualms in the contemporary mathematical community. In a report on the theory of sets published in 1900, Schoenflies was critical of this as well as other aspects of Borel's theory of measure. With regard to countable additivity, he writes that "the question of whether a property is extendable from finite to infinite sums cannot be settled by positing it but requires further investigation" (quoted in Hawkins, 1970, p. 107).

The further investigations were successfully carried out by Lebesgue in 1902. Lebesgue generalized the notions of inner and outer content of Peano and Jordan in such a way that the countable additivity of measure could be demonstrated; Lebesgue's definition of *outer measure* considers *denumerable* coverings. For each countable covering consider the limiting sum of the lengths of its constituent (open) intervals. The greatest lower bound of these numbers is the *outer measure* of the set in question. (Notice that the Lebesgue outer measure of the set of rationals in [0, 1] is 0.) For the inner measure of a bounded set, S, consider the closed intervals [a, b] which contain it. For each take its length, b – a, minus the outer measure of the set of points in it which are not in S. We can define the *inner measure* of S as the least upper bound of these numbers. (Of course, these numbers are all really the same. Any closed interval containing S will give the same result.)

Lebesgue was able to prove on the basis of these definitions, that the Lebesgue-measurable sets include both the Borel measurable sets and the sets measurable in the sense of Peano and Jordan; that Lebesgue measure agrees with each of these measures on the sets for which those measures were defined, and Lebesgue measure is countably additive. Furthermore, he showed that Lebesgue measure has the intuitively correct property of *translation invariance*. For a set S, and a real number a, let the set $S + a$ contain just the points $x + a$ for every x in S. The Lebesgue measure of any measurable set S, equals the measure of $S + a$ for any real number a.

Lebesgue's theory showed how the virtues of earlier theories could be combined and extended to provide an intuitive treatment for a very rich domain of measurable sets. In fact, at the time, it was not immediately apparent whether there were any bounded sets which were not measurable in the sense of Lebesgue.

The Vitali Paradox

In 1905 Vitali produced the first example of a non-Lebesgue measurable set. The argument is in many ways strikingly similar to that used in Zeno's paradox of measure. Since Lebesgue measure is only countably additive, rather than ultra-additive, one following the path of Zeno would have to seek a *countable* partition of the unit line segment into parts which by some symmetry consideration should have the same magnitude. With such a partition in hand, he could argue that if the members of the partition have zero measure, then the unit interval must have zero measure; if they have equal positive measure, the unit interval must have infinite measure. Both alternatives contradict the fact that the unit interval has Lebesgue measure one, so the members of the partition are not Lebesgue measurable.

Vitali found such a partition. To simplify matters slightly, we will construct the partition of the half-open interval [0,1). We can visualize this as wrapped around to form a unit circle. The relevant symmetry property of Lebesgue measure was mentioned in the preceding section. It is *translation-invariance*. Translation invariance implies translation invariance modulo 1, which in terms of our visualization means that if any Lebesgue measurable set of points is displaced a fixed distance around the circle, the resulting set will have the same Lebesgue measure. Consider the equivalence relation: $x - y$ is rational. This partitions [0, 1) into equivalence classes. Choose one member from each of these classes to form the choice set C. For each rational, r, in [0, 1) let C_r be the set gotten by adding (modulo 1) r to each member of C (i.e. by displacing C the distance r around the circle). The C_rs form a denumerable partition of [0, 1). Any one can be gotten by translation from any other. Since Lebesgue measure is translation-invariant, if they are Lebesgue measurable, they have the same measure. If so, the measure of [0, 1) must be either 0 or infinity. So the C_rs are not Lebesgue measurable. Such non-measurable sets are ubiquitous. It can be shown that every Lebesgue measurable set with non-zero measure contains a non-measurable set. Zeno would have been delighted.

Vitali's construction requires stronger mathematical methods than Zeno's. The crucial step involves the axiom of choice. This proves to be essential in the construction of a non-measurable set (Solovay, 1970).

The only facts about Lebesgue measure used in Vitali's argument other than translation invariance are that it is *countably additive* and *real valued* (the latter being used for the Archimedean property of the real numbers). Thus, the argument establishes a more general result: any translation-invariant, countably additive, real-valued measure defined on all the subsets of [0, 1)[9] must give [0, 1) either infinite measure or measure zero.

Must we, with Aristotle, concede that intervals (areas, volumes) of positive magnitude are made up of parts to which the concept of magnitude does not apply? Or can we plausibly weaken the foregoing set of three conditions which generate the Vitali paradox?

Finite Additivity and Non-Archimedean Measure

Lebesgue measure escaped Zeno's paradox by virtue of a weaker form of additivity. This suggests that a weakening of countable additivity to finite additivity might allow us to define a finitely additive measure on richer domain of sets. Of course, such a possibility would only be of interest if some of the virtues of Lebesgue measure could be retained; e.g. we would like each interval to have its length as its measure.

In fact, we can have this and more. There is a finitely additive, real-valued translation invariant measure defined on *all* subsets of [0, 1],[10] which agrees with Lebesgue measure on all the Lebesgue measurable sets.[11]

Returning to Vitali's example, it is clear that such a measure must give the sets C_r measure zero, for if it gave them positive measure, *finite* additivity and translation invariance would contradict the measure of [0, 1) being one. The C_rs can thus be accommodated by a finitely additive measure in the way in which the singletons were accommodated by Lebesgue measure; they have measure zero but the additivity properties of the measure are not strong enough for that to cause problems.

Some philosophers may, despite all of this, feel nagging Zenonian intuitions to the effect that a whole of positive magnitude simply should not be made up parts of measure zero. This is the intuition that measure should be *regular*; that only the null set should receive measure zero. We have seen that not even a *finitely* additive translation invariant measure can accommodate this intuition if it is real valued. But what if the values that the measure takes on lie in a domain which is non-Archimedean? Couldn't we get away with giving both the singletons and the C_rs infinitesimal measure in some way in which everything works out nicely?

Such speculations may be very old, but it has only been possible to give them substance since Abraham Robinson's creation of non-standard analysis (Robinson, 1966). Leibniz thought of infinitesimals as ideal elements which nevertheless obey the same laws as the numbers. But which laws? The answer cannot be "All" in too strong a sense; otherwise we would not be able to distinguish a theory which admits infinitesimals from one which doesn't. This question had to wait for the development of model theory for its proper answer. Robinson showed how a non-standard model of analysis could incorporate infinitesimals, which consequently must obey the *first-order laws* which govern the real numbers.

The crucial logical property of first-order languages that Robinson's construction uses is compactness: if a set of sentences is such that every finite subset of it has a model, then the whole set in question has a model. Compactness of first-order languages depends on their limited logical resources: the logical constants being limited to truth functions, identity, and first order quantifiers and their sentences being of only finite length. It does not depend on the languages being denumerable. Thus we could (and will) imagine first order languages with names for every real number, which are nevertheless compact. Compactness fails for second order logic given the 'natural' interpretation of second order quantifiers having as their domain the power set of the domain of the first order quantifiers. However, if we allow Henkin's *general* models in which higher order quantifiers are allowed to have as their domain subsets of their natural domain, higher order quantification theory is also compact.

Here then, is how we get a non-standard model of analysis which contains infinitesimal elements: Consider a rich first-order language which for every real number, r, contains a name o_r; a relational symbol for every relation on the reals; and an operation symbol for every operation on the reals. Let the theory ANALYSIS consist of all the true sentences of this language, and consider the theory which is the union of ANALYSIS with the set of all sentences of the form $o_r < y$ for each real r. Each finite subset of this theory has a model in the reals, so by compactness this theory does too. This is a non-standard model of the reals. The function which maps each real, r, onto o_r^*, the denotation in the non-standard model of its name, is an isomorphism. Each non-standard model contains an isomorphic copy of the reals. Working within the non-standard model, we will simply call these the *standard reals*. The denotation of the less-than relation totally orders the non-standard reals since the axioms of total order are first-order. According to that order, the element which the model assigns as the denotation of y is a infinite element; it is greater than any of the standard reals. There is a first order sentence which says that every number has a reciprocal and one which says that if x is greater that y, then the reciprocal of x is less than the reciprocal of y. Since the model makes these sentences true, there must be an element of the model that is the reciprocal of the infinite element and less than any positive standard real. This is an infinitesimal element. A great deal of knowledge about the structure of the infinitesimals follows from the fact that they obey all first order generalizations about the reals.

The question as to how such infinitesimals can be incorporated into non-standard measure theory is a bit more complicated, involving non-standard (general) models for a higher order language of analysis. (For details see Bernstein and Wattenberg, 1969.) They show that one can construct a measure defined for all subsets of the unit interval, which takes its values in a non-standard of the reals, which is *finitely additive, translation invariant up to an infinitesimal* which is *infinitesimally close to Lebesgue measure* on the Lebesgue measurable sets, and which is *regular* (i.e. only the null set gets measure zero). The Vitali sets of the last section, and the sets containing exactly one point will then both have infinitesimal measure.

It can be shown that in non-standard models of analysis every non-standard real is infinitesimally close to a unique standard real. Call the second the *standard part* of the first. Then if we have a non-standard measure of the kind described here, and derive a real-valued measure by considering only the standard parts of the values assigned by the non-standard measure, we get the sort of measure discussed at the beginning of this section: a real-valued, finitely additive, translation invariant measure defined on all subsets of [0, 1] which agrees with Lebesgue measure on the Lebesgue-measurable sets. What we gain by allowing our measure to take values in richer range – the non-standard reals – is *regularity*.

The Hausdorff Paradox

We seem to have finally seen how to get rid of non-measurability. Banach showed that at the cost of weakening additivity to finite additivity on the non-Lebesgue-measurable sets, we can make every bounded set of points on the real line measurable. This does not quite lay Zeno to rest, for he was ultimately concerned with magnitudes of volumes in three-dimensional space. We have been confining ourselves to one dimension for the sake of simplicity. To complete the story, we should show that Banach's result can be extended to three-dimensional space. It is not so. A construction due to Hausdorff (1914) and further generalized by Banach and Tarski (1924) shows that one cannot in three and higher dimensional Euclidean spaces have a finitely-additive measure, which assigns the unit cube measure 1, assigns congruent point sets equal measure, and assigns a measure to all subsets of the unit cube.

Here the appropriate invariance property is congruence-invariance. Points here are to be thought of as triples of real numbers. The *Euclidean distance* between two points, (x, y, z) and (x', y', z'), is given by the Pythagorean formula:

$$[(x - x')^2 + (y - y')^2 + (z - z')^2]^{\frac{1}{2}}.$$

Two sets of points in Euclidean three-dimensional space are congruent just in case there is a 1-to-1 function mapping the one onto the other which preserves Euclidean distance (i.e. the distance between any two points in the first set is equal to the distance between their images in the second set). Congruence invariance in one dimension is just translation invariance. The theory of Lebesgue measure for n-dimensional Euclidean space, developed analogously to the theory for one dimension, has the consequence that Lebesgue measure is congruence-invariant on the Lebesgue measurable sets. Banach actually showed that a finitely additive congruence-invariant extension of Lebesgue measure to all bounded sets is possible in both one- and two-dimensional Euclidean space. It is only in Euclidean spaces of three and higher dimensions where the theorem fails.

In an extended note to *Grundzüge der Mengenlehre* (1914) headed 'Unsolvability of the Measure Problem,' Hausdorff sets out to show that it is impossible to assign to all point sets on the surface of a sphere a finitely additive, congruence-invariant measure which assigns the whole surface a positive measure. To this end, he proves the following theorem:

> The spherical surface, K, can be decomposed into disjoint sets: A, B, C, Q, where Q is countable; A, B, C are congruent to each other; the union of B and C is congruent with each of the sets A, B, C.

Since Hausdorff is here considering real-valued measure, congruence invariance together with a finite measure for the surface entails that each countable point set has measure zero. (For one can by appropriate choice of rotations generate an infinite number of disjoint congruent point sets to any given denumerable point set. If the given set has positive measure, the surface of the sphere by finite additivity could not have finite measure.) Thus, under the stated assumptions, the measure of the surface, $m(K)$, would equal $m(A) + m(B) + m(C)$. Since A, B, and C are congruent with each other, $m(A) = 1/3\ m(K)$. Since A is congruent with $B \cup C$, $m(A) = 1/2\ m(K)$.

Hausdorff's theorem again depends on the axiom of choice (as does Banach's positive result for 1 and 2 dimensions). Hausdorff works with a group of rotations about two appropriately chosen[12] axes; the group generated by 1/2 rotation about the first axis, ϕ, and 1/3 rotation about the second, ψ. Hausdorff shows how this group of rotations can be decomposed into three disjoint sets: $G = A \cup B \cup C$, such that $A \cdot \phi = B \cup C; A \cdot \psi = B$; $A \cdot \psi^2 = C$.[13] Let Q be the countable set of fixed points of members of G. The set of points on the surface less this denumerable set, $S - Q$, is the disjoint union of the orbits of the group G. The axiom of choice comes into the picture to assure the existence of a choice set, M, containing exactly one member from each orbit. $S - Q$ consists of the union of the point sets that M is carried into by members of G. Let the *point* set A be the set of points that M is carried into by the rotations in the *set of rotations* A, likewise for B and C. Then A, B, C and Q are the requisite point sets for Hausdorff's theorem.

Hausdorff concludes: "A determination of measure for all bounded sets, which satisfies conditions . . . [congruence-invariance, unit cube has measure 1, finite additivity] is therefore impossible in three and higher dimensional Euclidean space, since otherwise it would also be possible on the sphere (where one would assign to a set on the sphere the volume of the corresponding conical body as its measure)."

The paradoxical results of Hausdorff and Vitali are analyzed and generalized in a celebrated paper of Banach and Tarski (1924). There they introduce the notion of *equivalence of sets of points by finite* (*and alternatively by denumerable*) *decomposition*. Two sets of points (in a metric space) are equivalent by finite decomposition iff there exist finite partitions $[p_1, \ldots p_n]$, $[q_1, \ldots q_n]$ of them respectively, whose respective members are congruent (p_1 congruent with q_1 & . . . & p_n congruent with q_n); analogously for equivalence by denumerable decomposition. Then generalizing Hausdorff's argument: "In a Euclidean space of $n \geqslant 3$ dimensions, two arbitrary sets, bounded and containing interior points (e.g. two spheres of different radius) are equivalent through finite decomposition" (Banach and Tarski, 1924, p. 244). In the form in which they develop it, Hausdorff's paradox is perhaps better known as the Banach–Tarski paradox. The analogous theorem holds for the surface of the sphere but fails for Euclidean spaces of 1 and 2 dimensions. For these spaces, however, we have a generalization of the Vitali paradox. For Euclidean spaces of dimension 1 and higher "two arbitrary sets (bounded or not) containing interior points are equivalent by *denumerable* decomposition" (Banach and Tarski, 1924, p. 244).

These rather surprising facts about *congruence* are at the heart of the restrictions on measurability that we have been discussing for the last three sections. They might be taken as calling into question the status of congruence-invariance as a desideratum for measure. Our intuitions in this regard are based on consideration of far simpler point sets than the ones involved in the Vitali and Hausdorff paradoxes. Before even raising questions of measure, we see that our intuitions regarding congruence of simple bodies in three-dimensional Euclidean space cannot be projected to arbitrary point sets.

One can extend Lebesgue measure to a finitely additive, *non-congruence invariant* measure on all the bounded subsets of Euclidean three-dimensional space. So measurability of all bounded sets can be achieved, but at an unexpected cost. One might wonder, however, whether if one pays the price of giving up congruence-invariance, one can avoid weakening countable additivity to finite additivity. The Vitali paradox and its generalizations, after all, used congruence invariance essentially. Things, however, are not quite so simple. Non-measurability has roots that go deeper than the metric structure of the underlying space.

Non-measurable Sets Without Congruence Invariance

Let us recall, for a moment, Zeno's two principles from Section I. Suppose that a whole can be partitioned into an infinite number of parts. Then Zeno thought:

(I) If the parts had positive (real) magnitude, then the whole would have infinite magnitude.

(II) If the parts had zero magnitude, then the whole would have zero magnitude.

Let us by 'magnitude' understand a *countably additive* measure. Then (II) is correct for an infinite partition into a denumerable number of parts, but fails for partitions into a non-denumerably infinite number of parts. On the other hand, (I) can fail for a denumerable partition unless some extra assumptions about the magnitudes of the parts are present (e.g. that they must all be the same by some invariance argument). A fact that we have not taken explicit notice of yet, is that (I) holds without restriction if the infinite partition is non-denumerable. *If a set has finite measure, it can contain at most a denumerable infinity of disjoint sets of positive measure*. Consider any partition of the set in question. Consider the set of members of this partition with measure greater than or equal to $1/2$. It must be finite. Otherwise by *finite* additivity of measure, the measure of S could not be finite. Likewise for the set of members of the partition with measure greater than or equal to $1/2^n$ and less than or equal to $1/2^{n-1}$, for each natural number n. Each member of the partition with non-zero measure is in one of these finite collections of sets. The number of such collections is denumerable, so the number of members of the original partition with positive measure is denumerable. Non-denumerable partitions make (I) true and (II) false;

denumerable partitions make (II) true [assuming countable additivity of the measure] and (I) false.

These facts make it possible to show, under the assumption of Cantor's continuum hypothesis, that there is no non-trivial countably additive measure on [0, 1] which gives all the unit point sets measure zero. The result is due to Banach and Kuratowski (1929) and was strengthened and generalized by Ulam (1930). No assumption of translation invariance is used; there is no appeal to metric considerations. Only the cardinality of the set in question plays a role. A set is of power aleph$_1$ iff it can be put into one-to-one correspondence with the ordinal numbers less than the first uncountable ordinal. The stated theorem is proved for arbitrary sets of power aleph$_1$. Cantor's continuum hypothesis enters to assure that [0, 1] is such a set.

I give the proof so that the reader can appreciate the Zenonian counterpoint: *Suppose that a countably additive (real-valued) measure is defined on a set, Z, of cardinality aleph*$_1$*, such that every one element subset receives measure zero. Then the measure of Z must be either infinite or zero.* Since Z is, by hypothesis, of power aleph$_1$, there is a well-ordering such that each element of Z is preceded by only countably many elements, i.e. for each y in Z, the set $\{x: x < y\}$ is countable. For each y, let $f_y(x)$ be a one-to-one mapping of this set into the positive integers. We can then consider $f(x, y)$ as a mapping from pairs (x, y) of elements of Z such that $x < y$, to integers. Now, for each x in Z and each positive integer, n, let the A_x^n be $\{y: x < y$ and $f(x, y) = n\}$. We can picture these sets as arranged in an infinite matrix with denumerably many rows and uncountably (aleph$_1$) many columns:

$$\begin{array}{l} A_1^1, A_2^1, \ldots A_n^1 \ldots A_\alpha^1 \ldots \\ A_1^2, A_2^2, \ldots A_n^2 \ldots A_\alpha^2 \ldots \\ \ldots\ldots\ldots\ldots\ldots\ldots\ldots \\ A_1^n, A_2^n, \ldots A_n^n, \ldots A_\alpha^n \ldots \\ \ldots\ldots\ldots\ldots\ldots\ldots\ldots \end{array}$$

The sets have been constructed so that: (a) The sets in any row are disjoint. (b) The union of the sets in any column is equal to the whole set Z minus a countable set. [(a) follows from the 1-to-1 nature of f considered as a function of x. For (b), any y greater than the x of the column belongs to the set in the column for which $n = f(x, y)$. The union of the sets in the column then differs from Z by the set of elements less than or equal to x, which, by hypothesis, is countable.]

If, in any row there is a non-denumerable number of sets of positive measure then by the correct form of Zeno's (I), Z cannot have finite measure.

If, on the other hand, in every row only a denumerable number of sets have non-zero measure, then only a denumerable number of sets in the whole matrix have non-zero measure since there are only a denumerable number of rows. Then there must be some column which contains all sets of measure zero, since there are a non-denumerable number of columns. The set of elements in Z not contained in the union of the sets in that column must also have measure zero by the correct form of Zeno's (II) since it is the union of a denumerable number of singletons, each of which, by hypothesis, has measure zero.[14]

At this stage of the game, the elimination of non-measurable sets may appear a rather quixotic goal. Lebesgue measure has extended measurability to a far richer domain then Zeno and Aristotle imagined possible. It meshes with an elegant and powerful theory of integration adequate to the needs of the physical sciences. Perhaps Lebesgue measure should be taken as the theory of measure for physical space, and the existence of non-measurable sets should be viewed as just a mildly surprising consequence of the theory rather than as a real difficulty. This is, I believe, the dominant view among mathematicians and mathematical physicists. The real bite of non-measurability comes not in physics or metaphysics, but in epistemology.

Measures of Degree of Belief

Let us turn our attention to probability measures which are meant to represent rational degrees of belief. In this area, questions of measurability take on a new pungency. It is one thing to say that some widely scattered set of points in Euclidean three-dimensional space does not have a natural volume associated with it; another to say that there must be propositions to which there cannot be a degree of belief.

Some of the assumptions used in demonstrating non-measurability also appear in a new light. Translation and congruence invariance appear now not as falsifiable claims about the structure of measure on physical space, but rather as the result of the exercise of someone's epistemological freedom. I wonder which point on a wheel of fortune will be the lowest point when it comes to rest. I come to degrees of belief which are invariant under translation about the circumference. Can it be denied that it is reasonable and proper for me to do so? Again, shouldn't we be able to have rational degrees of belief defined over the subsets of some set of power aleph_1, whether or not $c = \text{aleph}_1$? Furthermore, the Zenonian intuition that only the empty set (here the null proposition) should receive measure zero is supported by a kind of betting argument. Shimony (1955) showed that *regularity* of probability measure is entailed by *strict coherence*: One should reject systems of bets such that one could in no possible circumstance achieve a net gain although one could suffer a net loss.

It is for reasons such as the foregoing that interest in finitely additive, and non-Archimedean measure has largely been generated by the theory of personal probability. De Finetti has consistently rejected countable additivity as a postulate for the theory of personal probability. For a recent spirited defense of finite additivity, see de Finetti (1972 and 1974). Considerations of strict coherence, and of having conditional probabilities well defined in a natural way, can be used to motivate the move to non-Archimedean valued probabilities. For example, see Bernstein and Wattenberg (1969).

Let me back up, and put these questions in their proper setting. Suppose that we have a set, U, whose elements represent mutually exclusive and jointly exhaustive states of affairs. If we think of such states of affairs as individuated in a maximally specific way, we might call the constituents of U 'possible worlds' (but this raises questions which cannot be discussed here). The subsets of this set can be thought of as statements or 'propositions' if propositions are only individuated up to necessary equivalence relative to the original set of possibilities. A set of such 'propositions' closed under negation, conjunction and disjunction is a Boolean algebra of propositions (under countable conjunction and disjunction, and negation, a Boolean sigma algebra). A (finitely additive) measure defined on such a Boolean algebra of propositions which takes values in $[0, 1]$ and which gives a tautology measure 1 and a contradiction measure 0 is a *probability measure*. (I leave open the questions as to whether the algebra need also be a sigma algebra, whether the measure need also be countably additive, and whether $[0, 1]$ is to be taken as a set of standard reals or whether a non-standard model of the reals can also be utilized, these being material to the issues in question.) The question of measurability then is whether the set of all subsets of U can be taken as the appropriate sigma algebra of propositions, or whether we are forced to restrict our probability assignments to some smaller Boolean algebra.[15]

Suppose that degrees of belief are represented (obviously with some idealization) by a numerical-valued function from a Boolean algebra of propositions. There are well-known pragmatic virtues associated with that function being a probability measure. If it is not, and degrees of belief are used in the standard way in determining the fairness of bets, then the agent in question leaves himself open to a Dutch Book: a finite system of bets each of which he considers fair or favorable, such that the net result is a loss no matter what happens. A belief function which leaves one open to a Dutch Book is said to be incoherent. Coherent belief evaluation functions must be (finitely additive) probability measures. De Finetti argues that there is no comparable coherence argument for countable additivity; and that the imposition of countable additivity as a postulate has the undesirable consequences of (1) creating unmeasurable sets and (2) precluding probability assignments which are perfectly acceptable from a personalistic point of view, e.g. a uniform distribution on a denumerable set of possibilities. Consequently, he develops his theory of personal probability only under the assumption of finite additivity. Savage (1954) does likewise.

The question of the relation of additivity to coherence is not, however, quite so simple. Consider de Finetti's example of the uniform distribution on a denumerable set of possibilities (e.g. what ticket will win in a denumerable lottery). Finite additivity allows the uniform distribution which gives each ticket exactly zero chance of winning, while maintaining probability one that some ticket wins. I would love to have the chance of betting against someone having such a probability assignment. For each ticket, I will bet him \$100 against nothing that it wins; he will consider each of these bets fair. After the lottery, I collect my \$100. If he declines fair bets on the grounds that not betting is just as good, I can do as well offering favorable bets. I will bet \$101 against \$1/2 that the first ticket wins; \$101 against $\$1/2^n$ that the nth ticket wins. After the lottery, I am assured a net winning of at least \$100. The second example reveals clearly what the first may not; that in each case I am assuming sigma-additivity of the *payoff-values* in totaling up my net gain in the infinite system of bets. In fact, if we make these two assumptions: that a denumerable set of bets is permissible and that the payoff-values are sigma-additive, then one can show that the correlative notion of coherence implies countable additivity of the probability measure. The first notice of this fact of which I am aware is in Spielman (1977).

Let a *betting system* be a function from possible states of affairs to payoff values. A *bet* on a proposition p, is a betting system which has a gain, a, associated with every state of affairs in p, and a loss, b, associated with every state of affairs in the negation of p. The *aggregate* of two betting systems, $B_1 \# B_2$ is the betting system which has at each possible state of affairs w, the sum of the payoffs associated with B_1 and B_2:

$$B_1 \# B_2 (w) = B_1 (w) + B_2 (w).$$

Probability is to perform the practical function of placing a value, *expected value*, on bets and betting arrangements when the agent is uncertain as to the state of the world. It would be an attractive property for such evaluations to have that they are *extensional* in the sense that the valuation depend only on the betting system (the function from possible states of affairs to payoff values), and not on how it is described. Otherwise the agent would regard two different prices fair for the same arrangement, and could be systematically exploited by someone who repeatedly bought an arrangement from him cheap and resold it to him dear. Similar considerations support the contention that valuation under uncertainty, *expected value*, should be additive over aggregation:

$$EV(B_1 \# B_2) = EV(B_1) + EV(B_2).$$

For the agent would presumably sell a betting arrangement with expected value of X for X or more, and buy it for X or less. If payoff value is additive, expected value had better be! Now if $p;q$ are mutually exclusive propositions; B_1 and B_2 are bets on p and q respectively at the same stakes, then $B_1 \# B_2$ is a bet on their disjunction $p \vee q$. In particular, let B_1 be the bet that gains a dollar if p, loses nothing otherwise; B_2 be the bet that gains a dollar if q, loses nothing otherwise; B_3 be the bet that gains a dollar if $p \vee q$, loses nothing otherwise. Then $B_1 \# B_2 = B_3$, so by extensionality $EV(B_1 \# B_2) = EV(B_3)$ and by additivity of expected value over aggregation, $EV(B_1) + EV(B_2) = EV(B_3)$. Since these expected values equal by definition the respective probabilities of p, q, and $p \vee q$, we have finite additivity, $\mathrm{pr}(p) + \mathrm{pr}(q) = \mathrm{pr}(p \vee q)$. Now the point of going through all this, is to call attention to the fact that if payoff value is countably additive, then we can consider denumerable aggregates of bets, whose payoffs at each possible state of affairs, w, is the denumerable sum of the payoffs of its constituents:

$$\#_i B_i(w) = \Sigma_i B_i(w),$$

and run the analogous argument for countable additivity.

All the considerations that came into play regarding non-measurable sets in previous sections are now again on the table: finite and countable additivity, invariance, regularity, Archimedean and non-Archimedean values for the measure; domains of various cardinality on which the measure is to be defined. This is not the place to attempt to sort them out. Perhaps enough has been said to show that the truly deep issues first raised by Zeno still deserve to engage our interest.

University of California-Irvine

NOTES

[1] This view of the paradoxes is vigorously advocated by Owen (1957–8).
[2] e.g. Grünbaum (1952; 1963; 1968).
[3] See Luria (1933), Fränkel (1942), Owen (1957–8), Furley (1967 and 1969), Vlastos (1971).
[4] See also Aristotle's long discussion in the *Physics*, Bk. VIII, Ch. 8, 263^a4–264^a6, where Aristotle wrestles with the question of the midpoint and related questions having to do with open and closed intervals, and their relation to change.
[5] "The infinite by way of addition is in a manner the same as the infinite by way of division. Within a finite magnitude the infinite by way of addition is realized in an inverse way (to that by way of division); for, as we see the magnitude being divided *ad infinitum*, so, in the same way the sum of successive fractions when added to one another (continually) will be found to tend towards a determinate limit. For if, in a finite magnitude, you take a determinate fraction of it, and then add to that fraction in the same ratio, and so on [i.e., so that each part has to the preceding part the same ratio as the part first taken has to the whole], but *not* each time including (in the part taken) on and the same amount of the original whole, you will not traverse (i.e., exhaust) the finite magnitude. But if you increase the ratio so that it always includes one and the same magnitude, whatever it is, you will traverse it, because any finite magnitude can be exhausted by taking away from it continually any definite magnitude however small" (Aristotle, *Physics*, Bk. III, 6, 206_b, tr. Heath in Heath, 1949, p. 106).
[6] Let me put Zeno's statement in context of the fragment which contains it, using Furley's translation: "[Simplicius first summarizes this step in the following words – 'If a thing has no magnitude or bulk (πάχος) or mass, it would not exist.' Then he gives the reasoning in full]. For if it were added to something else that does exist, it would make it no greater; for if it were of no magnitude, and were added, it would not contribute anything to that magnitude. So it would follow that what was added was nothing. If when it is taken away, the other thing is to be no smaller, and is to be no bigger when it is added, it is clear that what was added or taken away was nothing" (Furley, 1967, p. 64). It may help to juxtapose this passage with one from *De generatione et corruptione* where Aristotle is explaining an argument which supposedly led Democritus to a doctrine of indivisible bodies: "Similarly, if it is made out of points, it will not be a quantity. For when they were in contact and there was one magnitude and they were together, they did not increase the magnitude of the whole; for when it was divided into two or more, the whole was no larger or smaller than formerly. So if they are all put together, they will not make a magnitude" (Furley, 1967, p. 84). Furley interprets this passage as arguing "that if a given line is divided in two, the sum of its two parts remains the same as the length of the original whole; yet there are now two points, at the inner end of each of the two half lines, where formerly there was only one; hence the extra point made no difference to the length – and so any number of points will make no difference to the length" (Furley, 1967, p. 85). That is, the argument is that the length of the (closed) line segment [0, 1/2] is exactly 1/2, as is the magnitude of [1/2, 1]. But the magnitude of [0, 1] is exactly 1, so the point 1/2 which is included in both [0, 1/2] and [1/2, 1] must be exactly zero. Whether or not this is its main purpose, such an argument could certainly be directed at someone who held that the midpoint (and points in general) have infinitesimal magnitude. One cannot help but wonder whether

it is being so used in the pseudo-Aristotelian polemic 'Concerning Indivisible Lines', 970^a21: "Further, the addition of the line will not (on the theory) make the whole line any longer than the original line to which the addition was made: For Simples will not, when added together, produce an increased total magnitude" (see also 970_b23–25 and 972^a12–14).

[7] "A thing that is in succession and touches is 'contiguous.' The 'continuous' is a subdivision of the contiguous: Things are called continuous when the touching limits of each become one and the same and are, as the word implies, contained in each other: continuity is impossible if these extremities are two. This definition makes it plain that continuity belongs to things that naturally in virtue of their mutual contact form a unity. And in whatever way that which holds them together is one, so too will the whole be one, e.g. by a rivet or glue or contact or organic union" (*Physics*, Book V, Ch. 3, 227^a6–16).

[8] Open intervals (a, b) and half-open ones $[a, b)$; $(a, b]$ are also assigned measure $b - a$. The endpoint makes things no bigger when added and no smaller when taken away.

[9] Or indeed on any sigma algebra containing the C_rs.

[10] And indeed on all bounded subsets of the reals.

[11] Banach (1923) and Banach and Tarski (1924).

[12] The axes are chosen so that distinct members of the group represent distinct rotations. Hausdorff proves that this is possible.

[13] A, B, C are constructed by recursion on the length of elements in G. 1 is in A; ϕ, ψ in B; ψ^2 in C. Continue as follows:

	x in A	x in B	x in C
x ends in ψ, ψ^2:	$x\phi$ in B	$x\phi$ in A	$x\phi$ in A
x ends in ϕ:	$x\psi$ in B and $x\psi^2$ in C	$x\psi$ in C and $x\psi^2$ in A	$x\psi$ in A and $x\psi^2$ in B

[14] The proof, essentially as I have given it is in Ulam (1930). He then strengthens it by showing that it holds for any set, Z, such that there is no weakly inaccessible cardinal less than or equal in power to Z.

[15] One way to do this would be first to assign a finitely additive probability measure to the sentences of a first-order language, and then extend it to a countably additive probability measure on the sigma algebra generated by the sets of models which satisfy sentences of the language (see Fenstad, 1980).

REFERENCES

Archimedes. *The Works of Archimedes*. Tr. T. L. Heath. (With a supplement, *The Method of Archimedes*.) New York: Dover, n.d.

Aristotle. *The Basic Works of Aristotle*. Ed. Richard McKeon. N.Y.: Random House, 1941. (Unless otherwise noted, all of the translations of Aristotle's works referred to in this paper are to be found in this edition, which is a selection of translations originally published in the *Works*, translated under the editorship of W. D. Ross. 12 vols. London: Oxford University Press, 1908–1931.)

pseudo-Aristotle. 'Concerning Indivisible Lines.' Tr. H. H. Joachim. In *The Works of Aristotle*, vol. VI. Ed. W. D. Ross. Oxford: Clarendon Press, 1913.
Banach, S. 1923. 'Sur le problème de la mesure,' *Fundamenta Mathematicae* **4**, 30–31.
Banach, S. and Kuratowski, C. 1929. 'Sur une généralisation du problème de la mesure,' V. 114, 127ff.
Banach, S. and Tarski, A. 1924. 'Sur la décomposition des ensembles de points en parties respectivement congruentes,' *Fundamenta Mathematicae* **6**, 244–277.
Bernstein, A. and Wattenberg, F. 1969. 'Non-Standard Measure Theory.' In W. A. J. Luxemberg (ed.), *Applications of Model Theory Algebra, Analysis and Probability*. pp. 171–185. New York: Holt Rinchart and Winston.
Boyer, C. B. 1968. *A History of Mathematics*. New York: Wiley.
Cajori, F. 1915. 'The History of Zeno's Arguments on Motion,' *American Mathematical Monthly* **22**, 1–6, 77–82, 109–115, 143–149, 179–186, 215–220, 253–258, 292–297.
Cajori, F. 1919. *A History of Mathematics*. 2nd ed. New York: Macmillan.
de Finetti, B. 1972. *Probability, Induction and Statistics*. New York: Wiley.
de Finetti, B. 1974. *Theory of Probability*. Tr. A. Machi. 2 vols. New York: Wiley.
Fenstad, J. E. 1980. 'The Structure of Probabilities Defined on First-Order Languages.' In *Studies in Inductive Logic and Probability II*. Ed. Jeffrey. pp. 251–262. Berkeley: University of California Press.
Fränkel, H. 1942. 'Zeno of Elea's Attacks on Plurality,' *American Journal of Philology* **63**, 1–25, 193–206. Revised version in Furley and Allen (1970–75), vol. 2, pp. 102–142.
Fritz, K. von 1945. 'The Discovery of Incommensurability by Hippasus of Metapontum,' *Annals of Mathematics* **46**, 242–264.
Furley, D. 1969. 'Aristotle and the Atomists on Infinity.' In *Naturphilosophie bei Aristoteles und Theophrast*, pp. 85–96. Heidelberg: Lothar Stiehm Verlag.
Furley, D. J. 1967. 'Indivisible Magnitudes.' In *Two Studies in the Greek Atomists*, by D. J. Furley. Princeton: Princeton University Press.
Furley, D. J. and Allen, R. E. 1970–75. *Studies in Presocratic Philosophy*. 2 vols. London: Routledge and Kegan Paul.
Grünbaum, A. 1952. 'A Consistent Conception of the Extended Linear Continuum as an Aggregate of Unextended Elements,' *Philosophy of Science* **19**, 290–95.
Grünbaum, A. 1963. *Philosophical Problems of Space and Time*. Chap. 6. New York: Knopf.
Grünbaum, A. 1968. *Modern Science and Zeno's Paradoxes*. London: Allen and Unwin.
Hausdorff, F. 1914. *Grundzüge der Mengenlehre*. Leipzig: Veit.
Hawkins, T. 1970. *Lebesgue's Theory of Integration*. New York: Chelsea.
Heath, T. L. 1949. *Mathematics in Aristotle*. London: Oxford University Press.
Lee, H. D. P. 1936. *Zeno of Elea*. Cambridge: Cambridge University Press.
Luria, S. 1933. 'Die Infinitesimallehre der antiken Atomisten,' *Quellen und Studien zur Geschichte der Mathematik* **2**, 106–195.
Mau, J. 1954. *Zum Problem des Infinitesimalen bei den antiken Atomisten*. Berlin: Akademie Verlag.
Owen, G. E. L. 1957–8. 'Zeno and the Mathematicians,' *Proceedings of the Aristotelian Society* **58**, 199–222. Reprinted in Furley and Allen (1970–75).
Robinson, A. 1966. *Non-Standard Analysis*. Amsterdam: North Holland.

Savage, L. J. 1954. *The Foundations of Statistics*. New York: Wiley.

Spielman, S. 1977. 'Physical Probability and Bayesian Statistics,' *Synthese* **36**, 235–269.

Shimony, A. 1955. 'Coherence and the Axioms of Confirmation,' *Journal of Symbolic Logic* **20**, 1–28.

Solovay, R. M. 1970. 'A Model of Set Theory in which Every Set of Reals Is Lebesgue-Measurable,' *Annals of Mathematics* **92**, 1–56.

Tannery, P. 1885. 'Le concept scientifique du continu: Zenon d'Elée et George Cantor,' *Revue Philosophique* **20**, no. 2.

Tarski, A. 1956. 'Foundations of the Geometry of Solids.' In *Logic, Semantics, Metamathematics*. Tr. and ed. by J. H. Woodger. Oxford: Clarendon Press.

Ulam, S. 1930. 'Zur Masstheorie in der allgemeinen Mengenlehre,' *Fundamenta Mathematicae* **16**, 140–150.

Vitali, G. 1905. *Sul problema della misura dei gruppi di punti di una retta*. Bologna:

Vlastos, G. 1971. 'A Zenonian Argument against Plurality.' In *Essays in Ancient Greek Philosophy*. Ed. by J. P. Anton with G. L. Kustas, pp. 119–144. Albany N. Y.: Suny Press.

JOHN STACHEL

SPECIAL RELATIVITY FROM MEASURING RODS

The mathematical structures associated with a space-time theory, such as the special theory of relativity (SRT) – or the general theory (GRT) for that matter – are numerous and interrelated in complex ways.[1] One may start their analysis with the concept of a point set, the elements of which are identified with events in space-time.[2] Imposing a continuity structure on this set leads to the concept of space-time as a four-dimensional topological manifold. Restriction to a differentiable structure then leads to the concept of space-time as a differentiable manifold. Various additional mathematical structures may now be introduced on this manifold: projective, affine, conformal and pseudo-metrical (a metrical structure with Minkowski signature). Each of these mathematical structures is closely associated with the behavior of some idealized physical entity in space-time. The projective structure is associated with the trajectories of structureless free test particles. If each particle carries some intrinsic measure of duration along its trajectory, it reflects the affine structure. The conformal structure is associated with the wave fronts of massless fields, such as the electromagnetic. A pseudo-metrical structure with Minkowski signature implies the existence of two fundamentally distinct types of interval which cannot be transformed into one another by any operation of the symmetry group defining the geometry (the inhomogeneous Lorentz group for SRT).[3] These two distinct types of interval are called spacelike and timelike, and physically quite distinct entities – measuring rods and clocks – are associated with their respective measurement.

While the various mathematical structures introduced in the analysis of space-time are conceptually quite distinct, in both SRT and GRT the projective, affine, conformal and pseudo-metrical structures are inextricably intertwined. One can, for example, mathematically derive all the other structures from the pseudo-metrical structure. Conversely, the pseudo-metrical structure may be derived from compatible projective and conformal structures.[4] In the case of SRT, one may even go a step further and derive the pseudo-metric structure (and thence the projective, of course) from the conformal structure alone, provided certain global assumptions about the entire space-time are added.[5]

This intertwining of projective, affine, conformal and pseudo-metrical

R. S. Cohen and L. Laudan (eds.), Physics, Philosophy and Psychoanalysis, 255–272.

structures makes possible numerous alternative ways of physically characterizing the space-time structure in terms of the behavior of various idealized physical entities. For example, one may use free test particles (associated with the projective structure) and null wave fronts – or the null rays (bicharacteristics) defined by these wave fronts (associated with the conformal structure) to characterize the metrical structure of space-time.[6] Or one may use measuring rods and clocks to characterize the metrical structure directly. In addition to rods and clocks, Einstein used light rays in his derivation of the Lorentz transformations; but it was soon realized that the light rays were superfluous.[7] Still other combinations are possible; for example, clocks and the paths of free test particles.[8] Of course, postulated properties of a set of physical entities in one method of characterizing the space-time structure become testable (in principle) deductions of another method. This circumstance has sometimes given rise to misunderstandings and even passionate attacks on the honor of one method or another.

The problem is compounded by the empiricist, operationalist or instrumentalist spirit in which these foundational exercises are often carried out, giving rise to (at least) two fundamental confusions. First of all, it is felt that the ideal entities initially introduced (rods, clocks, test particles, null rays or what have you) must be immediately (i.e., without any mediation) identifiable with objects used in laboratory tests of the theory. Unless the conceptual entities introduced into the foundations of the theory can be directly identified with such real objects the empirical foundations of the theory are thought to be insecure. Secondly, the order in which definitions and axioms characterizing these ideal entities are presented in the development of the theory is felt to be the order in which they must be empirically tested in order to have an empirically well-founded theory. The logical order of exposition of the foundations of the theory is identified with the sequence of operations to be followed in empirically testing the resulting theory.

But the entities introduced in the course of the logical exposition of a theory are theoretical, abstract, ideal, conceptual entities; the question of their relation to the actual, physical, concrete, real entities whose behavior the theory is intended to help understand is a complex one. It cannot be understood as a simple relation of direct correspondence. Nor is there any reason why the order of exposition of the relationships between the basic concepts used to build up a theory need parallel the empirical procedures (to say nothing of the other considerations) which give us confidence in the resulting theory. Again the relationship between a theory and the evidence for it is not one of simple parallelism.

Of course, there are some restrictions on the introduction of ideal entities in the development of a theory. For example, properties may not be initially attributed to such entities which are inconsistent with the resulting theory. To give an example relevant to the main subject of this paper: it is quite proper to introduce the concept of a perfectly rigid body into the foundations of Newtonian space-time structure, even though no real object exists which remains unaffected by sufficiently large stresses, because the existence of such a perfectly rigid body is not inconsistent with the principles of Newtonian kinematics. It is inconsistent, however, to introduce such a body into the foundations of SRT since it could be used to transmit signals with arbitrarily high velocities.

Another requirement for the introduction of ideal entities into the foundations of SRT arises from its character as a theory of principle.[9] Since SRT specifies the space-time structure independently of any dynamical considerations, properties attributed to ideal objects should similarly be restricted to kinematical ones. Once the special-relativistic space-time structure has been established, constructive theories describing the dynamical behavior of such things as particles and fields may be set up which accord with this structure. A special-relativistic theory of particle mechanics, electromagnetism, elasticity, etc., must be developed (it would be better to say special-relativistic theories, since the choice is not unique). One should require that solutions to the appropriate dynamical equations exist exhibiting the postulated kinematical behavior of the ideal entities.[10] One would thus be led to reject an ideal entity if it could be demonstrated that no dynamical theory consistent with SRT could have solutions which at least arbitrarily closely approximated the postulated kinematical behavior. To the extent that such solutions of the dynamical theory in question take us a step closer to understanding the behavior of certain real objects we may say that we have established a link between our ideal entities and those objects. Further refinement and complication of the constructive dynamical theories and/or their solutions may then lead to a more and more adequate understanding of the behavior of real objects. It is this complex process which constitutes the empirical testing of the theory – which is thereby tested as a whole, rather than foundation stone by foundation stone. Some parts of this testing procedure may involve less of the dynamical theory than others. And all of them will generally involve various ancillary items of testing apparatus, the theory of which may become more complicated than the theory we are testing. If there is circularity in this procedure, it is a healthy kind, inherent in the testing of an empirical theory, and not a vicious, logical kind. This is all I shall say here about the extremely complex question of the relationship between ideal entities and real objects, between the abstract and the concrete.

Different foundational approaches to SRT, carried out in the spirit indicated here, should not be looked upon as competing with but rather as supplementing each other by highlighting different facets of the total space-time structure. It is this structure as a whole, with its peculiar interrelationships between the projective, affine, conformal and pseudo-metrical structures, which constitutes the essential feature of SRT; and not the sequence in which these structures are introduced in a particular logical development of the theory.

One may attempt to directly characterize one or more of these structures, with the aim of ultimately providing a full characterization of the pseudo-metrical structure. Indeed, in the case of GRT this is all that can be done. But in the case of SRT an alternative but equivalent method exists: the pseudo-metrical structure may be fully characterized by the group of space-time symmetry transformations under which it is invariant, the inhomogeneous Lorentz group.[11] This is the route taken by Einstein (1905), before it was even clear that a space-time structure was thereby being characterized.[12] By physically characterizing a set of preferred space and time coordinates in an inertial frame and investigating the group of transformations between such coordinates in different inertial frames, Einstein was characterizing Minkowski space-time whether he knew it or not.

RIGID INERTIAL MOTIONS

With these lengthy preliminaries behind me, I can now state the purpose of this paper. I shall show – sometimes only in sketchy outline – that one may use the concept of a set of ideal entities in rigid inertial motion (RIM) with respect to each other as a sufficient foundation for SRT. Once this is done, one may then set up a special relativistic dynamical theory of elastic bodies and show that the motion of such bodies, when unstressed and acted on by no external forces, satisfies all the kinematic conditions initially imposed on entities in RIM. In accord with the criteria outlined in the previous section, this fully justifies the original introduction of the concept of entities in RIM into the foundations of SRT. An entity in RIM, with a straight line drawn on it and a unit of length indicated by the choice of two points on the line, constitutes a measuring rod. Thus, I call this approach SRT with only measuring rods.

Measuring rods have often been treated as objects of ill repute in SRT.

Two main reasons have been given for assigning pariah status to them. It has been objected that rods are complicated atomic structures, explanation of whose behavior requires the introduction of complicated dynamical (and even quantum-dynamical) laws. This is contrasted, for example, with particles and light rays, which are presumed to be much simpler entities to understand. As will hopefully be clear from the previous discussion, I regard this objection as based on confusion of an ideal entity, introduced in the development of a theory, with a real object which the theory is supposed to help us understand.

As theoretical entities, rods, clocks, particles and light rays are all on the same level of abstraction – or simplicity if you prefer. As for real objects whose behavior can be approximately understood by sufficiently complicated solutions to dynamical equations, the explanation of real rods, clocks, particles or light rays presents various complexities and difficulties. I am not arguing here for one and against another, of course. As indicated above, I favor peaceful coexistence at the foundational level.

One other main objection has been raised against measuring rods as theoretical entities: it is inconsistent to postulate the existence of a perfectly rigid entity in SRT. This is quite true, as mentioned above. But there is no need to postulate such entities. All that is needed is the postulation of a class of rigid inertial *motions*, and there is nothing in SRT which forbids the construction of dynamical systems capable of executing RIMs under appropriate conditions.

Having demonstrated, I hope, the legitimacy of introducing measuring rods (i.e., RIMs) as foundational elements, I now propose a counterattack in defense of the honor of the rod – not, however, in an aggressive imperialistic spirit, of course – by showing that no other foundational entities need be introduced to develop SRT. I do not claim to be doing anything very original thereby. All of the elements of my story can be found elsewhere. But I have never seen them put together to tell quite this tale in quite this spirit.[13] This constitutes my only justification for telling it here in honor of Adolf Grünbaum who, I believe, shares my affection for measuring rods.

The attentive reader should be puzzled by this point. I claim to be able to derive the pseudo-metrical structure of space-time with only measuring rods. Yet I emphasized at the outset that the pseudo-metrical structure involves the existence of two qualitatively quite distinct types of interval – spacelike and timelike – which are not intercomparable, and therefore require two distinct physical entities for their measurement: rods and clocks. The only way to avoid the apparent paradox is to construct a clock from measuring rods. Lest this seem an arithmetical miracle (two from one) let me hasten to point out that I really do have two distinct primitive concepts: measuring rods and their relative motion with constant velocity. But one must introduce the concept of constant or uniform relative motion as soon as one wants to consider the relationship between two or more inertial frames. So it is interesting to see that, if one takes it as a primitive concept, one can construct clocks, as well as define clocks running at the same rate at different points of an inertial frame and distant synchronization of two such clocks, all in terms of RIMs of measuring rods.

Before beginning to outline the development of SRT based on the concept of RIMs, I shall briefly discuss the nature of rigid and inertial motions in Minkowski space-time. A congruence of timelike worldlines defines a rigid motion if the orthogonal spacelike interval between neighboring worldlines remains constant, i.e., independent of where along the worldlines this distance is evaluated. This definition is not only applicable in SRT but also in GRT and even in Newtonian theory, if "orthogonal spacelike interval" is replaced by "spatial distance at constant absolute time" in the definition. A ponderable body is said to execute a rigid motion if the velocity four-vector of each point of the body is always tangent to some worldline of such a congruence. In Newtonian space-time there is a six-parameter class of such motions, but in SRT the class is much more strictly restricted (in GRT, the existence of even *one* rigid motion is a severe restriction on the metric tensor). A congruence of timelike worldlines defines an inertial or geodesic motion if the acceleration four-vector vanishes for each worldline of the congruence. A ponderable body is said to execute an inertial or geodesic motion if the velocity four-vector of each point of the body is tangent to some worldline of such a congruence. In Minkowski space-time the only inertial or geodesic worldlines are straight lines, but a family of such lines need not be parallel to define an inertial motion. However, if a congruence is both rigid and inertial (RIM) this cannot happen; the worldlines of the congruence must be parallel timelike straight lines, and thus define an inertial frame of reference (IFR) in the region of space-time which they cover.

ADAPTED SPATIAL AND TEMPORAL COORDINATES

I shall follow the traditional method of developing SRT, characterizing the pseudo-metrical space-time structure through its group of symmetries. An adapted system of spatial and temporal coordinates with respect to an IFR will be defined, as will the meaning of equal spatial and equal temporal intervals with respect to different IFRs. Then the group of transformations between the adapted coordinates of an event with respect to two IFRs will be investigated, yielding the inhomogeneous Lorentz group. All results will be local, in the sense that they will hold for some finite region of space-time but do not require the assumption that an IFR covers the entire Minkowski space-time.

My starting point is postulation of the existence of a sufficiently large class of bodies in RIM to satisfy all further assumptions. Being in RIM is a property of a single body. Two bodies in RIM either have a certain relation – being at relative rest – or they do not, in which case we say they are in relative motion. The relation of relative rest is reflexive (a body is at rest relative to itself), symmetric (if one body is at rest relative to a second, the second is at rest relative to the first) and transitive (if the second is at rest relative to a third, then the first is at rest relative to the third). Such a relation divides the class of bodies in RIM into equivalence classes: each member of an equivalence class being at rest relative to any other member of that class and in relative motion with respect to any member of another class. We now define an IFR as such an equivalence class of bodies in RIM.

Next assume that Euclidean geometry holds for each body in RIM, and hence for each IFR. Choosing such a body from the equivalence class defining an IFR (a body in the IFR for short), mark two points on it. This defines a unit of length for the body, and hence the IFR. Since the geometry has been postulated to be Euclidean, we may draw the straight line through the two points, prolong it and lay off the unit along it as often as needed, subdivide the unit as needed, etc. Such a straight line with a unit of length given on it will be called a measuring rod (or rod for short). Since I shall only refer to such rods when they are in RIM I shall not always add this qualification. Since the space of an IFR is Euclidean, it is homogeneous and isotropic; so rods may be used to map out its geometry and in particular to set up a system of Cartesian coordinates on any body in the IFR.

The principle of inertia (Newton's first law) is postulated as follows: Consider two distinct IFRs. Then any two bodies, one in each IFR, are in relative motion. A point A on one body successively coincides with a set of points on the other whose locus is a straight line. Two points A and B on the first body will then give rise to a pair of lines on the second body. If the two lines coincide, then the line AB on the first body defines the direction of motion of the second body with respect to the first. If the two lines do not coincide, then they are parallel. We still have not defined the speed of the second body with respect to the first, which will require a discussion of time measurement.

Before turning to that, however, I shall define equal spatial intervals (lengths, distances) with respect to two distinct IFRs.[14] Consider a case where points A and B on the first body give rise to a pair of distinct parallel lines on the second. Let a be a point on the second body on the line arising from the motion of A. Then a will in turn give rise to a straight line on the first body which contains A. Similarly, a point b on the other line yields a parallel straight line on the first body containing B. We thus have a pair of parallel lines on the first body which is correlated with a pair of parallel lines on the second. The perpendicular distances between such pairs of correlated parallel lines in different IFRs are defined as equal intervals. An equality relation must be symmetric and transitive. The given definition clearly defines a symmetric relation. We now assume that it also defines a transitive relation. Later, transitivity can be shown to follow from the properties of the Lorentz transformations. This establishes the consistency of the transitivity assumption at this point.

Next I turn to time measurement. Consider a point P at rest in some IFR, and a measuring rod in relative motion in the direction of its length. It follows from the principle of inertia that if *one* point of the rod ever coincides with P, then *all* the points of the rod will do so successively. Define a unit time interval at the point P of the IFR as the interval between the coincidence with P of the initial and final points of a unit spatial interval on the measuring rod. Such a rod thus constitutes a clock, measuring time intervals at the point P of the IFR. Assume that any two such clocks keep the same time except for units. That is, the ratio of the time intervals between two events at P as measured by two clocks is a constant depending only on the IFRs of the two rods defining the clocks at P and the IFR in which P is at rest. This assumption clearly requires that time with respect to any point in an IFR be homogeneous.

If the ratio between two clock rates is one, and if the rods defining the two clocks use equal spatial units, then we say that the two rods are travelling at the same speed with respect to P's IFR (they may be travelling in different directions, of course). To attach a numerical value to the speed, we need synchronized clocks a different points of the IFR, which I now proceed to define.

We may check whether clocks at two different points of an IFR are going at the same rate by means of a measuring rod moving along the straight line connecting the two points; or more generally, by means of two such rods moving at the same speed with respect to the IFR, one passing through each point. To define synchronous settings on two distant clocks in an IFR, we use two rods of unit length moving at the same speed with respect to the IFR, along the same straight line but in opposite senses. There will be two distinct events when the front of one rod coincides with the rear of the other rod (front and rear being defined with respect to the sense of motion along the line). Two clocks at rest in the IFR at the positions of the two events are synchronized if their readings are the same at the two moments of coincidence.[15]

The speed of one body in RIM with respect to a second is now defined as the reciprocal of the time it takes a point of the first body to pass between two points of the second which are a unit distance apart. The numerical value clearly depends on the units of length and time adopted, even if these units are equal for the two bodies. It does not follow directly that the speed of the second body with respect to the first is numerically equal to the speed of the first with respect to the second. This reciprocity principle, as it is called, will be proved in the course of the derivation of the Lorentz transformations.[16]

We must still define equal time intervals with respect to different IFRs. If two rods with equal unit lengths are in relative motion along the same straight line, then the time intervals for a unit length of one rod to pass a fixed point of the other are equal. More explicitly: Let t_1 be the time interval relative to the first rod's IFR between the coincidence with some fixed point of the first rod of the initial and final points of a unit length of the second rod. Let t_2 be the time interval relative to the second rod's IFR between the coincidence with some fixed point of the second of the initial and final points of a unit length of the first rod. Then we define t_1 to be equal to t_2. This is clearly a symmetric relation; the consistency of assuming it to be transitive again follows from the Lorentz transformations.

Assuming the homogeneity and isotropy of space and the homogeneity of time with respect to each IFR, it follows that the definitions of equal spatial and temporal intervals with respect to different IFRs is independent of times, places and directions of motion of rods occurring in the definitions.

LORENTZ TRANSFORMATIONS

Now I shall outline the derivation of the Lorentz transformations – or rather the one-parameter family that reduces to the usual Lorentz transformations when this parameter is set equal to c^2. The assumptions going into the derivation are that:

(1) Space with respect to any body in RIM, and hence any IFR, is Euclidean, i.e., homogeneous and isotropic with vanishing curvature; time with respect to any IFR is homogeneous.

(2) The law of inertia holds for bodies in RIM, and hence for IFRs.

(3) Cartesian spatial coordinates and temporal coordinates for each body in RIM, and hence any IFR, are fixed as described in the previous section, with equal unit spatial and temporal intervals (as there defined) employed in each IFR.

(4) The relativity principle holds.

All of these assumptions except the relativity principle have already been discussed. The relativity principle is only needed here in its kinematic form: no IFR may be distinguished from any other by any kinematical properties.[17] In particular, the relationship between kinematic properties of two IFRs can only depend on the relative velocity of one IFR with respect to the other.

The most important implication of the kinematic form of the relativity principle is that the Lorentz transformations form a group. First we must characterize the transformations to be studied. Any event may be located by its three Cartesian spatial coordinates and its time coordinate with respect to some IFR. The same event may be located by a similar set of coordinates with respect to a second IFR. Following tradition, I shall often refer to the two frames, and their coordinates, as the primed and unprimed frames, coordinates, etc. The relativity principle, together with the homogeneity and isotropy of space and time, implies that the primed coordinates of an event can only be functions of the unprimed coordinates of that event and of the relative velocity of the primed IFR with respect to the unprimed IFR. A sequence of two such transformations must be equivalent to a single transformation; and a similar transformation from the primed coordinates back to the unprimed coordinates must exist. Thus, the set of such transformations form a group.

The law of inertia implies that the transformation equations must be linear: If a sequence of events defines a straight line motion traversed with uniform speed with respect to one IFR, it must do so with respect to any other IFR; and indeed parallel straight lines traversed at the same speed must remain such under the transformation equations. The only transformations with this property are the affine transformations, i.e., linear transformations with non-vanishing determinant of the transformation coefficients. Thus, the primed coordinates must be linear functions of the unprimed coordinates. From our previous discussion, it follows that the coefficients can only depend on the relative velocity of the primed IFR with respect to the unprimed.

By a judicious use of coordinate freedom we can simplify the derivation of the transformation equations. Choose the x-axis in the direction of motion of the unprimed IFR with respect to the primed. It then follows from our definition of equality of spatial intervals with respect to different IFRs that equal coordinate differences in directions orthogonal to the x-axis and x'-axis respectively measure equal spatial intervals. By picking the x-axis and x'-axis to have an event in common and rotating the y'- and z'-axes if necessary, we can then assure that for all events $y' = y$, $z' = z$. So we need only consider linear transformation of the x and t coordinates:

$$x' = \gamma(x - wt) + x_0$$
$$t' = \Gamma(t - mx) + t_0.$$

Here γ, Γ, w, m, x_0, t_0, may depend on the relative velocity of the primed system with respect to the unprimed. With our choice of axes, this relative velocity only has an x-component, which I shall call v. Consider the point at rest in the primed system which passes through the origin of the unprimed system at the time $t = 0$. It must have x-coordinate $x = vt$ at any time t, and a constant x' coordinate at all times. This shows that $w = v$ and that x_0 must be independent of v in the first equation. t_0 is also a constant independent of v, expressing the freedom to pick the origin of the t' coordinate inherent in the homogeneity of time with respect to any IFR. If we require the event with vanishing unprimed coordinates to also have vanishing primed coordinates then $x_0 = t_0 = 0$, and we need only consider the homogeneous transformations:

$$x' = \gamma(v)(x - vt)$$
$$t' = \Gamma(v)[t - m(v)x].$$

It follows from the isotropy of space that if these transformation equations hold for an IFR moving with speed v along the x-axis of another IFR in one sense of motion, then transformation equations of the same form must hold for an IFR moving with the same speed in the opposite sense of motion. That is, if we reverse the signs of v, x, and x' in the above transformation equations, their form must remain unchanged.[18] Such a reversal gives:

$$x' = \gamma(-v)(x - vt)$$
$$t' = \Gamma(-v)[t + m(-v)x],$$

and comparison with the previous transformation equations shows they are identical provided $\gamma(v)$ and $\Gamma(v)$ are even functions of v, while $m(v)$ is an odd function.

Another condition on these three functions follows from the fact that the identity transformation must belong to the group. The identity transformation is obviously the transformation for which $v = 0$, showing that $\gamma(0) = \Gamma(0) = 1$, while $m(0) = 0$. We shall assume that they are all continuous functions of v.

Next we investigate the consequences of the group property of the transformations. Introduce a third IFR with coordinates x'', t'' moving with speed v' with respect to the primed IFR, and speed v'' with respect to the unprimed IFR along their common x- and x'-axes. Then the following two sets of transformation equations must hold:

$$\begin{aligned} x'' &= \gamma(v')(x' - v't') \qquad & x'' &= \gamma(v'')(x - v''t) \\ t'' &= \Gamma(v')[t' - m(v')x'] \qquad & t'' &= \Gamma(v'')[x - m(v'')t]\,. \end{aligned}$$

Replacing the coordinates x' and t' in the first set of equations by their expressions in terms of x and t, and comparing the coefficients of x and t in the resulting equations with the coefficients of x and t in the second set of equations gives the following four equations:

$$\begin{aligned} &\gamma(v'') = \gamma(v')[\gamma(v) + v'\Gamma(v)m(v)] \\ &v''\Gamma(v'') = \gamma(v')[v\gamma(v) + v'\Gamma(v)] \\ &\Gamma(v'') = \Gamma(v')[\Gamma(v) + m(v')\gamma(v)v] \\ &m(v'')\Gamma(v'') = \Gamma(v')[\Gamma(v)m(v) + m(v')\gamma(v)]\,. \end{aligned}$$

Dividing the second equation by the first, and the fourth by the third gives:

$$v'' = \frac{v\gamma(v) + v'\Gamma(v)}{\gamma(v) + v'\Gamma(v)m(v)}$$

$$m(v'') = \frac{\Gamma(v)m(v) + m(v')\gamma(v)}{\Gamma(v) + m(v')\gamma(v)}.$$

We may use these equations to prove the reciprocity principle. If we choose the double-primed IFR to coincide with the unprimed IFR, then $v'' = 0$, and v' becomes the inverse or reciprocal velocity to v, i.e., the velocity of the unprimed IFR relative to the primed. Letting $v'' = 0$ in the preceding two equations, and remembering $m(0) = 0$, we find that:

$$v' = -v\gamma(v)\Gamma^{-1}(v), \qquad m(v') = -m(v)\Gamma(v)\gamma^{-1}(v)$$

Multiplying these two equations shows that:

$$v'm(v') = vm(v).$$

Now v is a parameter which can very continuously from a value of zero and $vm(v)$ is a continuous, even function of v (since $m(v)$ is odd). There will be some range of positive values of v around zero for which this function is either monotonic or constant. If it is constant, its value must be zero, since the function vanishes for $v = 0$. This case gives the Galilei transformations, which can also be derived as a degenerate limit of the Lorentz transformations as we shall indicate later. So we need only investigate the monotonic case. Then the function $vm(v)$ may be inverted in this range of values, with the possibility $v' = v$; when we take negative values of v and the evenness of the function into account, we see that $v' = -v$ is also possible. The first case $v' = v$ is ruled out, since it implies that $\gamma(v) = -\Gamma(v)$ by the first of the above pair of equations, while $\gamma(0) = \Gamma(0) = 1$. So we have proved the reciprocity principle, that $v' = -v$. It now follows from the first of the above pair of equations that $\gamma(v) = \Gamma(v)$.

Using the latter result in the first and third of the set of four equations deduced earlier, and comparing the two resulting expressions for $\gamma(v'')$, shows that (remember v and v' are now unrelated):

$$vm(v') = v'm(v).$$

In other words $m(v)/v$ is independent of v. Let its value be k. Then $m(v) = v/k$, and the transformation equations take the form:

$$x' = \gamma(x - vt)$$
$$t' = \Gamma(t - vx/k),$$

where $\gamma = (1 - v^2/k)^{1/2}$, as may be found from several of the relations given above. The law of addition for parallel relative velocities results at once from substitution of the expression for $m(v)$ into the equation for v'' derived above:

$$v'' = \frac{v + v'}{1 + vv'/k}.$$

What about the value of k? *A priori*, it could have any non-zero value. If it were negative, the set of transformations would not form a group unless time inversions are admitted as well. So if we demand that the orthochronous transformations form a group, this restricts k to positive values.[19] Negative values of k also lead to curious results for the addition of velocities, such as the possibility of obtaining an infinite relative velocity by the compounding of two finite relative velocities. So I shall confine further discussion to the case $k > 0$.

If k is infinite, we get the Galilei transformation laws of Newtonian kinematics, which are thus a degenerate case. If k is finite, set $k = V^2$ giving a one-parameter family of groups of transformations each of which is formally similar to the usual Lorentz transformations, except that V plays the role of the fundamental velocity. V may be evaluated by any of several kinematical methods. Perhaps the simplest is comparison of the proper length of a rod (in its rest IFR) with its length relative to another IFR. If the rod is moving with respect to an IFR with a speed $v < V$, its relative length may be defined in either of two equivalent ways: the simultaneous distance between the positions of its end points with respect to an IFR; or the time interval with respect to an IFR for the rod to pass a fixed point of the IFR times its relative speed v. It may be deduced (in the well-known way) from the Lorentz transformations that this relative length is $L_0(1 - v^2/V^2)^{1/2}$, where L_0 is its proper length. One such measurement thus suffices to fix V.

In addition to the special homogeneous Lorentz transformations derived above and the additional ones arising from spatial rotations of axes, spatial and temporal translations of the coordinate axes with respect to any IFR are also symmetry transformations of the space-time. The homogeneous and inhomogeneous transformations together form the inhomogeneous Lorentz group (also often called the Poincaré group). This symmetry group fully characterizes the pseudo-metrical structure of Minkowski space-time.

It may now be checked that all assumptions made earlier, such as the transitivity of the equality definition for spatial and temporal intervals with respect to different IFRs, are indeed satisfied by the Lorentz transformed quantities.

CONCLUSION

I shall now very briefly outline how one could proceed to show that the kinematical assumptions made about ideal entities in RIM are consistent with special-relativistic dynamics, by setting up a special-relativistic theory of continuous media.[20] The very simplest such theory would suffice; for example, the theory of a perfectly elastic medium or even a perfect fluid (a fluid is quite capable of executing a RIM).[21] For any finite sample of such a medium, there will always be a solution to the dynamical equations which represents a state of the sample such that it is acted upon by no external forces, is unstressed, and each point of the medium moves inertially. Such a solution represents the sample in RIM. A requirement of the dynamical extension of the relativity principle is the invariance of all dynamical equations under the inhomogeneous Lorentz group. This implies that an inhomogeneous Lorentz transformation applied to the original solution representing a RIM yields a solution representing another such RIM (active interpretation of the transformation). Thus, a sufficiently large class of RIMs can be generated to satisfy all the kinematical conditions originally imposed in bodies in RIM. The dynamical consistency of the original kinematical assumptions is thus demonstrated.

Finally, I shall briefly mention the relevance of this derivation of SRT to the claims of the protophysicists. 'Protophysik' is the name given by a group of German-speaking philosophers of science to their program of founding all physics on criteria for the measurement of lengths (geometry), times (chronometry) and masses (hylometry).[22] The program lays great stress on the fundamental importance of starting from what is called a purely haptic basis (roughly, from mechanics), not using any non-mechanical means (such as optical) in constructing the protophysical foundations. Whatever one may think of this program as a basis for physics, it follows from the previous discussion that a purely haptic protophysics can be set up based on motions of 'rigid bodies', (which the proto-physicists use in founding geometry) and employing a definition of simultaneity with respect to an IFR which is equally applicable to non-relativistic and relativistic kinematics. The distinction between the two cases can also be made by purely haptic methods (whether or not the relative length of a moving rod is less than its rest length). Although it has been claimed that such purely haptic methods lead naturally to the concept of absolute time and the Galilei transformations, it is hard to see what objections a proto-physicist could have to the suggested procedures. In particular, which of the four assumptions on p. 264, leading to the space-time transformations, could be reject?

Boston University

ACKNOWLEDGEMENT

It is a pleasure to thank Dr. John Norton for a critical reading of the entire manuscript and in particular for suggesting the definition of equal spatial intervals with respect to different IFRs used here, and Abner Shimony and Roberto Torretti for valuable discussions and suggestions.

NOTES

[1] See Weyl (1923), Ehlers (1973a) for surveys of mathematic structures used in space-time theories.

[2] One need not start from an ontology of events. One could, for example, start with an ontology of processes, identified with finite regions of space-time. One would postulate (classically, at least) the possibility of indefinitely subdividing any process into smaller sub-procedure. Events would be introduced as ideal limiting elements of such a subdivision process, thus avoiding the need to postulate the existence of physical processes which occupy no space and take no time.

[3] The possibility of such a two-dimensional geometry was actually discussed by Poincaré (1887; 1952) two decades before Minkowski. Apparently he never recalled the connection of this geometry with Minkowski's (1909) "*Welt.*" Jon Dorling (1976) has recently based a very elegant axiomatic treatment of flat space-time geometry on the existence of distinct types of interval. Of course, there is also a third type of interval (null) separating timelike and spacelike intervals.

[4] See Ehlers *et al.* (1972) for the definition of compatible structures, and Weyl (1923) and Ehlers *et al.* (1972) for such a derivation.

[5] This was first done in Robb (1914). Also see Robb (1921; 1936). Weyl (1923) contains a brief discussion emphasizing the group-theoretical reasons for the success of Robb's approach.

[6] See Ehlers *et al.* (1972) for this approach.

[7] Einstein (1905) contains his original derivation. Ignatowsky (1910) was the first to investigate the possibility of eliminating the light axiom, followed by Frank and Rothe (1911). For further historical references, together with a brief outline of this approach, see Jammer (1979). Other derivations and references may be found in Schwartz (1962), Berzi and Gorini (1969), Süssmann (1969), Lee and Kalotas (1975), and Lévy-Leblond (1976).

[8] See Synge (1960; 1964) for this approach.

[9] In 1919 Einstein described "principle-theories" as follows: "The elements which form their basis . . . are . . . general characteristics of natural processes, principles that give rise to mathematically formulated criteria which the separate processes or the theoretical representation of them have to satisfy." He contrasted them with "constructive" theories which "attempt to build up a picture of the more complex phenomena out of the materials of a relatively simple formal scheme . . . " He gave thermodynamics as an example of a theory of principle, the kinetic theory of gases of a constructive theory, and stated that the theory of relativity was a "principle theory." See Einstein (1954), p. 228).

[10] This requirement is suggested by the comment in Einstein (1979), pp. 55–56: "It is striking that the theory . . . introduces two kinds of physical things, i.e., (1) measuring rods and clocks, (2) all other things, e.g., the electromagnetic field, the material point, etc. This, in a certain sense is inconsistent; strictly speaking, measuring rods and clocks should emerge as solutions of the basic equations . . . , not, as it were, as theoretically self-sufficient entities." Einstein had in mind a unified field theory strong enough to allow for the construction of such stable entities as singularly-free solutions. More modest consistency requirements are all that may be imposed on SRT and GRT.

[11] This method is in the spirit of Felix Klein's Erlangen Program, which characterizes a geometry by the invariants of some fundamental group of transformations.

[12] This only became clear after the work of Minkowski (1909).
[13] I am especially indebted to Dixon (1978), as well as the references in Note 7.
[14] Equality of spatial intervals with respect to different IFRs is often defined by means of the transport of a rigid rod from one IFR to the other. Such a definition is unproblematical in the non-relativistic case, since an initially RIM can remain a rigid motion while being continuously accelerated to another RIM. However, this is not possible in the relativistic case. Thus, the transfer of an extended body from one IFR to another cannot be discussed without entering into dynamical considerations, based on some model of the body (e.g., elastic solid), of the effects of acceleration on its shape. It seems preferable – and is essential to my approach – to give definitions of equality of intervals – both spatial and temporal – with respect to different IFRs based exclusively on kinematical considerations.

I shall not discuss here the question of how one verifies equality of Euclidean spatial intervals within an IFR, nor of conventionalism in geometry, to the discussion of which Adolf Grünbaum has made such vital contributions.

[15] This definition is inspired by the example given in Einstein (1911), slightly amplified in Pauli (1958). Other equivalent kinematic definitions of simultaneity with respect to an IFR are possible. One due to Thoma (see Feenberg, 1979, pp. 333–4) is based on a construction similar to that in the definition of equal spatial intervals given above.
[16] See Berzi and Gorini (1969) for a fuller discussion of the reciprocity principle.
[17] For relativistic constructive theories the relativity principle must be extended to exclude the possibility of distinguishing between IFRs by means of any proposed dynamical laws. Of course, particular solutions to such equations may pick out preferred IFRs. Confusion sometimes results when this distinction is not observed.
[18] In one spatial dimension, this reversal appears to be a discontinuous transformation (i.e., unconnected with the identity transformation). But in three dimensions it may be obtained from the identity by continuous spatial rotation – by 180° rotation about the y-axis for example. So we are really considering a transformation which is proper and orthochronous.
[19] See Berzi and Gorini (1969) for a fuller discussion of this point.
[20] Ehlers (1973b), pp. 89–98, contains an excellent brief discussion of the foundations of continuum mechanics and thermodynamics.
[21] A special relativistic theory of perfect elasticity was first set up in Herglotz (1911). For recent discussion see Soper (1976), Carter and Quintana (1972).
[22] Pfarr (1981) is an anthology of papers on protophysics, expository and critical, centered on the protophysical attack on special relativity.

REFERENCES

Berzi, V. and Gorini, V. 1969. 'Reciprocity Principle and the Lorentz Transformations,' *J. Math Phys.* **10**, 1518–1524.

Carter, B. and Quintana, H. 1972. 'Foundations of General Relativistic High-Pressure Elasticity Theory,' *Proc. Roy. Soc. Lond.* **A 331**, 57–83.

Dixon, W. G. 1978. *Special Relativity: The Foundation of Macroscopic Physics*. Cambridge: Cambridge University Press.

Dorling, J. 1976. 'Special Relativity Out of Euclidean Geometry,' unpublished.

Ehlers, J. 1973a. 'The Nature and Structure of Spacetime.' In J. Mehra (ed.), *The Physicist's Conception of Nature*, pp. 71–91. Dordrecht and Boston: D. Reidel.

Ehlers, J. 1973b. 'Survey of General Relativity Theory.' In W. Israel (ed.), *Relativity, Astrophysics and Cosmology*, pp. 1–125. Dordrecht: D. Reidel.

Ehlers, J., Pirani, F. A. E., and Schild, A. 1972. 'The Geometry of Free Fall and Light Propagation.' In L. O'Raifeartaigh (ed.), *General Relativity*. Oxford: Clarendon Press.

Einstein, A. 1905. 'Zur Elektrodynamik bewegter Körper,' *Ann. Phys.* **17**, 891–921.

Einstein, A. 1911. 'Zum Ehrenfestchen Paradoxon,' *Physik. Zeitschr.* **12**, 509–510.

Einstein, A. 1954. 'What Is the Theory of Relativity?' In his *Ideas and Opinions*, pp. 227–232. New York: Crown.

Einstein, A. 1979. *Autobiographical Notes: A Centennial Edition*. LaSalle/Chicago: Open Court.

Feenberg, E. 1979. 'Distant Synchrony and the One-Way Velocity of Light,' *Found. Phys.* **9**, 329–337.

Frank, P. and Rothe, H. 1911. 'Ueber die Transformation der Raumzeitkoordinaten von ruhenden auf bewegte Systeme,' *Ann. der Phys.* **34**, 825–855.

Herglotz, G. 1911. 'Ueber die Mechanik des deformierbaren Körpers vom Standpunkte der Relativitätstheorie,' *Ann. Phys.* **36**, 493–533.

Ignatowsky, W. 1910. 'Einige allgemeine Bemerkungen zum Relativitätsprinzip,' *Physik, Zeitschr.* **10**, 972–975.

Jammer, M. 1979. 'Some Foundational Problems in the Special Theory of Relativity.' In G. Toraldo di Francia (ed.), *Problems in the Foundations of Physics*, pp. 202–236. Amsterdam/New York/Oxford: North-Holland.

Lee, A. R. and Kalotas, T. M. 1975. 'Lorentz Transformations from the First Postulate,' *Am. J. Phys.* **43**, 434–437.

Lévy-Leblond, J. M. 1976. 'One More Derivation of the Lorentz Transformation,' *Am. J. Phys.* **44**, 271–277.

Minkowski, H. 1909. *Raum und Zeit*. Leipzig/Berlin: B. G. Teubner.

Pauli, W. 1958. *Theory of Relativity*. New York: Pergamon.

Pfarr, J. (ed.). 1981. *Protophysik und Relativitätstheorie*. Mannheim/Vienna/Zürich: B. I. Wissenschaftsverlag.

Poincaré, H. 1887. 'Sur les hypothèses fondamentales de la géometrie,' *Bull. Soc. Math. France* **15**, 203–216.

Poincaré, H. 1952. 'Non-Euclidean Geometries.' In his *Science and Hypothesis*, pp. 35–50. New York: Dover.

Robb, A. A. 1914. *A Theory of Time and Space*. Cambridge: Cambridge University Press.

Robb, A. A. 1921. *The Absolute Relations of Time and Space*. Cambridge: Cambridge University Press.

Robb, A. A. 1936. *The Geometry of Space and Time*. Cambridge: Cambridge University Press.

Schwartz, H. M. 1962. 'Axiomatic Deduction of the General Lorentz Transformations,' *Am. J. Phys.* **30**, 697–707.

Soper, D. E. 1976. *Classical Field Theory*. New York: Wiley-Interscience.

Süssmann, G. 1969. 'Begründung der Lorentz-Gruppe mit Symmetrie-und Relativitäts-Annahmen,' *Zeitschr. Naturf.* **24a**, 495–498.

Synge, J. L. 1960. *Relativity: The General Theory*. Amsterdam: North-Holland.

Synge, J. L. 1964. *Relativity: The Special Theory*. 2nd ed. Amsterdam: North-Holland.

Weyl, H. 1923. *Mathematische Analyse des Raumproblems*. Berlin: Springer.

ROBERTO TORRETTI

CAUSALITY AND SPACETIME STRUCTURE IN RELATIVITY

1

The modern idea of causality, dominant since the seventeenth century, contrasts sharply with the earlier acceptation of the term *cause* (*causa, aitia*) in the Aristotelian tradition. While the Aristotelian physicist sought for the causes of *things*, and admitted as such anything, be it stuffs or structures, drives or goals, that might reasonably account for the object under study, the modern scientist inquires after the causes of *processes* or *states of affairs*, meaning the agents that bring them about. This drastic change in the sense and scope of one of the fundamental categories of scientific thought has often been said to stem from a change of purpose: modern science, it is said, pursues the domination, not the contemplation of nature. With such an end in mind, the scientist would tend of course to ignore nature's own ends – if indeed it has any – and to view each natural stuff as nothing but an aggregate of passive and active dispositions to aid or resist man's aims. However, though the prospective "master and lord of nature"[1] might thus free himself from the Aristotelian preoccupation with the so-called final and material causes, it is unlikely that he could direct all his attention to "sources of change" (*arkhai tes metaboles*) or efficient causes, in utter disregard of form or structure. For efficient causes can only be grasped as such in a context that must be structurally conceived. It is true that the experimental method characteristic of modern scientific inquiry is primarily fit for disclosing the presence and mode of operation of natural agents. But the results of experiment can only make sense if they and the experimental setting that yields them are described and understood in structural terms. Why then has 'cause', in modern parlance, become synonymous with 'efficient cause', while the idea of 'formal causation' and explanation by structure, though alive and strong in actual scientific practice, has all but vanished from 'metascientific' discouse? A brief reflection on the research program sketched in the Preface to Newton's *Principia* can throw some light on this question. Newton expected natural philosophy to find out the forces of nature by studying the phenomena of motion, and then, from those forces, to infer the other phenomena.[2] The program's ultimate goal is therefore

R. S. Cohen and L. Laudan (ed.), Physics, Philosophy and Psychoanalysis, 273–293.

to derive the course of nature from the acting sources of change. Natural philosophy is thus clearly understood as a quest not just for causes in general, in the Aristotelian and medieval sense, but specifically for *efficient* causes, powers, forces. The latter, however, must be gathered from a particular type of changes, namely, motions, i.e. changes of place in time. But motions can only be described in terms of the twofold mathematical structure of Newtonian time and space (which are assumed to be isomorphic, respectively, to the real number field $\mathbf{R}$, and to $\mathbf{R}^3$, endowed with the standard metric) – or, as we would rather put it nowadays, in terms of the single mathematical structure of Newtonian spacetime (homeomorphic with $\mathbf{R}^4 = \mathbf{R} \times \mathbf{R}^3$). Hence, the new natural philosophy does not neglect what Newton thought was the structural background of physical events, but rather takes it for granted as a prerequisite for grasping the causal relations which Newton's program aims at discovering. The close connection between causality and spacetime is a fundamental premise of the physics of central forces that prevailed in the heyday of Newtonianism in the eighteenth and early nineteenth centuries. According to this view – illustrated, for instance, by Coulomb's electrostatics and Ampère's electrodyanmics – the forces of nature depend on the changing positions of their sources in space, so that causality is subordinate to geometry.

In an important manuscript 'On the Gravity and Equilibrium of Fluids', published by the Halls in 1962, Newton expressly characterized space and time as pure structures or relational systems: "It is only through their mutual order and positions that the parts of time and space can be seen to be the very same which they are in fact. They do not have any other individuation principle besides the said order and positions, which therefore cannot change."[3] Such structures have their own peculiar mode of existing (*quendam sibi proprium existendi modum*) and do not fall under the categories of substance and accident into which traditional ontology classified all beings. They are not substances, for they do not underlie any actions, such as a mind's thoughts or a body's motions. Nor are they accidents, for we can clearly conceive them as existing without any substrate, "as when we imagine spaces outside the world or places empty of body".[4]

Philosophers have seldom felt at ease with Newtonian space and time. Being for the most part less daring and open-minded than Newton, they have been unwilling to accept that something can exist which is neither a substance, nor an attribute or relation of substances, and which therefore cannot be described in the established Aristotelian terminology. Even such a bold thinker as Kant could only reconcile himself with Newtonian space and time by regarding them as "mere forms" of human sensibility and degrading everything that is contaminated with them to the status of second-rank, phenomenal being. We need not recall here the numerous attempts to develop a 'relational' theory of physical space and time, whereby these structures or relational systems would no longer be viewed as self-sustained – or as sustained by God alone – as in Newton, but as an abstract reflection of the actual network of changing relations between material things. Such attempts were revitalized in this century by Einstein's profound revision of our ideas of space and time. It was generally thought that the relativistic conception of spacetime lent strong support to the relationist school, at least until Adolf Grünbaum (1957) issued a warning about "the philosophical retention of absolute space" in General Relativity. More precisely, it was believed that spacetime structure, as countenanced by Einstein's theory, was no more than an outline – so to speak, an intellectual shadow – of the system of causal relations between real events. According to Hans Reichenbach, this "causal theory of space and time" is "the philosophical result of the theory of relativity".[5] "The combined space-time order reveals itself as the ordering schema of causal chains (*Ordnungsschema der Kausalreihen*), and thus as the expression of the causal structure of the universe . . . The order of causal chains is ultimately reflected in all space-time determinations."[6]

I intend here to analyze and evaluate these claims in connection with Special and General Relativity. As we shall see, they cannot mean quite the same in both contexts. But before proceeding with this task, let us take a look at Reichenbach's statements in their abstract generality.[7] To a traditional empiricist, who believes that all scientific knowledge is to be attained by induction from experimental results, they must sound inherently plausible. He might argue as follows: Relativity does not take space and time for granted; the Special Theory criticizes the Newtonian view of them; the General Theory binds the spacetime structure of the world to the dynamics of gravity; thus spacetime geometry becomes a matter of experimental inquiry; but experiments are properly directed at disclosing causal relations; hence, the structure of spacetime – insofar as it is not a factitious and freely disposable convention – ought to reflect the causal system of nature ('causal' being of course understood in its restricted modern sense). In the light of the foregoing argument we can readily see why the causal theory of spacetime is favored by inductivist empiricists. We can also surmise why its advocates are wont to season it with a dash or two of chronogeometric conventionalism – namely, in order to cover those features of spacetime geometry which apparently cannot be accounted for by the causal theory. But the argument also implies that, if the causal view of relativistic spacetime is untenable or can only be held in a Pickwickian sense, Relativity is incompatible with inductivism.

2

We turn now to the causal interpretation of flat relativistic spacetime, i.e. the spacetime of Special Relativity. To understand its claims and motivation we must recall the main features of Minkowski's geometrical formulation of the theory. The standard practice of labelling physical events with real number quadruples that vary smoothly as the argument ranges over space and time presupposes that events themselves or their 'locus' or 'arena' constitute a four-dimensional differentiable manifold. Minkowski called this manifold "*die Welt*" – "the world" – but we prefer to call it "spacetime". The (special) Principle of Relativity, stated by Einstein in 1905, implies that the spacetime manifold admits an atlas of global charts with the following property: the laws of physics take the same mathematical form when referred to any of these charts. The charts, which we shall call *Lorentz charts*, combine a set of Cartesian space coordinates bound to an inertial rigid frame, with the Einstein time coordinate defined by means of bouncing light signals, as explained in Section 1 of 'Zur Elektrodynamik bewegter Körper'.[8] The postulated form invariance of physical laws under coordinate transformations from one Lorentz chart to another obtains of course only if the same standard units of time and length are employed in the construction of all Lorentz charts. The general form of such coordinate transformations can be readily ascertained if we assume with Einstein that they preserve in particular the two following laws: the Law of Inertia and the Principle of the Constancy of the Speed of Light (regardless of the state of motion of its source). If x and y are two Lorentz charts with the same origin (i.e. if x and y send the same spacetime point to $0 \in \mathbf{R}^4$), the coordinate transformation $x \cdot y^{-1}$ is demonstrably a Lorentz transformation, i.e. a linear permutation of $\mathbf{R}^4$ which can be characterized as follows: choose, for simplicity's sake, the standard units employed in the definition of Lorentz charts so that the constant speed of light in vacuo equals 1; then, the matrix A of any Lorentz transformation satisfies the equation $A^T HA = H$, where A^T is the transpose of A and H is the diagonal matrix $(-1, 1, 1, 1)$. If x and y do not share the same origin, $x \cdot y^{-1}$ is a so-called Poincaré transformation, i.e. the product of a Lorentz transformation and a translation. Lorentz transformations obviously form a group.[9] Consequently, the Poincaré transformations, being generated by the Lorentz group and the group of translations, form a group too. The Poincaré group acts transitively and effectively on $\mathbf{R}^4$ through the coordinate transformations between Lorentz charts. Since the latter are global, each coordinate transformation $x \cdot y^{-1}$ between two arbitrary Lorentz charts x and y is matched by a unique point transformation $x^{-1} \cdot y$ of spacetime onto itself. Through these point transformations the Poincaré group acts transitively and effectively on the spacetime manifold (as conceived by Special Relativity). This is the key to Minkowski's geometrical reading of Einstein's theory. According to Felix Klein's Erlangen Program, the action of a group on a manifold endows the latter with a geometric structure characterized by the group invariants. The Minkowski geometry induced on the spacetime manifold by the stated action of the Poincaré group is contained, so to speak, in a nutshell in a single two-point invariant that I shall call the *separation* between spacetime points. Let P and Q be two such points and let (P^i) and (Q^i) denote their respective coordinates in any given Lorentz chart; their separation is then given by

$$|P - Q| = \Sigma_i \Sigma_j \eta_{ij} (P^i - Q^i)\ (P^j - Q^j),$$

where i and j range over $\{0, 1, 2, 3\}$ and the η_{ij} are the elements of the matrix H (i.e. $\eta_{00} = -1$; $\eta_{ii} = 1$ if $i \neq 0$; $\eta_{ij} = 0$ if $i \neq j$).[10] We say that P and Q are *separate* if $|P - Q| > 0$, and *connected* if $|P - Q| \leqslant 0$. It will be noted that both relations are symmetric (since $|P - Q| = |Q - P|$). All points separate from a given point P lie outside a two-sheet (hyper-)cone with its vertex at P. For any point Q on this cone $|P - Q| = 0$, which is why it is called the *null-cone* at P (*Null* being the German word for *zero*). Poincaré point-transformations map null-cones onto null-cones. We can choose one of the two sheets of the null-cone at a point P and call it the 'future' null-cone at P. Our choice is then propagated unambiguously by translation to all other null-cones. A point transformation that preserves the choice of a 'future' is said to be orthochronous. Orthochronous Poincaré (resp. Lorentz) transformations form a subgroup of the Poincaré (resp. Lorentz) group.

The causal interpretation of Minkowski geometry stems from two simple facts concerning separateness and connectedness between spacetime points:

(i) The relation 'greater than' between the time coordinates of two spacetime points P and Q is preserved by all orthochronous Poincaré transformations if P and Q are connected, but it is not generally preserved by such transformations if P and Q are separate. If, as was commonly assumed before the advent of General Relativity, admissible time coordinates reflect the 'real', 'objective' time order between events, it is clear that such a time order can subsist only between events at connected points, but not between events at separate points.

(ii) The physical conditions governing Lorentz charts imply that two separate spacetime points P and Q can only be joined by a signal travelling faster than light (relatively to inertial reference frames endowed with Einstein time).

Combining these two facts, we see that any faster-than-light signal, i.e. any transfer of energy and momentum with speed $|dr/dt| > 1$ (where t and r respectively denote the time and the length of the position vector in an arbitrary Lorentz chart), must be undetermined as to its origin and destination, for its emission relative to a Lorentz chart turns out to be its reception relative to another one, and vice versa. If such indeterminateness is repugnant to our metaphysical insights or feelings we are bound to conclude that all transfers of energy and momentum occur at a speed equal to or less than that of light, so that physical influence can only be exerted between connected spacetime points. It follows that *events at separate points are always causally unrelated*, while *causally connected events must take place at connected points*.

The preceding considerations bestow a clear causal meaning on the Minkowski geometry but do not by themselves provide it with a purely causal foundation. There is apparently no reason to expect that the spacetime relations of separateness and connectedness, which have thus been given a causal meaning, will encompass or generate the entire system of chronogeometric predicates. However, this important and seemingly unlikely result can be extracted, after some conceptual reshuffling, from the axiomatic formulation of Minkowski geometry published by A. A. Robb in 1914. Robb's work was overlooked by Reichenbach in the 1920's, but was put to good use in Henryk Mehlberg's 'Essai sur la théorie causale du temps' (1935/37). Mehlberg was able to derive the spacetime geometry of Special Relativity from a set of axioms with a single primitive term, namely, a binary predicate whose intended meaning is precisely that of our "connectedness", but which Mehlberg unabashedly called "*le rapport causal*". We ought not to set much store by such feats of syntactic economy. Usually, the complexity of unproven propositions does more than compensate for the paucity of undefined terms. For instance, Mario Pieri (1899) built Euclidean geometry from the sole concept of 'motion', but the motions he speaks about are so restricted by his twenty axioms that they turn out to be precisely the Euclidean motions. Similarly, Mehlberg's 'causal relation' is not the general, fairly neutral notion that physicists and philosophers associate with this expression, for Mehlberg's axiom system has loaded it with the full connotation of connectedness in Minkowski spacetime. There is, however, a deep and powerful reason why Mehlberg's – and Robb's – achievement is more substantial than Pieri's. If we compare Mehlberg's "*rapport causal*", i.e. connectedness, with the fundamental invariant of Minkowski geometry, namely, the numerical separation between spacetime points, it will look as if the latter's invariance were a much stiffer requirement than the former's. After all, connectedness holds between two spacetime points when their separation takes *any* non-positive value, and there are uncountably many such values. And yet the group of point transformations that preserve connectedness – the so-called "causal group" – is only slightly (one is tempted to say, insignificantly) larger than the Poincaré group – which is a subgroup of it. As E. C. Zeeman (1964) has shown, the causal group of Minkowski spacetime is generated by the Poincaré group and the group of dilatations (i.e. the spacetime permutations that multiply all separations by a positive constant factor). There are of course uncountably many dilatations too – they add a whole dimension to the Poincaré group. But one would tend to agree that this addition is trivial and lacks a genuine geometrical import. Thus, Zeeman's Theorem, together with the aforesaid causal interpretation of chronogeometrical connectedness, apparently suffice to establish that, in a universe governed by Special Relativity, the spacetime structure does no more than reflect the system of causal relations. And yet, as I shall now show, several arguments can be put forward which undermine the strength or curtail the scope of this contention.

In the first place, let me note that, while the causal group of Minkowski spacetime does not differ *geometrically* from the Poincaré group in a significant way, this does not mean, however, that their difference is negligible also from a *physical* point of view. Let F and $T(F)$ be two configurations in Minkowski spacetime, such that $T(F)$ is the image of F by the Poincaré point transformation T. Then, according to the (special) Principle of Relativity, two deterministic experiments whose initial conditions are represented by F and $T(F)$, respectively, must have similar outcomes – more precisely, the outcome of the latter must be the transform by T of the outcome of the former. But if T is a typical element of the causal group, and hence the product of a Poincaré transformation and a dilatation, there need not be such a neat correspondence between both experiments. (A *coordinate* dilatation amounts indeed to a conventional change of units; but a *point* dilatation involves a real rescaling of physical things and processes, which, given the variety of ways in which times and lengths occur in the laws of nature, obviously cannot be trivial.)

In the second place, we ought to bear in mind that in the proposed causal interpretation separateness and connectedness are *modalized* causal predicates: two causally related events *necessarily* take place at connected points; their spacetime locations *cannot* be separate.[11] This does not mean, however, that every connected pair of points is actually joined by a causal chain; indeed, for all we know, they might not even be the site of any events at all. (Minkowski avoided the latter alternative by conceiving spacetime as a densely packed plenum of events;[12] under the field laws of classical physics each event would then influence or be influenced by whatever happened at locations connected with its own.) Thus, chronogeometrical connectedness is not an abstract distillation of concrete causal connections, but rather a prerequisite which the latter must fulfil. Spacetime structure, more than a reflection of actual causal relations, turns out to be a norm by which such relations are constrained.

In the third place, it must be emphasized that the interpretation of spacetime connectedness and separateness as modalized causal concepts does not logically follow from Special Relativity and is not supported by its empirical evidence, but is motivated by philosophical sentiments. Special Relativity does indeed entail that any object moving with the speed of light in an inertial frame travels with the same speed relative to every inertial frame; and that no material object can be accelerated from rest in such a frame to a velocity equal to or greater than that of light. This implies, in turn, that only connected point pairs can be joined by the worldlines (i.e. the spacetime trajectories) of ordinary matter or radiation. But Special Relativity does not preclude the existence of physical objects that always move faster than light in every inertial frame. Hypothetical particles with this property were extensively discussed in the 1960's under the name of 'tachyons', but have not been detected hitherto. According to Special Relativity, a tachyon would be in many ways peculiar, but not inherently absurd. (For instance, if a particle moves with velocity $v > 1$, the Lorentz factor $\gamma = 1/\sqrt{1 - v^2}$ is an imaginary number. Consequently, such a particle can only be assigned a measurable, and hence real-valued, energy $\mu\gamma$ and momentum $\mu v\gamma$ if its 'rest mass' μ is imaginary-valued. This is strange, but consistent with the fact that nobody will every find the particle at rest in his lab.) Since all events in a tachyon's history occur at mutually separate spacetime points they are subject to the sort of temporal ambiguity that I mentioned earlier: the tachyon's emission in an inertial frame may be its absorption relative to another one, and vice versa. Such ambiguity would mean indeed that Einstein time is more deeply at variance with Newtonian time than was initially believed, but is otherwise credible. After sixty or more years of General Relativity we know enough about the physical meaning of coordinates to accept even wilder changes of description as a consequence of coordinate transformations. One may still feel inclined to ask which is the 'real' time order of tachyonic events; but can we be so sure that this category is universally applicable – that it is not solely appropriate to the like of us, slow and heavy creatures, with only a modest power of chronometric resolution? Genuine paradoxes would no doubt ensue if we could use tachyons to change our past. Let A and B be two connected spacetime points, e.g. in my own life history, and let C be another point, separate from A and B. If energy and momentum can be transferred through tachyons there could be a closed causal chain from A to B to C to A. Might I not then attempt to change events at A from B even though B lies to A's future? In my view, such an attempt would not only run against common sense, but also against the very notion of a closed causal chain. If someone is caught in a closed causal chain he is thereby deprived of every chance of altering it. This may be beneath our presumed human dignity but it is certainly compatible with commonly accepted ideas about physical existence.

The fact that tachyons are compatible with Special Relativity, even if they do not happen to exist, implies that the distinction between connected and separate spacetime points does not by itself provide a classification of event pairs into those that are causally connectible and those that are not; but corresponds rather to a distinction between two conceivable forms of causal connectibility, namely, through tachyons and through massive matter and radiation, of which the former, as far as we can tell, is not realized in nature. This conclusion agrees well with the empirical foundations on which Special Relativity was originally based. They did not include a statistical sampling of event pairs, showing which were and which were not causally connected; but consisted of evidence for the validity of the Maxwell–Lorentz equations of electrodynamics in every inertial frame ('coordinatized' by a Lorentz chart), and in particular for the sameness of the speed of light *in vacuo* in all such frames. The invariant speed of light was then demonstrably the least upper bound of the speed of ordinary, massive matter relative to inertial frames, which implied, in turn, the said distinction between two modes of propagation of causal influence.

Finally, I should point out that the causal theory of spacetime cannot account for what is perhaps the most important physical manifestation of the Minkowski geometry in Special Relativity, namely, the behavior of free massive particles and free light pulses, i.e. ordinary matter and radiation exempt from the causal interference of other matter and radiation. The worldline of such an isolated object is a straight, i.e. a path in Minkowski spacetime whose image by a Lorentz chart satisfies a linear equation. Any two points, P and Q, of this path satisfy the inequality $|P - Q| < 0$, if the object is a free massive particle, and the equation $|P - Q| = 0$, if it is a light pulse. Presumably, free tachyons, if they existed, would follow the same law (with $|P - Q| > 0$). Thus, the spacetime does not only regulate the several avenues open (or closed) to causal connection. It also fixes the course to be followed by whatever is causally disconnected from everything else. The causalist cannot find much comfort in Special Relativity.

3

The efficacy of spacetime structure in guiding the force-free motion of matter and radiation was later cited by Einstein as a strong reason for replacing Special by General Relativity:

> The principle of inertia ... seems to compel us to ascribe physically objective properties to the space-time continuum. Just as it was consistent from the Newtonian standpoint to make both the statements, *tempus est absolutum, spatium est absolutum*, so from the standpoint of the special theory of relativity we must say, *continuum spatii et temporis est absolutum*. In this latter statement *absolutum* means not only 'physically real', but also 'independent in its physical properties, having a physical effect, but not itself influenced by physical conditions' .... But ... it is contrary to the mode of thinking in science to conceive of a thing (the space-time continuum) which acts itself, but which cannot be acted upon.[13]

In the General Theory, the spacetime geometry still provides a "guiding field" (*Führungsfeld*) for (now) freely-falling matter and radiation, but is in turn responsive to the actual distribution of matter and energy. Thus, the relation between geometry and the system of causal interactions in nature takes a new twist: the former is not merely the abstract reflection of the latter, or the norm with which the latter must comply, but is engaged in it as a full-fledged participant.

To see how this conception works, let us recall that General Relativity regards spacetime as a 4-dimensional Riemannian manifold whose metric field induces a Minkowski geometry on the tangent space at each point. The metric components g_{ij} (relative to an arbitrary spacetime chart x)[14] are the solutions of the Einstein field equations,

$$R_{ij} - \tfrac{1}{2}Rg_{ij} = -kT_{ij},$$

where the R_{ij} are the components (relative to x) of the Ricci tensor constructed from the metric (R denotes the trace $\Sigma_k R^k_k$ of the 'mixed' Ricci components), while the T_{ij} are the components of a suitable generalization of Laue's (1911) stress-energy tensor, representing the distribution of matter and (non-gravitational) energy. Let g denote the metric and consider a vector field V, defined on an open set in spacetime. $\mathbf{g}(V, V)$ is then a real-valued function (or scalar field) on the domain of V. V is said to be *timelike, null*, or *spacelike* depending on whether $\mathbf{g}(V, V)$ is everywhere negative, zero, or positive. It is usually assumed that spacetime admits a smooth global timelike vector field, whose value at each point can be used for distinguishing 'the past' from 'the future'. (This does not follow from the preceding assumptions.) A smooth curve in spacetime is *timelike* (*null, spacelike*) if it is everywhere tangent to a timelike (resp. null, spacelike) vector field. General Relativity originally postulated – and subsequently proved, under special conditions, from the Einstein field equations – that the possible worldlines of freely-falling non-spinning uncharged massive particles are the timelike geodesics of spacetime. The theory implies moreover, as shown by Laue (1920) and Whittaker (1928), that the possible worldlines of light pulses in vacuo are precisely the null geodesics. The transmission of gravitational pulses also takes place along null geodesics (Einstein, 1918). Due to the local validity of Special Relativity, massive particles must always describe timelike curves, while tachyons, if they existed, would describe spacelike curves.

Let us say that two spacetime points are *connected* if they are joined by a null or a timelike curve, and *separate* if they cannot be so joined. When thus extended to General Relativity, the relations of separateness and connectedness obviously preserve the same causal meaning they had in the Special Theory. But there can be no question of building on them a general causal theory of relativistic spacetime. Except in the special Minkowskian case, the metric of a relativistic spacetime is not determined merely by assorting its points into connected and separate pairs. A theorem proved in the early 1920's by Hermann Weyl tells us that in order to fix the metric up to a constant scale factor, something else is needed besides the classification of spacetime point-pairs according to their separateness or connectedness; namely, to single out all the geodesic paths (i.e. the ranges of geodesics).[15] The fact that the stated information contains essentially all that there is to know about the Riemannian geometry of relativistic spacetime says much about the latter's true physical import. For our two sets of data jointly specify the geodesic paths that consist exclusively of mutually connected points, i.e. the spacetime tracks reserved for radiation and the simplest forms of ordinary freely falling matter. Indeed, the specification of these tracks is by itself sufficient – as well as necessary – for determining the metric (*modulo* a constant scale factor).[16] Thus, the relativistic spacetime structure turns out to be richer in the general case than the abstract scheme of viable causal links articulated in the relations of separateness and connectedness; but it is just as strong as is required for it to play its appointed role as a guiding field of matter and radiation. Therefore – we may conclude – it is this role that is the *raison d'être* of spacetime geometry.

The idea that a geometric structure might contribute to the actual shaping of physical events is so alien to our received modes of thought that, from the advent of General Relativity, there has been a tendency to explain this idea away, to reduce spacetime geometry to the role of a mere mediator between interacting material things, perhaps no more than an aid to calculation. Even Einstein himself may have yielded to this tendency in the earlier days of the theory, when he still bowed to Mach's authority. This interpretation is apparently inbuilt in the Einstein field equations, for the matter distribution represented by the T_{ij} on the right-hand side is obviously meant to be the source of the field represented by the g_{ij}, through which gravitational action is exerted on test particles. However, as de Sitter showed already in 1917, there are sourceless solutions (with the T_{ij} identically zero) of the field equations – apart from the Minkowski solution, which may be dismissed as trivial. It can indeed be argued that such solutions are merely a consequence of the formal mathematical properties of the Einstein equations, but have nothing to do with the real meaning of their terms in actual physical applications. But even if we grant this allegation,[17] the fact remains that an equation is a two-sided affair, and it is therefore more sensible to say that the Einstein equations express a mutual dependence of matter and geometry, rather than a total subordination of the latter to the former. Indeed, there are physically significant solutions of the Einstein equations in which the metric field can hardly be said to mediate between its sources and a test particle subjected to its guidance, for the prevailing conditions do not allow any gravitational action of the sources on that particle, or permit only such action as would not be sufficient to account for the particle's predicted behavior.

Consider, for example, the Schwarzschild (1916) vacuum solution, applied in all classical tests of the theory. The solution does not follow from a detailed knowledge of the matter distribution, but is worked out from global assumptions concerning the metric field – namely, that it can be decomposed into isometric, asymptotically flat, spherically symmetric spacelike slices.[18] One naturally assumes that the field's source surrounds the center of symmetry. But there is no indication as to how the source might have produced the field, which, being static, bears of course no trace of its origin. Indeed, if the source is entirely contained within the so-called Schwarzschild radius, no gravitational action can ever irradiate from it into the field beyond this radius, so that the field and any test particle traversing it ought to be regarded as causally independent from the source. Matter falling into the 'black hole' determined by the Schwarzschild radius becomes dynamically idle with regard to the space outside. (This is apt to be overlooked if one still thinks in terms of Newtonian instantaneous action-at-a-distance. It shows, by the way, that the fact that a Schwarzschild field will persist unchanged if its source collapses isotropically under its own weight does not imply that the source remains in hiding inside the singularity at the center of symmetry, instead of vanishing there right out of existence. In classical General Relativity, anything that sinks into such a singularity is withdrawn forever from the economy of nature.)

It may be objected to the preceding discussion that the Schwarzschild field is no less unreal than the sourceless solutions that we chose to ignore. It is valuable only as a first approximation to the field surrounding a big star, which is, of course, virtually static and spherically symmetric, and goes over to near flatness in the vast interstellar spaces. But all embodiments of the Schwarzschild solution in the real world must have been generated at some time or other by dynamical processes issuing from their respective material sources. To avoid such objections we must turn to the cosmological solutions of the Einstein equations, which embrace all events within their scope. Let us take a look at the earliest and simplest: the family of solutions discovered by Alexander Friedmann (1922; 1924), of which the 1917 Einstein (static) and de Sitter (sourceless) solutions are limiting cases. As is well known, the Friedmann solutions provide a framework within which one can readily account for the systematic shift towards the red of the spectra of distant galaxies first observed by Slipher, and for the isotropic microwave black-body radiation discovered by Penzias and Wilson. To achieve his results, Friedmann postulated that matter can be represented as a pressureless fluid and made the following assumptions concerning, as he put it, "the general, so to speak geometrical character of the world":[19]

(i) Spacetime admits a global time coordinate whose parametric lines are the integral curves of the worldvelocity field of matter (the time coordinate measures proper time along each worldline of matter).

(ii) The hypersurfaces orthogonal to the said integral curves are maximally symmetric 3-dimensional subspaces of spacetime, and hence Riemannian spaces of constant curvature (relative to the positive definite metric induced in each of them by the spacetime metric).

Einstein's field equations rigorously imply that the particles of a pressureless fluid describe timelike geodesics.[20] These geodesics are, of course, the integral curves of the worldvelocity field of matter mentioned under (i), and therefore they constitute a congruence that fills the entire Friedmann universe. Except in some special cases (which include the static Einstein and the de Sitter solution), the geodesics of the said congruence are incomplete,[21] and are focussed in the directions of decreasing and increasing time (expanding and recontracting Friedmann universe), or, at least, in one of these directions (the universe then expands from a point to infinity, or contracts from infinity to a point). Since timelike geodesics can be parametrized by proper time, matter in a typical Friedmann universe has only a finite past and/or a finite future. The feature of the Friedmann universes that is most relevant to our present discussion is the existence of horizons.[22] Let p denote a particle and $S(p)$ the set of spacetime points from which a light signal – or a gravitational pulse – can be sent to p. The *event horizon* of p is the boundary of $S(p)$. Let E denote an event in the history of p, and let $S(E)$ be the set of matter worldlines from which a light signal – or a gravitational pulse – can be sent to p so that it arrives before E. The boundary of $S(E)$ – in the quotient space of spacetime by the congruence of matter worldlines – is the *particle horizon* of E. If $S(p)$ is not empty there are spacetime points from which no electromagnetic or gravitational influence can ever be received by p; if $S(E)$ is not empty there are particles of matter from which no such influence can arrive in p before E. Friedmann universes, and other more realistic world models related to them, have non-empty particle horizons.[23] Consider a particle p in a Friedmann universe that expands from a point. Let $t(E)$ be the time of an event E at p, as given by the standard Friedmann global time coordinate. As $t(E)$ decreases, approaching the greatest lower bound of time, fewer and fewer particles are found within E's particle horizon, until, in the limit, there remains p alone. In the earliest times of cosmic evolution each particle has only had opportunity to suffer gravitational influence from or exert it on its nearest neighbors. Therefore, it makes little sense to explain the global structure of the Friedmann field, which shapes the worldlines of matter, by the gravitational interaction of each material particle with all the rest. Mutual communication between the several parts of matter is gradually attained, within the framework laid down by the metric field.

To ascribe such a dramatic dynamic effect as the expansion of the universe to the inherent evolutionary tendencies of the spacetime geometry will probably be distasteful to many philosophers. However, in the context of relativistic cosmology, I do not see how one could avoid it. When John Earman wrote some time ago that in General Relativity "the deviation of the world lines of a system of test particles is caused by the curvature of space-time",[24] Bas van Fraassen remarked that "the assertion that space-time causes deviations in world lines is a category mistake."[25] Van Fraassen would doubtless be right if the spacetime curvature were expected to deviate world-lines in the way in which a policeman might deviate traffic. But all that Earman meant to say was that, given the mathematical fact of Jacobi's equation, a non-zero spacetime curvature at a point entails a change in the rate at which a timelike geodesic through that point approaches or recedes from nearby timelike geodesics. Now, physicists ordinarily say, in an altogether similar sense, that the electromagnetic field deflects a beam of charged particles, and nobody has ever accused them of making a category mistake when they say so. Indeed, as soon as Riemann succeeded in characterizing a broad family of geometric structures – including the familiar Euclidean and the then still unknown Minkowski geometry – by means of a tensor field, i.e. by the same type of mathematical object usually employed for representing a field of force, the way was open for incorporating physical geometry into the dynamical transactions of nature. However, even if the tensor field representation of metric relations in spacetime does furnish us the means of understanding and accurately formulating their interdependence with matter and the forces of nature, we ought not to jump to the conclusion that General Relativity has overcome the duality of cause and structure, power and form, that we encountered in Newtonian dynamics. The very concept of a tensor field presupposes a manifold structure on which the field is defined. The dynamical interplay of fields governed by the Einstein equations requires a smooth, topologically shaped arena in which to evolve.

Several attempts have been made to construct the topology of relativistic spacetime from that of causal chains, which is assumed to agree at least locally with the topology of the linear continuum. (The latter can, of course, be defined in terms of the relation of causal precedence, if we introduce the necessary denseness and continuity postulates which many believe to be empirically vindicated by our direct awareness of our own lives.) Very interesting in this connection is the *path topology* of spacetime, defined by Hawking, King and McCarthy (1976).

In order to define it I must introduce a few terms. Let M be an arbitrary relativistic spacetime. A smooth curve in M is *future-directed* if the tangent vector at each point of the curve lies within the future lobe of the local null cone. Consider a point $P \in M$ and a subset $U \subset M$. The *future* $I^+(P, U)$ of P *relative* to U is the set of all points in U that can be reached from P by a future-directed smooth curve of finite extent whose range is wholly contained in U. We abbreviate $I^+(P, M)$ to $I^+(P)$ and call it simply the future of P. Past-directed smooth curves, the past $I^-(P)$ of a point $P \in M$ and its past $I^-(P, U)$ relative to a subset $U \subset M$ are similarly defined. A neighborhood U of a point $P \in M$ is said to be *normal* if the exponential mapping Exp_p maps a neighborhood of the zero vector in the tangent space at P diffeomorphically onto U. U is *convex* if any two points $Q_1, Q_2 \in U$ are joined by a geodesic, unique up to reparametrization, whose range lies wholly within U. A continuous curve $\gamma: J \longrightarrow M$ is future-directed at $t_0 \in J$ if there is a connected neighborhood H of t_0 in J and an open normal convex neighborhood U of γt_0 in M such that, for any $t \in H$, $\gamma t \in I^-(\gamma t_0, U)$ if $t < t_0$ and $t \in I^+(\gamma t_0, U)$ if $t_0 < t$. The continuous curve γ is future-directed if it is future-directed at every $t \in J$. Past-directed continuous curves are defined analogously. A continuous curve that is either past or future directed is said to be *timelike*. It is not hard to see that smooth timelike curves as defined earlier are also timelike in this extended sense. Consider now the collection S of all timelike paths (i.e. all ranges of timelike continuous curves) in M.[26] Each element of S is a one-dimensional subspace of M. This implies that if U is a 'coordinate patch' (the domain of a chart) of M and $P \in S$, $U \cap P$ is open in P. The *path topology* of M is the strongest topology that induces in each timelike path the same subspace topology as the standard manifold topology of M. In contrast with the latter (which is generated by the coordinate patches) the path topology is uniquely determined – through the requirement of maximal strength – by the one-dimensional topologies of the paths themselves, i.e. by neighborhood relations on each possible worldline of matter. David Malament (1977) proved that a permutation f of M that preserves the path topology (i.e. such that both f and its inverse f^{-1} map open sets onto open sets) will also preserve the standard manifold topology of M, its differentiable structure and the distinction between connected and separate point-pairs.[27]

Two more theorems, also due to Malament (1977), further illustrate the links that can be established between topology and causality in a relativistic spacetime. If f is any permutation of the spacetime M such that both it and its inverse f^{-1} map timelike paths onto timelike paths, f preserves the standard manifold topology of M. Moreover, if M is both past- and future-distinguishing, i.e. if no two distinct points of M share the same future or the same past, then any permutation f of M which maps the future and the past of each $P \in M$ respectively onto the future and the past of its image $f(P)$ will also preserve the standard manifold topology of M. The latter is therefore determined, in a vast family of cases, by the scheme of possible worldlines of matter.[28]

Malament's results testify to close bonds between the fundamental neighborhood relations of spacetime points and the possible causal connections between events at them. But one can hardly claim that the former have thereby been reduced to the latter. It is just as well that they haven't. For when the turn comes to explain what is meant by causality, philosophers of the most diverse persuasion mention spatial and temporal contiguity as its surest and least problematic ingredient. Thus, David Hume's first "rule by which to judge causes and effects" is that "the cause and the effect must be contiguous in space and time" (*Treatise*, I. iii. 15, p. 173). And C. J. Ducasse's celebrated anti-Humean analysis of the causal relation is wholly stated in spatial and temporal terms.[29] Obviously both Hume and Ducasse had a feeling that our ideas of time and space were clearer and less in need of explication than our idea of cause. If, as I sense, they had the right feeling, the causal theory of spacetime would be a notorious example of *elucidatio per obscurius.*[30]

Universidad de Puerto Rico

NOTES

[1] Descartes, *AT*, VI, 62 (*Discours de la méthode*).
[2] "Omnis enim philosophiae difficultas in eo versari videtur, ut a phaenomenis motuum investigemus vires naturae, deinde ab his viribus demonstremus phaenomena reliqua" (Newton, *Principia*, p. 16).
[3] Newton (1962), p. 103.
[4] *Ibid.*, p. 99.
[5] Reichenbach (1928), p. 307; (1958), p. 269.
[6] Reichenbach (1928), pp. 307 f.; (1958), p. 268.
[7] This implies, in particular, that I am paying no heed to the special context in which Reichenbach made these statements, in connection with his astounding claim that "in the most general gravitational fields" countenanced by General Relativity "all of the

metrical properties of the space-time continuum are destroyed" (because all rigid rods and clocks are shattered where the spacetime curvature is large enough!).
[8] Einstein (1905), pp. 893f.
[9] Let A and B be the matrices of two Lorentz transformations. Then $(AB)^T H(AB) = B^T(A^T HA)B = B^T HB = H$. Recalling that the transpose of the inverse matrix A^{-1} is the inverse of the transpose A^T, we verify that $(A^{-1})^T HA^{-1} = (A^T)^{-1} A^T HAA^{-1} = H$, so that the inverse of A is also the matrix of a Lorentz transformation. It is obvious, moreover, that the identity matrix I meets the condition $I^T HI = H$.
[10] It can be readily seen that $|P - Q|$ does not depend on the choice of the particular Lorentz chart in which the coordinates of P and Q are (P^i) and (Q^i). For let $(\bar{P}^i)$ and $(\bar{Q}^i)$ be their respective coordinates in another Lorentz chart related to the former by the transformation $x^i \mapsto \Sigma_j A_{ij} x^j + k^i$, where the k^i are constants and (A_{ij}) is the matrix of a Lorentz transformation. We have then that

$$\begin{aligned}
&\Sigma_{ij}\eta_{ij}(\bar{P}^i - \bar{Q}^i)(\bar{P}^j - \bar{Q}^j)\\
&= \Sigma_{ijkh}\eta_{ij}A_{ik}A_{jh}(P^k - Q^k)(P^h - Q^h)\\
&= \Sigma_{ijkh}A^T_{ki}\eta_{ij}A_{jh}(P^k - Q^k)(P^h - Q^h)\\
&= \Sigma_{kh}\eta_{kh}(P^k - Q^k)(P^h - Q^h) = |P - Q|.
\end{aligned}$$

[11] Since, according to some authors, causal relations themselves are modal too, one might be tempted to speak here of 'second modalization'.
[12] Minkowski (1909), p. 104: "Um nirgends eine gähnende Leere zu lassen, wollen wir uns vorstellen, dass aller Orten und zu jeder Zeit etwas Wahrnehmbares vorhanden ist."
[13] Einstein (1956), pp. 55f.
[14] Let x^i be the ith coordinate function of x; denote by $\partial/\partial x^i$ the vector field whose integral curves are the parametric lines of x^i. Then, if g stands for the metric tensor, $g_{ij} = \mathrm{g}(\partial/\partial x^i, \partial/\partial x^j)$. (A (0, 2)-tensor field such as g assigns to each pair of vector fields defined on an open subset U of spacetime a smooth real-valued function on U.) The tensor components R_{ij} and T_{ij} that occur in the Einstein field equations are defined analogously.
[15] Weyl (1921).
[16] Ehlers, Pirani and Schild (1972).
[17] Such philosophically motivated restrictions on the possible physical meaning of mathematical objects are often unquestionable (as when an electrical engineer discards the imaginary part of the complex numbers that turn up in the end result of his calculations); but they can also severely impair the heuristic power of a theory. (Just think of the consequences of dismissing the negative energy solutions of Dirac's equation as 'physically meaningless'.)
[18] We now know that the Schwarzschild metric is essentially the one and only spherically symmetric solution to the Einstein vacuum field equations, so that Schwarzschild's further assumptions are redundant. See Birkhoff (1923), p. 255; Petrov (1969), p. 360; Hawking and Ellis (1973), App. B; cf. Adolf Grünbaum's apt reminder (1973, p. 840)
[19] Friedmann (1922), p. 378.
[20] Einstein and Grossmann (1913), p. 10. For a proof, see Rindler (1977), p. 182.
[21] Like any (parametrized) curve, a geodesic is a continuous mapping of an interval of the real number field into spacetime. A timelike geodesic γ is said to be incomplete

if its domain is a proper subset of the real number field and γ is not the restriction to that domain of a timelike geodesic defined on the entire real number field.

22 Rindler (1956).

23 MacCallum (1971).

24 Earman (1972), p. 84.

25 van Fraassen (1972), p. 93.

26 Hawking *et al.* (1976) actually defined the path topology for so-called strongly causal spacetimes (i.e. spacetimes in which every point has a neighborhood that no timelike or null curve penetrates more than once). I should note also that they use 'curve' for what I call 'path', and vice versa.

27 Strictly speaking, what Malament showed was that, if $f: M \longrightarrow M$ is a homeomorphism of the path topology, then f is smooth and the spacetime metric g is conformal with its pull-back f^*g (in other words: $f^*\mathrm{g} = \omega^2\mathrm{g}$, where ω is a smooth real-valued function on M; this condition evidently implies that both g and f^*g determine the same null-cone in the tangent space at each point of M).

28 More precisely: $P \ll Q$ if there is a timelike future directed curve γ: $[a, b] \longrightarrow M$, such that $\gamma(a) = P$ and $\gamma(b) = Q$.

29 Ducasse (1926) defined causality as follows:

Considering two changes, C and K (which may be either of the same or of different objects), the change C is said to have been sufficient to, i.e. to have caused, the change K, if: 1. The change C occurred during a time and through a space terminating at the instant I at the surface S. 2. The change K occurred during a time and through a space beginning at the instant I at the surface S. 3. No change other than C occurred during the time and through the space of C, and no change other than K during the time and through the space of K.

30 I am very grateful to David Malament and John Stachel for reading an earlier version of this paper and suggesting several important improvements.

REFERENCES

Birkhoff, G. D. 1923. *Relativity and Modern Physics*. Cambridge, Mass.: Harvard University Press.

Descartes, R. (*AT*), *Oeuvres*. Edited by C. Adam and P. Tannéry. 12 vols. Paris: Cerf, 1897–1910.

Ducasse, C. J. 1926. 'On the Nature and the Observability of the Causal Relation,' *Journal of Philosophy* **23**: reprinted in Ducasse, *Truth, Knowledge and Causation*, pp. 1–14. London: Routledge, 1968.

Earman, J. 1972. 'Notes on the Causal Theory of Time,' *Synthese* **24**, 74–86.

Ehlers, J., F. A. E. Pirani and A. Schild. 1972. 'The Geometry of Free Fall and Light Propagation.' In O'Raifeartaigh (ed.), *General Relativity. Papers in Honour of J. L. Synge*. Oxford: Clarendon Press.

Einstein, A. 1905. 'Zur Elektrodynamik bewegter Körper,' *Annalen der Physik* (4) **17**, 891–921.

Einstein, A. 1917, 'Kosmologische Betrachtungen zur allgemeinen Relativitätstheorie,' *Sitzungsberichte d. Preuss.Akad.der Wiss.*, pp. 142–152.

Einstein, A. 1918. 'Ueber Gravitationswellen,' *Sitzungsber. d. Preuss.Akad.der Wiss.*, pp. 154–167.

Einstein, A. 1956. *The Meaning of Relativity*. 5th ed. Princeton: Princeton University Press.

Einstein, A. and M. Grossmann. 1913. *Entwurf einer verallgemeinerten Relativitätstheorie und einer Theorie der Gravitation*. Leipzig: Teubner.

Fraassen, B. C. van. 1972. 'Earman on the Causal Theory of Time,' *Synthese* **24**, 87–95.

Friedmann, A. 1922. 'Ueber die Krümmung des Raumes,' *Zeitschrift für Physik* **10**, 377–386.

Friedmann, A. 1924. 'Ueber die Möglichkeit einer Welt mit konstanter negativer Krümmung des Raumes,' *Zeitschrift für Physik* **21**, 326–332.

Grünbaum, A. 1957. 'The Philosophical Retention of Absolute Space in Einstein's General Theory of Relativity,' *Philosophical Review* **66**, 525–534.

Grünbaum, A. 1973. *Philosophical Problems of Space and Time*. 2nd enlarged edition. *Boston Studies in the Philosophy of Science*, vol. 12. Dordrecht: Reidel.

Hawking, S. W. and G. F. R. Ellis. 1973. *The Large Scale Structure of Space-Time*. Cambridge: The University Press.

Hawking, S. W., A. King, and P. McCarthy. 1976. 'A New Topology for Curved Space-Time which Incorporates the Causal, Differential and Conformal Structures,' *Journal of Mathematical Physics* **17**, 174–181.

Hume, D. 1888. *A Treatise of Human Nature*. Ed. by L. A. Selby-Bigge. Oxford: Clarendon Press.

Laue, M. Von. 1911. 'Zur Dynamik der Relativitätstheorie,' *Annalen der Physik* (4) **35**, 524–542.

Laue, M. von. 1920. 'Theoretisches über neuere optische Beobachtungen zur Relativitätstheorie,' *Physikalische Zeitschrift* **21**, 659–662.

MacCallum, M. A. H. 1971. 'Mixmaster Universe Problem,' *Nature, Phys. Sci.* **230**, 112–113.

Malament, D. 1977. 'The Class of Continuous Timelike Curves Determines the Topology of Spacetime,' *Journal of Mathematical Physics* **18**, 1399–1404.

Mehlberg, H. 1935/1937. 'Essai sur la théorie causale du temps,' *Studia philosophica* (Lemberg), **1**, 119–260; **2**, 111–231. See also Mehlberg (1980).

Mehlberg, H. 1980. *Essay on the Causal Theory of Time*. (Vol. I of *Time, Causality and the Quantum Theory*. Ed. by R. S. Cohen 2 vols.) *Boston Studies in the Philosophy of Science*, vol. 19. Dordrecht, Boston: D. Reidel. Translation of Mehlberg (1935/1937).

Minkowski, H. 1909. 'Raum und Zeit,' *Physikalische Zeitschrift* **10**, 104–111.

Newton, I. 1962. *Unpublished Scientific Papers of Isaac Newton*. Edited by A. R. and M. B. Hall. Cambridge: The University Press.

Newton, I. 1972. *Philosophiae naturalis principia mathematica*. Ed. by A. Koyré and I. B. Cohen. 2 vols. Cambridge, Mass.: Harvard University Press.

Petrov, A. Z. 1969. *Einstein Spaces*. Oxford: Pergamon.

Pieri, M. 1899. 'Della geometria elementare come sistema ipotetico-deduttivo; monografia del punto e del moto,' *Memorie della R.Acad. delle Scienze di Torino*, Cl.Sc.Fis.Mat.e Nat. (2) **49**, 173–222.

Reichenbach, H. 1928. *Philosophie der Raum-Zeit-Lehre*. Berlin: W.de Gruyter. See also Reichenbach (1958).

Reichenbach, H. 1958. *The Philosophy of Space and Time*. Transl. by M. Reichenbach and J. Freund. New York: Dover. Translation of Reichenbach (1928).

Rindler, W. 1956. 'Visual Horizons in World Models,' *Monthly Notices of the Royal Astronomical Society* **116**, 662–677.

Rindler, W. 1977. *Essential Relativity, Special, General and Cosmological*. 2nd ed. Berlin: Springer.

Robb, A. A. 1914. *A Theory of Time and Space*. Cambridge: The University Press.

Schwarzschild, K. 1916. 'Ueber das Gravitationsfeld eines Massenpunktes nach der Einsteinschen Theorie,' *Sitzungsber. d. Preuss. Akad. der Wiss.*, pp. 189–196.

Sitter, W. de. 1917. 'On Einstein's Theory of Gravitation, and Its Astronomical Consequences, III,' *Monthly Notices of the Royal Astronomical Society* 78, 3–28.

Weyl, H. 1921. 'Zur Infinitesimalgeometrie. Einordnung der projektiven und konformen Auffassung,' *Nachrichten der Gesellschaft der Wissenschaften zu Göttingen*, Math.-phys.Kl., pp. 99–112.

Whittaker, E. T. 1928. 'Note on the Law That Light-Rays are the Null-Geodesics of a Gravitational Field,' *Cambridge Philosophical Society Proceedings* **24**, 32–34.

Zeeman, E. C. 1964. 'Causality Implies the Lorentz Group,' *Journal of Mathematical Physics* **5**, 490–493.

The interested reader can also derive much instruction, as I have, from the following writings of Lawrence Sklar:

Sklar, L. 1974. *Space, Time and Spacetime*. Berkeley: University of California Press. Chapter IV.

Sklar, L. 1977. 'What Might Be Right about the Causal Theory of Time?,' *Synthese* **35**, 155–171.

Sklar, L. (Forthcoming.) 'Prospects for a Causal Theory of Space-Time,' Royal Institute of Philosophy Conference on Space, Time and Causality, held at the University of Keele, Summer 1981.

BAS C. VAN FRAASSEN

CALIBRATION: A FREQUENCY JUSTIFICATION FOR PERSONAL PROBABILITY *

If a physical theory states that the probability of some event, under certain conditions, is thus or so, we naturally take that to be a statement of objective fact, descriptive of the way the world is. And we expect that fact, if it is indeed the case, to be reflected in frequencies of occurrence among the described events. What is called the frequency interpretation of probability intends something more: namely, that such a probabilistic theory is really *only* about actual frequencies of occurrence.[1]

But the language of probability has uncontestably another use as well: it serves to formulate and express our opinion and the extent of our avowed ignorance concerning matters of fact. This use invites the epithets 'subjective' or 'personal' because it is keyed to the state of the user. When I say that it seems likely to me that it will rain today, or that rain seems as likely as (more likely than, twice as likely as) not, I express my very own opinion and judgment, I express some aspect of my own expectations for today. Any satisfactory view about probability must explicate this second use as well.

Here adherents to the frequency interpretation have fared very badly. And adherents of subjectivist or Bayesian views have done very well, on two counts. First, they have made an effort to show that within their own framework they can recapitulate the explanatory and explicatory successes of their objectivist rivals. Secondly, they have demonstrated that observance of the probability calculus in the expression of personal opinion or degree of belief is required, on their interpretation, by very minimal criteria of rationality ('coherence'). The paradigm example of the first is de Finetti's theorem in his 'Foresight: Its Logical Laws, Its Subjective Sources'; of the second, the well-known Dutch Book Theorem.[2] And finally, there appears to be a consensus in the literature that frequentists have never succeeded in meeting the major criticisms of their views as applied to this second use of probability language.

In this paper I shall attempt to redress the balance somewhat. I shall outline how the use of probability language to express personal opinion about a single event can be understood in a way that avoids the major problems with which frequentists have struggled. And I shall attempt to demonstrate

R. S. Cohen and L. Laudan (eds.), Physics, Philosophy and Psychoanalysis, 295–319.

that observance of the probability calculus in such expression of opinion is equivalent to satisfaction of a basic frequentist criterion of rationality (*frequency coherence*). Based on the idea of *scoring*, also a subject investigated by de Finetti and other Bayesians, this will be a frequency analogue of the Dutch Book Theorem.

1. THE PHENOMENON: PERSONAL PROBABILITY JUDGMENTS

As a form of speech, expressions of personal opinion are often easy to recognize. "It seems likely to me that it will rain" can not be equated with any precise probability evaluation, but "likely" is here surely synonymous with "very probable." And "He is twice as likely to win the race as is his brother" is a very exact statement of odds, which we equate in turn with a probability ratio. When the weather forecast on the radio says, finally, that the chance of precipitation equals 0.6, that sounds at once very precise and very objective, but it is an announcement of the metereologist's professional opinion, reached after conscientious consultation of the data. To say that the opinion is professional, does not even imply that all the professional colleagues he respects would have to reach the same estimate when given the same data, though it does imply a large measure of agreement among them.

How shall we understand this activity? We can perceive it in two ways, not perhaps mutually exclusive: as *expressing* attitudes or as *asserting* autobiographical facts. To bring out the difference, think of the somewhat parallel case of promising. Yesterday I said, "I promise to give you a horse." But I did not give you anything, and today you accuse me of the heinous immorality of breaking a promise. No, I reply, I am not guilty of that at all, but only of the much lesser offense of lying. All that happened was that yesterday I stated falsely that I was promising to give you a horse.

It is easy to see what is wrong with this story. In saying, "I promise . . .", I must (normally?) be taken to be doing something more than implying or stating an autobiographical fact. In just the same way, if I say, "It seems likely to me that . . . ", I may be implying or stating a fact about my own attitude or judgment; but I am first and foremost doing something else: expressing that attitude or judgment.

Attitudes, once expressed, are evaluated in two ways. The first question is one which it should, in principle, be possible to answer right away: is this attitude *reasonable*? The second concerns the future: is this attitude *vindicated*? Again an imperfect parallel may help: a practical decision to devote the evening to attending a certain play. Was this decision a reasonable one? That depends on the reviews you have read, the amount of money and time you have, the time and the alternatives contemplated at that time. Was it vindicated? That depends on factors not settled for you at the time of decision: how good the performance turns out to be, how much pleasure or insight you gained from it, and also on what else happened that evening that you missed or could have prevented or influenced if you had not gone to the play.

A morass in which frequentists have often sunk is their search for objective criteria of how reasonable a judgment is, in the light of available information. The most ambitious and most successful attempt along these lines is that of Kyburg. I will not say that he *is* stuck in a morass: his program of defining the right reference class and a recipe for determining the correct epistemic probabilities on the basis of available statistical information, may be successful. But we cannot yet say that it is. The Bayesian approach appears to eliminate this enterprise, and its problems, entirely. And still the subjectivist Bayesian is not silent on the question of reasonableness. How is that possible?

Looking again at the parallel of practical or moral decisions, we see one minimal criterion of reasonableness that connects it with vindication. A decision is unreasonable if vindication is *a priori* precluded. The Bayesian equates a probabilistic expression of opinion with an announcement of betting odds the person is willing to accept. Vindication consists clearly in gaining, or at least not losing, as a consequence of such bets. The Dutch Book Theorem says that such vindication is *a priori* precluded if and only if the probability calculus is violated. Thus the possibility of vindication is taken as a requirement of reasonableness.

This general insight and strategy are open to all contestants. Let the frequentist equate probabilistic expression of opinion with something else; and let him investigate the conditions under which such vindication is not *a priori* excluded.[3]

2. THE THEORY AND ITS PROBLEMS

The phenomenon to be addressed is the constant stream of judgment expressed in (vague) probabilistic language. A theory will propose models of what is going on, in which phenomena of this sort can fit. Because we, as philosophers, are interested in epistemology rather than psychology, we look to such theories only to find out two things: understanding of what this activity *could be*, and of the conditions under which this activity is *rational*. An answer to the first will suggest one to the second, for rationality consists largely in the suitability of chosen means to intended ends. What the activity *is* should determine its criteria of success. We will evaluate its rationality by seeing whether its aim is pursued in an optimal fashion, first with respect to its own criteria of success and secondly in view of other aims of the larger projects of which it is part.

John Venn, in his *Logic of Chance*, was perhaps the first to formulate explicitly the frequency interpretation as an answer to the first question. The activity of judgment, expressed in such utterances as "It seems to me as likely as not that it will rain today," "It seems 95% probable to me that it will snow today" is assigned two main underlying factors. The first is a *selection of a reference class* – a classification of the subject – and the second an *estimate of relative frequency* in that class – in these examples, frequency of rain or of snow. This sketches the very simplest model of the activity which is suggested by the idea that probability talk is centrally and essentially concerned with frequencies. In the Appendix, I shall discuss this further.

The basic objections to this theory were already – and perhaps best – formulated by John Maynard Keynes in his *Treatise on Probability* (especially Ch. VIII, Sections 7–13). They take the form of three questions. The first is: how is the reference class selected? The second: how or where does the person obtain his estimates of frequencies? And the most important: why should personal probabilities, arrived at in this fashion, either obey, or be rationally required to obey, the probability calculus?

We may take the first two questions to be a request for elaboration of the theory. Can we construct models in which all probabilistic judgments, including those concerning statistical frequencies, appear as the outcome of such a process? And in such models, what is the exact mechanism of reference class selection, et cetera? It is noteworthy that the most extensive and sophisticated attempt to construct such models, namely that of Henry Kyburg's *Logical Foundations of Statistical Inference*, is also an attempt to do so in the most constrained manner possible. In contrast, John Venn explicitly allowed for an element of subjective choice and volition in the selection of reference classes, differing from occasion to occasion.[4]

These first two questions, however, do not strike me as going to the heart of the matter at all. Why should we ask Reichenbach, for instance, for a recipe for arriving at a judgment (in the light of our own background beliefs and information) about which horse will win this specific race, when we certainly have no right to ask Ramsey or de Finetti how to arrive at a specific bet on this particular occasion? A presupposition that Kyburg gives the appearance of accepting, and Venn apparently rejected,[4] is that the judgment will have been arrived at in a rational manner, exactly if the input (background beliefs and information) determines via the dictates of rational deliberation, a uniquely right answer – the rationally compelled one. The alternative view, which I urge as the correct one, is that requirements of rationality can only go so far, and that what is rational is what stays within their bounds; thus allowing for an element of subjectivity and personal volition within rational choice. Rationality is only bridled irrationality.

The heart of the matter appears in Keynes' third question. Whether or not our judgments are reasonable should be determinable at the time we make them. But such underlying factors as statistical estimates and reference class selection are *hidden variables*, they do not belong to the surface phenomenon of judgment, at least in general, and are not (entirely) accessible to introspection either. (Consider the famous case of the chicken sexers, or any other sort of expertise in professional judgment where we speak of talent as well as of book learning.) The one paradigm rule of thumb for a preliminary evaluation of the reasonableness of judgment, which can indeed be applied at the time and without acceptance of any interpretation, is to see whether the axioms of probability are not violated. Let the frequentist either justify this rule or show why it should be rejected or restricted.

The frequentist cannot answer this challenge by pointing out that finite proportions in classes (or suitably chosen relative frequencies in sequences) obey those axioms. For the choice of reference class plays a crucial role as well. Suppose I am asked two questions about today: will there be any precipitation? Will there be any snow? And imagine that for the first question I consult the almanac, which says that here in Toronto approximately one in five days is marked by precipitation. The second question I answer after I have looked outside and taken account of the fact that today is a cold, overcast December day. Then I announce my probabilities: 1/5 for the first, 1/3 for the second. I chose different reference classes; now I have given a lower probability to the first proposition although it is entailed by the second, a violation of probability theory. Does it not seem that the frequentist must show why it is necessary to avoid this, and that he can only do it by formulating and defending intricate rules for the choice of reference classes?

But as I explained in the preceding section, there is a general strategy for answering this third question of Keynes. We can give a frequentist explication of the criteria of success for such judgments – *vindication*; then set down as a minimal requirement of rationality that the judgments not be such as to preclude *a priori* the very possibility of vindication; and finally, demonstrate that this requirement entails non-violation of the probability calculus.

Vindication I shall explicate in terms of *calibration*, a measure of how reliable one's judgments have been as indicators of actual frequencies. Possibility of vindication I shall then explicate as *potential calibration*. And the required demonstration will take the form of two adequacy theorems and the sketch of a third.

This is an alternative to the well-known strategy of laying down rules for the choice of reference classes. Quite apart from the morass of complexities which has beset *that* strategy, it leaves an obvious open question: why is it rational to follow *those* rules in selecting a reference class? Those rules also need justification, so they may still force us back to what I here propose: an analysis of the possibility of vindication, for the judgments which result. Hence I advocate the outlined alternative strategy.

3. VINDICATION: SCORING AND THE CALIBRATION LEMMA

After a metereologist announces, in the morning, a chance of 0.8 for rain, during the day it then rains or does not rain. In the first case, the meteorologist may look proud, but in the second he need not look ashamed – obviously what happens on that day does not make his forecast correct or incorrect. But he has announced these probabilities for a year – how good was his forecasting performance? The first problem to solve is that of devising measures to 'score' his performance. The second is to show, of some such measure, that it makes sense, that is, that it measures success with respect to the aim of his enterprise.

The first problem was given a solution, in 1950, which became generally accepted. Weather forecasters are evaluated by the *Brier Score.*[5] When given feedback on their cumulative Brier score, they also improve that score – which is lovely if it really measures their success, and regrettable if it does not. As analyzed afterward, the score actually combines two criteria. The first is *informativeness* or extremeness: the score tends to improve if the announced probabilities are closer to *one* and *zero*. The second is called *calibration*; its basic idea is that the forecasts fit the series of actual events perfectly, exactly if it rained on 60% of the days on which he said the probability of rain was 0.6, and so on for the other stated numerical values.

It is of course very typical to see this combination of two criteria, of just that sort. Of a traditional, non-statistical scientific theory, philosophers of every persuasion demand (in their various terminologies) both informativeness and truth, in certain respects. The two aims are in desperate tension, for the more informative we make our theories (the more audacious we are, the bolder our conjectures) the less sure we can be that they are true, the greater the chance they will be false (in the intended respect).[6] Calibration plays here the conceptual role that truth, or empirical adequacy, plays in other contexts of discussion.

Now it will be clear that calibration is meant to be a measure of how reliable the forecasts are, cumulatively, as indicators of actual frequencies of occurrence. Just what the frequentist would pose intuitively as the aim of the forecaster's activity.[7] But is the basic idea that motivates the proposed measure a good one? Can we say, from our chosen point of view, that this clever idea of perfect calibration marks correct execution of the judgmental activity? If that chosen point of view is the frequency interpretation, certainly.

This we can establish by means of a simple demonstration. Suppose the forecaster acts exactly as frequentists describe. Each morning he classifies the day x as belonging to a reference class $\beta(x, \mathit{rain})$. The classifications open to him here form a logical partition, that is, he has one and only one such reference class for any day with which he is presented. Suppose also that for each class Y of days that he ever uses as a reference class, he has an estimate $\alpha(\mathit{rain} \mid Y)$ of the relative frequency of rain in Y. So on the morning of day x he announces the number $\alpha(\mathit{rain} \mid \beta(x, \mathit{rain}))$ as his probability for rain on that day.

Now he could fare badly, even if he correctly classifies each day (x belongs to $\beta(x, \mathit{rain})$ in each case), and even if he has perfectly correct estimates of the frequency of rain in each of these classes. For the total set D of days with which he is presented may be an unrepresentative sample of days in general. That would be just plain unlucky for him. But if we assume that the world does not make him unlucky in that way, then the following little lemma shows that the correctness of his proportion estimates and reference class selection guarantees perfect calibration.

To state the lemma, let the correct proportions (equalling by assumption the estimated ones) be represented by an additive set function m defined on D. As usual define the conditionalization $m(A \mid B)$ as $m(A \cap B) \div m(B)$, where the denominator is not *zero*.

(3.1) CALIBRATION LEMMA. *If X is a finite partition of D, in the domain of m, and for each x in D the function P_x is defined by*

$$P_x(A) = m(A \mid B^x)$$

where B^x is the member of X to which x belongs, then

$$(3.2) \qquad m(A \mid \{x : P_x(A) = r\}) = r$$

wherever defined.

Note that if m is the proportion estimate, and X the set of reference classes for the question, then $P_X(A)$ is the forecast probability for A.

To prove this lemma, denote as $A(r)$ the set $\{x : P_x(A) = r\}$ of cases in which the announced probability of A was the number r. That set is also exactly the union of the sets B in X – the reference classes – for which the estimated proportion $m(A \mid B)$ equals r:

$$(3.3) \qquad \begin{aligned} A(r) &= \cup\, \{B \in X : m(A \mid B) = r\} \\ &= B_1 \cup \ldots \cup B_k \text{ (say)}, \end{aligned}$$

a union of disjoint members of the partition. Hence:

$$(3.4) \qquad \begin{aligned} m(A \cap A(r)) &= m(\cup \{A \cap B_i : i = 1, \ldots, k\}) \\ &= \sum_{i=1}^{k} m(B_i) m(A \mid B_i) \\ &= r \sum_{i=1}^{k} m(B_i) \\ &= rm(B_1 \cup \ldots \cup B_k) \\ &= rm(A(r)). \end{aligned}$$

Hence also $m(A \mid A(r)) = m(A \cap A(r)) \div m(A(r)) = r$, provided of course the denominator is not zero.

The argument extends at once to countable partitions if m is sigma-additive, but that seems a bit irrelevant for personal probabilities or rain forecasts. We may conclude in any case that the basic idea of perfect calibration is exactly the idea of complete correctness to be associated with the frequency interpretation: a selection of reference classes and estimate of proportions that happen to be exactly right for the presented sample.

4. REASONABLENESS: POTENTIAL APPROACH TO CALIBRATION

Let us now proceed slightly more abstractly: I am given a field or Boolean algebra F of attributes and a domain D of individuals, and asked to express a judgment concerning whether x has A, for various attributes A and various entities x in that domain. Let Q be a finite set of such propositions [x *has* A]

and let function P, defined on the whole family of these propositions, be used to represent my judgments. Call P a *scheme* for D and F. I shall assume that P assigns real numbers as 'grades' of personal probability.

The first notion we must define, for such a set Q, is the proportion of truths in it. But which are the true propositions? That depends on the state of the world, which must be represented too – by a model M. (The obvious form for such a model is a couple $\langle D, loc \rangle$ where D is the domain and *loc* some function that determines what attributes in F the members of D have, i.e., their 'location' in the possibility space determined by F. But that is a technical detail.) Each proposition is *true* or *false* in each model, and the Boolean operations on propositions cohere with the usual 'truth-table' assignments of truth values in a model. Denote by 'TRUE (M)' the set of propositions (for D and F) true in the model M. Then define the proportion of truths:

$$(4.1) \qquad \%MQ = \#(\text{TRUE}(M) \cap Q) \div \#Q$$

where # denotes the set's cardinality.

The next obvious step is to define the subset $Q(r)$ of propositions to which P assigns value r, and then to call P perfectly calibrated on Q exactly if $r = \%MQ(r)$ for each such assigned value. But because we are now dealing with questions that may relate to more than one attribute, that procedure is too rough and ready. Suppose for example I am asked about each of 100 days: will it rain? will it rain or snow?, will it rain or snow or hail? If the actual proportions were 0, 1/2, 6/10, and my announced probabilities were the same on each day, namely 3/10, 1/2, 3/10, the calibration would be perfect. (For Q contains 300 propositions, to 200 of which I have assigned 3/10; but of that 200, all of the first hundred (the ones of form "it rains on day x") are false while 60 of the remaining one hundred are true, and 60/200 equals 3/10.) This perfect calibration on the subsets $Q(r)$ hides the irrationality of assigning a lower value to one proposition than to a second which *implies* the first. But that irrationality would become readily apparent if we subdivided further in the obvious way, in terms of the attributes as well as the numbers assigned.

$$(4.2) \qquad Q_P(A, r) = \{E \in Q : (\exists z)(E = [z \text{ has } A]) \text{ and } P(E) = r\}.$$

When no confusion threatens, abbreviate $Q_P(A, r)$ to $Q(A, r)$. Then we call *P perfectly calibrated on Q with respect to model M* exactly if

(4.3) $r = \%MQ(A, r)$ for each value r and attribute A for which $Q(A, r)$ is not empty.

Such perfect calibration may, however, be precluded for trivial reasons.

If, for example, Q contains only a single member, then P must assign it either *zero* or *one* if (4.3) is to hold. Moreover, given that Q is finite, a look at (4.1) shows that P cannot be perfectly calibrated on Q at all unless P assigns only proportions of the finite number $\#Q$. Surely it cannot be a precondition of vindication that our personal probabilities come in rational fractions! Reflect especially on the fact that we do not generally know beforehand how many questions we shall be asked. Our first need here is for a measure of approximation, or distance from perfect calibration. The obvious measure to come to mind here (especially to readers of Brier's article) is the length of the vector $\langle r_i - \%MQ(A, r_i)\rangle$ where $r_1, \ldots, r_n$ are the numbers (in some order) which P assigns to members of Q. But because I shall be concerned with the measure only with respect to the *possibility* of its decrease toward zero, we can without loss of *finesse*, use a much cruder one.

(4.4) P is *calibrated to within* distance q, *on* set Q, *with respect to* model M, exactly if q is the supremum of the numbers $|r - \%MQ(A, r)|$ such that $Q(A, r)$ is not empty.

To be perfectly calibrated is then to be calibrated to within distance zero. Now this may be impossible to achieve, for stated reasons, even if Q is increased indefinitely. But with such increase we may hope for ever better approximation.

It is too early, though, to announce this hope as furnishing a criterion for reasonableness. For suppose that I first state my probability for rain as 1/6 and then you ask me about one thousand tosses of a fair die for the probability of ace and I say 1/6 each time. On the total set of 1001 questions, my personal probability will probably be quite well calibrated, but that reveals nothing about the reasonableness of my initial judgment about rain. To see the problem in acute form, let this first judgment be replaced by two: adding to it also the judgment that the probability of there being no rain equals 1/6 as well. Calibration on the total [illegible] of 1002 propositions will be quite good, whereas there is something drastically wrong with my probabilities for the first two.

So the possibility of ever better calibration which we require, must be on extensions of the initial set of propositions which are in a relevant sense *like* the original ones. A frequentist would say that optimally, the additional questions raised should be about the same attributes for entities for which the person selects the same reference classes. That selection being a 'hidden variable' of his judgment, however, we must make do with a relation of likeness reflected entirely in the personal probability function P, i.e., in the actual expression of the judgments.

(4.5) Entities x and y are *P-alike* exactly if $P[x \text{ has } A] = P[y \text{ has } A]$ for each attribute A.

(4.6) Q' is a *P-alike extension of Q* if and only if $Q \subseteq Q'$, P is defined for every member of Q', and if $[z \text{ has } A]$ is in Q' then there is an entity y such that y and z are P-alike, and for each attribute B, $[z \text{ has } B]$ is in Q' if and only if $[y \text{ has } B]$ is in Q.

Thus a typical P-alike extension of $Q = \{[y \text{ has } A], [y \text{ has } B]\}$ looks like $Q' = \{[y \text{ has } A], [y \text{ has } B], [z_1 \text{ has } A], [z_1 \text{ has } B], \ldots, [z_n \text{ has } A], [z_n \text{ has } B]\}$, where $z_1, \ldots, z_n$ are all P-alike to y. Having introduced the relevant notion of likeness, we can now define potential calibration in two steps.

(4.7) Let P be a scheme for D and F and P' a scheme for D' and F'. Then P' is an *extension* of P exactly if $D \subseteq D'$, $F \subseteq F'$ and $P[y \text{ has } A] = P'[y \text{ has } A]$ for each y in D and A in F.

(4.8) P is *potentially calibrated* on finite set Q of propositions on which it is defined exactly if for every real positive number q there exists an extension P' of P and P'-alike extension Q' of Q such that P' is calibrated to within q on Q', on some model.

As minimal criterion of rationality, from a frequentist point of view, I state the requirement that our body of judgments should be representable by at least one scheme which is potentially calibrated on every finite set of propositions for which it is defined. Note that since "calibrated to within q" has been defined with reference to proportions, and hence only for finite sets of propositions, the P-alike extensions that play a role in determining potential calibration on Q are also all finite.

5. FIRST ADEQUACY THEOREM

In this section and the next I shall address what seems at first sight to be a weaker criterion than I announced in the preceding paragraph:

(5.1) Scheme P is frequency coherent exactly if it is potentially calibrated on every finite set of propositions of form $Q = \{[x \textit{ has } A] : A \text{ in } FQ\}$ for which it is defined.

Note that here all members of Q are about the same, single subject. The scope and limits of this requirement will be discussed in Section 7.

(5.2) THEOREM. *If P is a scheme for D and F, and for each element x of D the function P_x defined by*

$$(5.2^*) \quad P_x(A) = P[x \textit{ has } A]$$

is a probability function on F, then P is frequency coherent.

To prove this theorem, we proceed in two stages. First of all, assuming P, D, F to be as described, consider the set $Q = \{[y \textit{ has } A_1], \ldots, [y \textit{ has } A_k]\}$. The attributes $A_1, \ldots, A_k$ generate a finite sub-algebra F^* of F. Let $B_1, \ldots, B_m$ be the atoms of F^*. Think of these atoms as boxes, and the other elements of F^* (which are finite joins of these atoms) as composite boxes. Place n_j items in box B_j, for $j = 1, \ldots, m$ with the total $n = n_1 + \ldots + n_m$. Select any positive number r you like; you can then choose those 'occupation numbers' for the boxes so that $n_j/n = P_x(B_j) \pm 1/r$. The reason is of course that the $P_x(B_j)$ are non-negative numbers that sum to *one*, by the hypothesis that P_x is a probability function. If now A is in F^*, say $A = B_1 \cup B_2 \cup B_3$, then $P_x(A)$ is determined by the additivity of P_x and the occupation number for A similarly:

$$(5.3) \quad \begin{aligned}(n_1 + n_2 + n_3)/n &= P_x(B_1) + P_x(B_2) + P_x(B_3) \pm 3/r \\ &= P_x(A) \pm 3/r\end{aligned}$$

In general, the divergence can be no more than m/r. And because the number m is fixed as the number of atoms in F^*, we can set m/r less than or equal to any pre-selected positive number q by appropriate choice of r. This reasoning establishes the unsurprising fact that a probability function on a finite domain can be arbitrarily closely approximated by proportion in an urn.

As second stage, we turn this demonstration into the construction of a model which shows the potential calibration of P on Q. To the original domain we add $n - 1$ new entities. We extend P to P' by setting P' equal to P where both are defined, and all the new entities P'-alike to y itself. Now we take a model M in which the set consisting of y itself plus the new entities is distributed in proportions n_j among the atoms B_j of F^*. Finally we consider the calibration of P' on the larger set $Q' = \{[z \text{ has } A_i] : i = 1, \ldots, k,$ and $z = y$ or z is a new entity$\}$ and find that P' is calibrated to within q on Q' with respect to model M. Hence we conclude, by generalizing on this construction, that P is potentially calibrated on Q itself.

6. SECOND ADEQUACY THEOREM

As converse to the first result, we find that obedience of the probability calculus is also a necessary condition for frequency coherence.

(6.1) THEOREM. *If P is a scheme for D and F, and is frequency coherent, then each function P_x defined by (5.2*), for x in D, is a probability function on F.*

The axioms of probability theory (for personal probability I consider only finitary constraints) are

(I) $\quad 0 = p(\Lambda) \leqslant p(A) \leqslant p(K) = 1$
(II) $\quad p(A \cup B) + p(A \cap B) = p(A) + p(B)$

where Λ and K are the minimal and maximal elements of Boolean algebra F and $\cup, \cap$ its join and meet operations.

Assuming now that P, D, F are as described in the antecedent of the theorem, it is clear first of all that $P_y(\Lambda)$ must equal to zero. For $[y \text{ has } \Lambda]$ is the impossible proposition, and so that proportion of truths in any subset of $Q' = \{[z_1 \text{ has } \Lambda], \ldots, [z_n \text{ has } \Lambda]\}$ equals zero. Hence no extension P' of P will be calibrated on Q' to within $q > 0$ unless $P'[z_i \text{ has } \Lambda]$ has an absolute value which is less than or equal to q. Thus $P[y \text{ has } \Lambda]$ must have an absolute value less than every positive number, if P is potentially calibrated on $\{[y \text{ has } \Lambda]\}$. Similarly $P[y \text{ has } K] = 1$ if P is potentially calibrated on $\{[y \text{ has } K]\}$.

It is just as easy to see that $P[y \text{ has } A]$ must be in the interval $[0, 1]$. For if the value assigned is a distance q outside that interval, then no extension P' of P can be calibrated to within less than q on any P'-alike extension of $\{[y \text{ has } A]\}$ – simply because all the relevant proportions are within it. Finally, we consider the four member set:

(6.2) $\quad Q = [y \text{ has } A \cup B], [y \text{ has } A \cap B], [y \text{ has } A], [y \text{ has } B]$

Let us suppose that we have a violation of Axiom II:

(6.3) $P_y(A \cup B) + P_y(A \cap B) = P_y(A) + P_y(B) + d$

where d may be either positive or negative. We now extend P to a scheme P' for a larger domain D' in which all new entities are P'-alike to y. And we consider the P'-alike extension Q' of Q in which the same propositions occur with not only y but also these new entities as subjects. Let us abbreviate:

$$\begin{array}{ll} a_2 = P_y(A \cup B) & b_1 = \%MQ'(A \cup B, a_1) \\ a_2 = P_y(A \cap B) & b_2 = \%MQ'(A \cap B, a_2) \\ a_3 = P_y(A) & b_3 = \%MQ'(A, a_3) \\ a_4 = P_y(B) & b_4 = \%MQ'(A, a_4) \end{array}$$

where M is some appropriate model.

Because the new entities are all P-alike to y, $Q'(A', r)$ will be empty for all cases not listed in the above table. Thus for example $Q'(A \cup B, a_1)$ is the set of all propositions of form [z *has* $A \cup B$] in Q', there are, let us say, m of these (and hence Q' has 4 m members exactly) of which m_1 are true, in which case $b_1 = m_1/m$. If we similarly set $b_i = m_i/m$ for $i = 2, 3, 4$ then it is clear that $m_1 + m_2 = m_3 + m_4$, so $b_1 + b_2 = b_3 + b_4$. We have $a_1 + a_2 = a_3 + a_4 + d$ and $b_1 + b_2 = b_3 + b_4$, and therefore:

(6.4) $(a_1 - b_1) + (a_2 - b_2) - (a_3 - b_3) - (a_4 - b_4) = d$

from which we conclude

(6.5.) $|a_1 - b_1| + |a_2 - b_2| + |a_3 - b_3| + |a_4 - b_4| \geqslant d$

which means that P' is not calibrated on q' to within less than d.

This argument being general with respect to extensions P' of P, P'-alike extensions Q' of Q, and relevant models M, we conclude that calibration to within less than d is impossible for these extensions, and so P is not potentially calibrated on Q.

7. ADEQUACY OF THE FREQUENCY COHERENCE CONCEPT

We have now established that a scheme P is frequency coherent if and only if each of its relativizations P_x is a probability function. But the reader may now have doubts about the significance of the notions used. Frequency coherence, as defined, relates only to calibration on sets of propositions that are all about the same subject. What about more diverse sets? This initial doubt, at least, can be put to rest.

(7.1) THEOREM. *P is frequency coherent if and only if P is potentially calibrated on all finite sets of propositions on which it is defined.*

The proof, which I shall sketch, relies on the simple lemma:

(7.2) LEMMA. *If P is calibrated to within q on disjoint sets* $Q_1, \ldots, Q_n$ *then P is also calibrated to within q on their union.*

For suppose Q is the union of those disjoint sets $Q_1, \ldots, Q_n$. The $Q(A, r) = Q_1(A, r) \cup \ldots \cup Q_n(A, r)$. Hence the proportion of M-truths in $Q(A, r)$ can neither be higher than all the numbers $\%MQ_i(A, r)$ nor lower than all of them. Hence the distance between that proportion and r cannot be larger than the supremum of all the numbers $|r - \%MQ_i(r)|$.

In the models, as we have conceived them so far, the questions whether [*xhas A*], [*y has B*] are true are totally independent. Hence we will be able to carry out the construction utilized in the first and second adequacy theorem simultaneously for any finite set of entities $y_1, \ldots, y_n$ in D. With respect to the set

$$Q' = \{[y_1 \textit{ has } A_1^1], [y_1 \textit{ has } A_2^1], \ldots, [y_n \textit{ has } A_{k_n}^n]\},$$

we can then find an appropriate model M such that the relevant extension P' is calibrated to within q on each subset

$$Q_j' = \{[y_j \textit{ has } A_1^j], \ldots, [y_j \textit{ has } A_{k_j}^j]\}$$

and therefore also on their union, i.e., Q' itself, by the above lemma.

But perhaps this is 'stonewalling'. For the uneasiness may lie exactly in the idea that there may be connections or relations among the entities. In that case, the questions whether x has A and whether y has B are *not* independent. Especially logic-minded readers, who want to see probabilities attached to all propositions expressed in a first-order predicate language, will be inclined to feel that the discussion so far has ignored relations among entities in the domain.

When Tarski reduced the problem of truth to the definition of *satisfaction*, he was showing, in effect, how questions about several entities can always be thought of as being about a single entity. For example, the following are equivalent:

(7.2) *x has A* and *y has B* and *x bears R to y*

(7.3) $\langle x, y\rangle$ *has* $A \otimes K$
$\langle x, y\rangle$ *has* $K \otimes B$
$\langle x, y\rangle$ *has* R

for an appropriately chosen product construction. The usual definition of satisfaction relates countably infinite sequences to open sentences. But it is quite possible to do the same job for, on the one hand, finite sequences, and on the other, the calculus of relations represented by sets of such sequences. In the second appendix I shall describe this construction somewhat more fully. Here I shall only state the conclusion that *if* there are significant relations among the entities in a domain, we should represent the person's judgments not simply by means of a scheme for that domain and a family of attributes pertaining to its members – but also by schemes for powers, or unions of powers, of that domain and pertinent relational attributes. All the schemes used to represent his judgments need to be frequency coherent; and that reflection should remove the uneasiness expressed above.

8. CONCLUSION: IS THERE A FUTURE FOR THE FREQUENCY INTERPRETATION?

Can we understand the activity of judgment, expressed in (vague) probability language, in a way that accords with the frequency interpretation of probability? I think we can, in two ways. The first is via the contention that the very aim of our judgment is to be a reliable indicator of actual frequencies of occurrence. The second is via a reflection on how that aim could be achieved, without essential recourse to deliberation about anything except the correct classification of the subjects and estimates of relative proportions among the classes involved.

As the central problem for this attempt I have selected Keynes' third question: how can the frequency interpretation justify our observance of the rules of the probability calculus, as intelligible and rational? It is clear that even with correct estimates of statistical frequencies, the selection of different reference classes on different occasions could easily lead to violations of those rules. Selection of the same reference class for all questions, on the other hand, would rob our judgments of all informative content.

My solution consisted in describing the expression of judgment as the expression of an epistemic attitude, and to discuss the proper evaluation of such an attitude under two headings: vindication and reasonableness. As a basic criterion of reasonableness (without any suggestion that it is the only criterion), I pointed to the requirement that vindication should not be *a priori* precluded. Now the main task at this point, for any interpretation of probability, is to explicate exactly what is vindication for a body of judgments. I argued that from the frequentist point of view, the notion of *calibration*, as it appears in the Brier score, is the core criterion of vindication.

After having refined this notion so as to allow for at least a crude measure of approximation, and to explicate the relevant sense of possibility when we consider whether a person's judgments have potentially good calibration, I could then formulate the correlate basic criterion of reasonableness. This was a special concept of *potential calibration* which I called *frequency coherence*. And it was possible to prove that satisfaction of this criterion is equivalent to non-violation of the probability calculus. Hence Keynes' challenge has been met.

Now I believe that this has far-reaching consequences for the frequentist program as a whole. I insisted, in my short discussion of frequency schemes (i.e., models of judgment formation) that we should reject the idea that we must provide a *recipe* – i.e., set of determinate rules – for the selection of reference classes and formation of frequency estimates. This was on the more general grounds that we should not identify rationality with being compelled by requirements of rationality, but rather with being within their bounds, allowed by them. Rationality is only bridled irrationality.

The demonstration that potential vindication requires obedience to the probability calculus can now take over much of the job that recipes for reference class selection were meant to do. For suppose we choose reference classes for some basic questions, and form corresponding judgments. The probability calculus will then constrain our further judgments to a large extent – and to that extent, we can be totally uninterested in a recipe for what reference classes are or should be chosen in those further cases. Suppose that you choose reference classes for *rain all day* and *dry all day* and announce your personal probabilities as 0.2 for today's having the first attribute and 0.3 for its having the second. Now I ask you about its having the attribute *rain all day or dry all day*. Why should you stop to consult a recipe for choosing a reference class? You know now that whatever one you choose, you will be irrational unless you come up with the answer 0.5. Hence if anyone is interested in building such frequency interpretation models for judgment formation, he should, I think, be counselled that he can now, *without loss to his program*, make the probability calculus part of the constraints on the selection of reference classes. For the use of that calculus has been justified on independent but frequentist grounds.

An unsympathetic reader may at this point ask why we should bother with the frequency interpretation at all. Certainly I am much more anxious that contemplation of 'objective' probabilities should not lead to a belief in propensities – anxious, that is, to maintain an empiricist view of probabilistic scientific theories – than I am to deny 'subjective' probability a status *sui generis*. But there is a philosophical question: why is the same name 'probability' appropriate for both? The wonder can presumably be removed only by a plausible explanation which entails either that the question is mistaken, and there is no connection at all, or else that there is a very intimate connection. Since the question can perhaps best be focussed on the special fact that the same axiomatic theory proves to be wonderfully useful in the explication of both uses of 'probability', we should especially ask why that should be. I hope to have shown that the frequency interpretation can remove this wonder by exhibiting an intimate connection that implies that the same, familiar axioms should cover both uses.[8]

Princeton University

APPENDIX I. FREQUENCY SCHEMES

In Kyburg's work, the literature contains an impressive, large-scale attempt to give a model for judgment which includes judgments concerning statistical frequencies (relative proportions in classes) and single case probabilities ('epistemic probabilities') based on these statistical judgments plus rule-governed selection of the right reference classes, via a generalization of the concept of the 'statistical syllogism'. As a result it is now difficult to stand back and canvass in an abstract way how frequentists *could*, in principle, go about constructing their models. Such a survey would nevertheless be of value, even if we came to see Kyburg's work as entirely succeeding in its aims, for it would be valuable to know whether the aims could be achieved some other way.

Given a domain of entities D and a Boolean algebra F of attributes (perhaps identified with subsets of a larger domain that includes D) I call a *scheme* any map P of the *propositions* [x *has* A] with x in D and A in F, into real numbers. This scheme is meant to represent the surface phenomena of judgment after initial regimentation into probabilistic form. The notion of *frequency scheme* is much vaguer: a structure, suggested by the frequency interpretation, one part of which is such a scheme (i.e., a theoretical model for the phenomena of judgment). I have not indicated what entity the proposition [x *has* A] is; the reader may choose a convenient identification, for example, with the ordered pair $\langle x, A\rangle$.

Suppose we ask the subject on a given occasion whether entity x has attribute A. I propose in general that his judgment is determined by four factors. The first is a partially defined scheme σ for D and an extension F' of F – his *initial scheme*. The second is his *estimate* α which is a binary function partially defined on F'; "$\alpha(A|B) = r$" is read as "the proportion of entities that have A among those that have B equals r." Note that α has nothing to do with domain D *per se*, at least at this general level of discussion. The third is his *selector* β, a function that selects for each x in D and each A in F a class $\beta(x, A)$ of attributes in the algebra F, and perhaps for some in $F' - F$ as well. Note that I have generalized the choice of a reference class to selection of a class of reference classes, for reasons made clear below. And the fourth is his *strategy* Σ which determines a numerical grade (*personal probability*) for each proposition on the basis of the foregoing. Let $M = \langle\sigma, \alpha, \beta, \Sigma\rangle$ be called a *frequency scheme*, and abbreviate

(I.1) $\quad P_M\,[x \textit{ has } A] = \Sigma(\sigma, \alpha, \beta)\,[x \textit{ has } A]$

which is the *scheme* of frequency scheme M. We may at once impose the requirement that Σ be entirely determined by σ where defined, that is

(I.2) $\quad P_M\,[x \textit{ has } A] = \sigma\,[x \textit{ has } A]$ when defined

The remainder of the structure represents the procedure of deliberation whereby the initial scheme is extended to other propositions in accordance with frequentist intuitions.

Now I will give some examples of what frequency schemes can be like. The first is the simplest. The initial scheme represents only something like initial full belief (or 'taking as evidence'), it just assigns zeroes and ones to some propositions. The selector β now acts as follows: for the couple $\langle x, A\rangle$ it selects a single attribute B such that (i) σ assigns 1 to $[x \textit{ has } B]$ and (ii) $\alpha(A|B)$ is defined. Denote the attribute selected, in general, as $\beta_{x,A}$. Finally, Σ then simply assigns that estimated proposition. Thus we have, for $M = \langle\sigma, \alpha, \beta, \Sigma\rangle$:

(I.3) $\quad P_M\,[x \textit{ has } A] = \alpha(A|\beta_{x,A})$.

This follows closely Venn's original idea that we classify the subject (with no account taken of doubts about the classification) and announce the statistical frequency (assumed known) in that reference class.

But this does not seem very realistic to me. Does it not seem more plausible that we base our opinion in part on classifications of the subject, for which we have only partial certainty? So we can envisage a slightly more elaborate frequency scheme in which σ assigns some numbers between *zero* and *one* as well. Here let β select as $\beta(x, A)$ a *partition* of attributes $B_1, \ldots,$ B_k for which σ is defined, that is, $\sigma(B_j \cap B_j) = 0$, $\sigma(B_1 \cup \ldots \cup B_k) = 1$, with $\alpha(A | B_i)$ defined for each. Then Σ should act so as to yield, for $M = \langle \sigma, \alpha, \beta, \Sigma \rangle$:

$$\text{(I.4)} \qquad P_M [x \textit{ has } A] = \Sigma \{\sigma [x \textit{ has } B] \, \alpha(A | B) : B \in \beta(x, A)\}$$

where this capital sigma is the summation sign.

But now, as $\Sigma(\sigma, \alpha, \beta)$ extends σ, the new propositions to which probabilities are assigned, can also begin to play a role in deliberation. So we can describe a third type of frequency scheme. In that larger algebra F', we may introduce a partial ordering. This has nothing to do with the Boolean operations *per se*, but it may have something to do with the subject x of the question at issue, so call it "x-precedes." In this type of scheme, σ is defined for [x *has* A] only if nothing x-precedes A. In that case principle (I.2) applies. Next we look at the case in which something x-precedes A; then β may select a partition of attributes that x-precede A, with the same conditions fulfilled for σ and α, so that (I.4) can apply. (Note that (I.3) is just a special case of (I.4) if σ assigns only zeroes and ones.) Finally, we come to the case where something x-precedes A but β does not act so that (I.4) can apply; then β must still select a partition $\beta(x, A)$ of attributes which x-precede A, and the following principle should be applicable:

$$\text{(I.5)} \qquad P_M [x \textit{ has } A] = \Sigma \{P_M [x \textit{ has } B] \, \alpha(A | B) : B \in \beta(x, A)\}.$$

The use of the partial ordering *x-precedes* allows these principles to govern the action of the strategy Σ without circularity. To the extent that P_M is defined on propositions [x *has* B] for attributes B that *x-precede* A, it takes over the role of initial scheme σ in the constraints on $\beta(x, A)$ and determination of P_M [x *has* A].

At this point we might even speculate again that restriction of σ to the assignment of zeroes and ones only, might not unduly impoverish the stock of frequency schemes of this third type. If the attributes A are sophisticated enough, a proposition [x *has* A] might well be the exact information that the proportion of Bs among Cs equals 0.75, say. In that case, α could be built up simultaneously with Σ. These two reflections are in the direction of what Kyburg's constructions are meant tc .hieve. As among Bayesians, we can see a divergence of inclination toward 'global models' and 'local models' ('small worlds') respectively, open to frequentists as well. I have not discussed severe constraints on the selector here; for that I refer back to the last section of the body of this paper.

APPENDIX II. RELATIONS AND PRODUCT CONSTRUCTIONS

In Section 7 I discussed calibration of a scheme for sets of propositions about several individuals, and the difficulties that could occur due to relations among these. For example, x might be Christmas day and y Christmas morning, so that rain on y and dry weather throughout x are not logically independent. I shall here describe in some more detail the kind of product construction in which questions about several individuals are reduced to ones about a single entity, in a way directly relevant to this paper.

For definiteness I shall take the algebra F of attributes to be a field of subsets of a given set K (the maximal element of F). A *model* in which propositions receive truth values is then a couple $M = \langle loc, D \rangle$, where *loc* maps D into K, and [x *has* A] is true in M exactly if $loc(x)$ is a member of set A.

To take account of relational attributes, we focus on domain D^∞, which is the class of all finite sequences of members of D. We let K be itself a set K_0^∞ and F a field of subsets thereof. Intuitively we identify the binary relation R with all the sequences $e = \langle e(1), \ldots, e(n) \rangle$ in K such that $e(1)$ bears R to $e(2)$. Thus R is identified with a set which contains all finite elongations of its members. In the present case we say that R nevertheless has degree 2, because there is a subset of K_0^2 which 'determines' R. Stated precisely:

(II.1) If Y is a subset of X^∞ then Y^+ is the set of all members of X^∞ which have some initial segment that is in Y; and the *degree* of Y (if any) is the least positive integer m for which there exists a set Y^* such that for all e in X^∞, e is in Y if and only if $\langle e(1), \ldots, e(m) \rangle$ is in Y^*.

(II.2) RESTRICTION. *Each attribute in F has a finite degree.*

It follows at once that if A is in F, then $A = A^+$ (the operation$^+$ understood contextually here with reference to K_0^∞). This restriction is compatible with the Boolean character of F as a field of sets, but keeps it from being a sigma-field. (For example, the degree of $A \cap B$ is the maximum of the degrees of A, B if any.) Note that the degrees of K and of Λ equal 1, since e is in K (respectively, Λ), if and only if $\langle e(1) \rangle$ is in K_0^1 (respectively, Λ), where we denote as X^n the set of all n-tuples of members of X.

A *model* must be restricted so as to observe the structural relations among the sequences:

(II.3) $M = \langle loc, D^\infty \rangle$ is a *model* for D and F exactly if *loc* maps D into K_0 and for x in D^∞, of length n, $loc\,(x) = \langle loc\,(x(1)), \ldots, loc\,(x(n))\rangle$.

We define the following operations on K and its subsets:

(II.4) If $A_1, \ldots, A_n$ have degrees $m(1), \ldots, m(n)$ respectively, then $A_1 \otimes \ldots \otimes A_n$ is the set whose members are all sequences e in K such that $\langle e(1), \ldots, e(m(1))\rangle$ is in $A_1, \ldots, \langle e(m(n-1)+1), \ldots, e(m(n-1)+m(n))\rangle$ is in A_n.

(II.5) $e(m+n)b = \langle e(1), \ldots, e(m), b(1), \ldots, b(n)\rangle$ and undefined if the lengths of e, b are less than m, n respectively.

(II.6) Where m, n are the degrees of A, B respectively,
$A \mathbin{\overline{\wedge}} B = \{b(m+n)d : b \text{ in } A \text{ and } d \text{ in } B\}^+$
$A \mathbin{\underline{\vee}} B = \{b(m+n)d : b \text{ in } A \text{ or } d \text{ in } B\}^+$.

I shall call $\overline{\wedge}$ and $\underline{\vee}$ the *directed meet* and *directed join*. Note that the degree of $A_1 \otimes \ldots \otimes A_n$ equals the sum of the degrees of $A_1, \ldots, A_n$, and similarly for $A_1 \mathbin{\overline{\wedge}} A_2$, $A_1 \mathbin{\underline{\vee}} A_2$. We impose on F also:

(II.7) RESTRICTION. *If $A_1, \ldots, A_n$ are in F so are $A_1 \otimes \ldots \otimes A_n$, $A_1 \mathbin{\overline{\wedge}} A_2$, $A_1 \mathbin{\underline{\vee}} A_2$, and each set K_0^n, $n = 1, 2, \ldots$.*

This is again compatible with the Boolean character of F and with (II.2).

It is clear that the directed meet and join are not commutative, but on the level of truths of propositions commutation is effectively restored. The $(m+n)$ operation makes sense for any sequences, hence can be used on D as well. Then we see

(II.4) $[x(m+n)y$ *has* $A \mathbin{\underline{\vee}} B]$ is true in model $M = \langle loc, D\rangle$ iff $\langle loc\,(x(1)), \ldots, loc\,(x(m))\rangle$ is in A or $\langle loc\,(y(1)), \ldots, loc\,(y(n))\rangle$ is in B, hence iff $[x$ *has* $A]$ or $[y$ *has* $B]$ is true in M

where it was assumed that A, B have degrees m and n respectively, and x, y appropriate lengths. Thus we see that *if we identify a proposition with the set of models in which it is true*, then

(II.5) $[x(m+n)y$ *has* $A \mathbin{\underline{\vee}} B] = [x$ *has* $A] \cup [y$ *has* $B]$,

and similarly for directed meet and intersection.

Let us now inspect the adequacy proofs for the special case of such a product construction. There are just two points that must especially be made. It may seem at first that we need to modify the notion of P-alike, by stipulating that not only $P[x \textit{ has } A] = P[y \textit{ has } A]$ for all A in F, but also that $P[\langle x(k), \ldots, x(k+m)\rangle \textit{ has } A] = P[\langle y(k), \ldots, y(k+m)\rangle \textit{ has } A]$ for all A in F as long as $k + m$ is not too long. Actually no such emendation is needed, because $[\langle x(k), \ldots, x(k+m)\rangle \textit{ has } A]$ is the same proposition as $[x \textit{ has } K_0{}^{k-1} \otimes A \otimes K_0{}^r]$ where r equals the length of x minus $(k + m)$. The second point relates to the justification of the additivity principle in the second Adequacy proof, the only place where we deal explicitly with an initial set containing more than one proposition. Consider:

$$P([x \textit{ has } A] \cup [y \textit{ has } B]) + P([x \textit{ has } A] \cap [y \textit{ has } B]) \\ = P([x \textit{ has } A]) + P([y \textit{ has } B]).$$

Let m and n be the degrees of A, B respectively. Then the four propositions can all be seen to be identical with propositions about the single entity $x(m+n)y$:

$$[x \textit{ has } A] \cup [y \textit{ has } B] = [x(m+n)y \textit{ has } A \veebar B] \\ [x \textit{ has } A] \cap [y \textit{ has } B] = [x(m+n)y \textit{ has } A \barwedge B] \\ [x \textit{ has } A] = [x(m+n)y \textit{ has } A \otimes K_0{}^n] \\ [y \textit{ has } B] = [x(m+n)y \textit{ has } K_0{}^m \otimes B].$$

After this the proof can proceed as before.

NOTES

* Through his writings and as my teacher, dissertation supervisor, and friend, Adolf Grünbaum has been my main guide into philosophy of science ever since I read his 1955 article on the foundations of special relativity, which I came across as a undergraduate. I dedicate this paper to him, with sincere gratitude and warm affection. Support for this research by the National Science Foundation is gratefully acknowledged. A preliminary version of this paper was circulated in September 1979. I have learned that results that appear to be similar to the theorems in this paper were stated in a public lecture by Abner Shimony in 1978. I also wish to thank J. Hellige and W. Edwards, of the University of Southern California, for helpful discussions.

[1] The equation is not a simple one; see my (1979).

[2] *Annales de l'Institut Henri Poincaré*, vol. 7 (1937), in English translation in Kyburg and Smokler (1964).

[3] Both Reichenbach and Salmon have discussed vindication (I owe the term to Salmon) of predictions in connection with probability and inductive strategy, and have been concerned to analyze the condition of *possible* vindication. Hence my strategy here continues *one* venerable strand in frequentist thinking.

[4] Venn (1888), p. 213.

[5] See Brier (1950); there is now a large body of literature on scoring in general. For the decomposition into calibration and extremeness, see Dickey (1974), and Murphy (1972). See also de Finetti (1965), Pickhardt and Wallace (1974), Shuford *et al.* (1966), Winkler and Murphy (1968); and see further Note 8 below.

[6] In my own view of theories the truth requirement is one of empirical adequacy (truth about what is both actual and observable) only. Information has several objective dimensions, such as logical strength and what I call empirical strength, but also plays an essential role in such pragmatic virtues as being explanatory (informative in 'relevant' respects). See my (1980), (1981), (1982).

[7] Reichenbach formulated a crude measure of vindication which he used at several places in his (1949), including in his discussion of "single case probability" (which he called a "pseudo-concept," that "must be replaced by a substitute constructed in terms of class probabilities.") That is, if a person assents to all propositions about individual events of sort B, when he believes the relative frequency of B to be $\geqslant r$, then he will be right in proportion $\geqslant r$ of the cases, if that belief is correct. By taking $r > 1/2$ he will thus be right more often than not. This refers to a choice of the same reference class for each question about an individual having attribute B. Reichenbach then points out that if we switch to a smaller reference class, in which the proportion of B is higher, the proportion of success in our predictions will also increase. He did not, as far as I know, investigate what proportion of success is possible in the general case in which the questions are about different attributes, and the reference classes chosen may vary, even for the same attributes from individual to individual. But although measurement by a division of this type (assent at level $\geqslant r$) is crude, it is the *sort* of measure of vindication that is needed here.

[8] The reader may well have wondered how Bayesians can or should approach the question of 'correct' scoring procedures. A good indication is found in the results proved by Shuford *et al.* (1966) who suggest as a basic criterion that a scoring procedure is admissible exactly if anyone can maximize his expected score if and only if he correctly reports his personal probabilities. (The expected score is of course the score's expectation value calculated by his own personal probability.)

REFERENCES

Brier, G. W. 1950. 'Verification of Forecasts Expressed in Terms of Probability,' *Monthly Weather Review* 78, 1–3.

de Finetti, B. 1965. 'Methods for Discriminating Levels of Partial Knowledge Concerning a Test Item,' *British Journal of Mathematical and Statistical Psychology* **18**, 87–123.

Dickey, J. M. 1974. 'Comments on Suppes,' *Journal of the Royal Statistical Society* **B36**, 179–180.

Keynes, J. M. 1921. *Treatise on Probability*. London: Macmillan.

Kyburg, H. E., Jr. 1974. *The Logical Foundations of Statistical Inference*. Dordrecht: D. Reidel.

Kyburg, H. E., Jr. and H. E. Smokler (eds.), 1964. *Studies in Subjective Probability*. New York: Wiley.

Murphy, A. H. 1972. 'Scalar and Vector Partitions of the Probability Score,' *Journal of Applied Metereology* 11, 273–282.

Pickhardt, R. C. and J. B. Wallace. 1974. 'A Study of the Performance of Subjective Probability Assessors,' *Decision Sciences* 5, 347–363.

Reichenbach, H. 1949. *The Theory of Probability*. 2nd ed. Berkeley: University of California Press.

Salmon, W. 1967. *The Foundations of Statistical Inference*. Pittsburgh: University of Pittsburgh Press.

Shuford, E. H., A Albert, and H. E. Massen. 1966. 'Admissible Probability Measurement Procedures,' *Psychometrika* 31, 125–145.

van Fraassen, B. C. 1979. 'Foundations of Probability Theory: A Modal Frequency Interpretation.' In G. Toraldo di Francia (ed.), *Problems in the Foundations of Physics*. Amsterdam: North-Holland.

van Fraassen, B. C. 1980. *The Scientific Image*. Oxford: Clarendon Press.

van Fraassen, B. C. 1981. 'Theory Construction and Experiment: An Empiricist View.' In P. Asquith and R. Giere (eds.), *PSA 1980*, vol. 2. East Lansing, Michigan: Philosophy of Science Association.

van Fraassen, B. C. 1982. 'Glymour on Evidence and Explanation.' In J. Earman (ed.) *Minnesota Studies in the Philosophy of Science*, vol. 10. Minneapolis: University of Minnesota Press, forthcoming.

Venn, J. 1888. *The Logic of Chance* (1886). London: Macmillan.

Winkler, R. L. and A. H. Murphy 1968. ' "Good" Probability Assessors,' *Journal of Applied Metereology* 7, 751–758.

BIBLIOGRAPHY OF ADOLF GRÜNBAUM

1948

[1] 'Critical Study of F. S. C. Northrop's *The Logic of the Sciences and the Humanities*,' *Yale Law Journal* 57, 1332–1338.

1950

[2] 'Realism and Neo-Kantianism in Professor Margenau's Philosophy of Quantum Mechanics,' *Philosophy of Science* 17, 26–34.
[3] 'Relativity and the Atomicity of Becoming,' *Review of Metaphysics* 4, 143–186.

1951

[4] 'Some Recent Writings in the Philosophy of Mathematics,' *Review of Metaphysics* 5, 281–292.
[5] 'The Philosophy of Continuity,' Ph.D. Thesis, Yale. (See also [110].)

1952

[6] 'Some Highlights of Modern Cosmology and Cosmogony,' *Review of Metaphysics* 5, 481–498.
[7] 'Messrs. Black and Taylor on Temporal Paradoxes,' *Analysis* 12, 144–148.
[8] 'Causality and the Science of Human Behavior,' *American Scientist* 40, 665–676. (This article was reprinted as item P–639 in the Bobbs-Merrill Reprint Series in the Social Sciences. (See also [11, 34, 43, 82, 86, 125].)
[9] 'A Consistent Conception of the Extended Linear Continuum as an Aggregate of Unextended Elements,' *Philosophy of Science* 19, 288–306.

1953

[10] 'Critique of H. Dingle's Objection to the Bondi-Gold Cosmology,' *Scientific American* (December 1953), pp. 6–8.
[11] A version of [8] appeared in H. Feigl and M. Brodbeck (eds.), *Readings in the Philosophy of Science*, pp. 766–778. New York: Appleton-Century-Crofts.
[12] 'Some Remarks on Professor Ushenko's Interpretation of Causal Law,' *Journal of Philosophy* 50, 115–120.
[13] Review article on Hans Reichenbach's *Rise of Scientific Philosophy, Scripta Mathematica* 19, 48–54.
[14] 'Comments on Power and Science,' *Review of Metaphysics* 6, 378–380.

R. S. Cohen and L. Laudan (eds.), Physics, Philosophy and Psychoanalysis, 321–331.

[15] 'Relativity, Causality and Weiss's Theory of Relations,' *Review of Metaphysics* 7, 115–123.

[16] 'Whitehead's Method of Extensive Abstraction,' *British Journal for the Philosophy of Science* 4, 215–226.

1954

[17] Review of R. Dugas' *Histoire de la mécanique, Philosophy and Phenomenological Research* **15**, 119–121.

[18] Review of H. Feigl and M. Brodbeck (eds.), *Readings in the Philosophy of Science, American Journal of Physics* **22**, 498–499.

[19] Review of R. von Mises' *Positivism, Scripta Mathematica* **20**, 75–76.

[20] 'E. A. Milne's Scales of Time,' *British Journal for the Philosophy of Science* **4**, 329–330.

[21] 'Science and Ideology,' *Scientific Monthly* **79**, 13–19. (See also [24].)

[22] 'The Clock Paradox in the Special Theory of Relativity,' *Philosophy of Science* **21**, 249–252. (See also [27].)

[23] 'Operationism and Relativity,' *Scientific Monthly* **79**, 228–231. (See also [40].)

[24] A version of [21] appeared in *Humanist* **14**, 161–173.

1955

[25] 'Time and Ethics,' *Scientific Monthly* **80**, 200–201.

[26] Review of E. Nagel's *Sovereign Reason, Science* **121**, 862.

[27] A version of [22] appeared in *Proceedings of the II International Congress of the IUHPS*, vol. 3, pp. 55–60. Neuchâtel: Editions du Griffon.

[28] 'Time and Entropy,' *American Scientist* **43**, 550–572. (See also [35, 41].)

[29] 'Logical and Philosophical Foundations of the Special Theory of Relativity,' *American Journal of Physics* **23**, 450–464. (See also [48].)

[30] 'Modern Science and Refutation of the Paradoxes of Zeno,' *Scientific Monthly* **81**, 234–239.

1956

[31] Review of H. Reichenbach's *The Direction of Time, American Scientist* **44**, 292A–300A.

[32] 'Relativity of Simultaneity within a Single Galilean Frame: A Rejoinder,' *American Journal of Physics* **24**, 588–590.

[33] 'Historical Determinism, Social Activism and Predictions in the Social Sciences,' *British Journal for the Philosophy of Science* 7, 236–240.

1957

[34] A version of [8] appeared in M. Mandelbaum, F. W. Gramlich, and A. R. Anderson (eds.), *Philosophic Problems*, pp. 328–338. New York: Macmillan.

[35] A German translation of [28] appeared as 'Das Zeitproblem,' *Archiv für Philosophie* 7, 165–208.

[36] 'Some Remarks on Moon and Spencer's "On the Establishment of a Universal Time",' *Philosophy of Science* **24**, 77–78.

[37] (With E. L. Hill.) 'Irreversible Processes in Physical Theory,' *Nature* **179**, 1296–1297.

[38] 'The Philosophical Retention of Absolute Space in Einstein's General Theory of Relativity,' *Philosophical Review* **66**, 525–534. (See also [76].)

[39] 'Complementarity in Quantum Physics and Its Philosophical Generalization,' *Journal of Philosophy* **54**, 713–727.

[40] A version of [23] appeared in P. Frank (ed.), *The Validation of Scientific Theories*, pp. 84–94. Boston: Beacon Press. (A paperback edition of this volume was published by Collier–Macmillan, New York, 1961.)

1958

[41] A Hebrew translation of [28] appeared in *IYYOUN* **9**, 249–62.

[42] A revised version of [29] appeared as 'Fundamental Philosophical Issues in the Special Theory of Relativity.' In K. Sapper (ed.), *Kritik und Fortbildung der Relativitätstheorie*, pp. 1–26. Graz: Akademische Druck u. Verlagsanstalt.

1959

[43] A version of [8] appeared in R. S. Daniel (ed.), *Readings in General Psychology*, pp. 328–336. Boston: Houghton Mifflin Co.

[44] 'Conventionalism in Geometry.' In *Symposium on the Axiomatic Method*, pp. 204–222. (*Studies in Logic and the Foundations of Mathematics*). Amsterdam: North Holland Publishing Co.

[45] 'Remarks on Dr. Kubie's Views.' In S. Hook (ed.), *Psychoanalysis, Scientific Method and Philosophy*, p. 225. New York: New York University Press.

[46] 'The Falsifiability of the Lorentz-Fitzgerald Contraction Hypothesis,' *British Journal for Philosophy of Science* **10**, 48–50.

1960

[47] 'The Role of *A Priori* Elements in Physical Theory.' In L. W. Friedrich (ed.), *The Nature of Physical Knowledge*, pp. 109–128. Bloomington: Indiana University Press.

[48] A version of [29] appeared in A. Danto and S. Morgenbesser (eds.), *Philosophy of Science*, pp. 399–434. New York: Meridian.

[49] A rejoinder to H. Dingle's critique of [46] appeared in *British Journal for Philosophy of Science* **11**, 143–145.

[50] 'The Duhemian Argument,' *Philosophy of Science* **27**, 75–87. (This article was reprinted as item Phil–90 in The Bobbs-Merrill Reprint Series in Philosophy. (See also [141].)

1961

[51] 'The Genesis of the Special Theory of Relativity.' In H. Feigl and G. Maxwell (eds.), *Current Issues in the Philosophy of Science*, pp. 43–53. New York: Holt, Rinehart and Winston. (See also [96, 126].)

[52] 'Law and Convention in Physical Theory,' *ibid.*, pp. 140–155.

[53] 'Discussion of Professor Feyerabend's Comments on "Law and Convention in Physical Theory",' *ibid.*, pp. 161–168.

[54] 'Professor Dingle on Falsifiability: A Second Rejoinder,' *British Journal for the Philosophy of Science* **12**, 153–156.

1962

[55] 'Prologue' and 'Epilogue' in P. W. Bridgman's posthumous *A Sophisticate's Primer of Relativity*, pp. vii–viii and 165–191. Middletown, Conn.: Wesleyan University Press. A British edition was published in London in 1963 by Routledge and Kegan Paul. (Some subsequent printings of Bridgman's book include neither the Prologue nor the Epilogue.)

[56] 'Geometry, Chronometry and Empiricism.' In H. Feigl and G. Maxwell (eds.), *Minnesota Studies in the Philosophy of Science*, vol. 3, pp. 405–526. Minneapolis: University of Minnesota Press. (See also [74].)

[57] 'The Nature of Time.' In R. G. Colodny (ed.), *Frontiers of Science and Philosophy*, pp. 147–188. Pittsburgh: University of Pittsburgh Press; London: Allen and Unwin.

[58] 'A. N. Whitehead's Philosophy of Science,' *Philosophical Review* **71**, 218–229.

[59] 'Temporally-Asymmetric Principles, Parity between Explanation and Prediction and Mechanism vs. Teleology,' *Philosophy of Science* **29**, 146–170. (See also [69, 71, 100].)

[60] 'The Special Theory of Relativity as a Case Study of the Importance of the Philosophy of Science for the History of Science,' *Annali di Matematica* **57**, 257–282. (See also [70].)

[61] 'The Structure of Science,' *Philosophy of Science* **29**, 294–305.

[62] 'Science and Man,' *Perspectives in Biology and Medicine* **5**, 483–502. (See also [84, 86].)

[63] 'The Falsifiability of Theories: Total or Partial? A Contemporary Evaluation of the Duhem–Quine Thesis,' *Synthese* **14**, 17–34. (See also [72, 83].)

[64] 'The Relevance of Philosophy to the History of the Special Theory of Relativity,' *Journal of Philosophy* **59**, 561–574.

1963

[65] 'Reflexive Predictions: Comments on Professor Roger Buck's Paper,' *Philosophy of Science* **30**, 370–372. (See also [101].)

[66] 'Space and Time.' A lecture broadcast by The Voice of America on July 1, and published by The Voice of America, U.S. Information Agency, Washington, D.C. (See also [91].)

[67] *Philosophical Problems of Space and Time*. New York: Knopf. (See also [73, 99, 130].)

[68] 'Carnap's Views on the Foundations of Geometry.' In P. A. Schilpp (ed.), *The Philosophy of Rudolf Carnap*, pp. 599–684. LaSalle, Ill.: Open Court Publishing Co. (See also [75, 131].)

[69] A version of [59] appeared in B. Baumrin (ed.), *Philosophy of Science*, pp. 57–96. (*The Delaware Seminar*, vol. I). New York: John Wiley and Sons.

[70] A version of [60] appeared in *ibid.*, vol. II, pp. 171–204.

[71] A version of [59] appeared in H. Kyburg and E. Nagel (eds.), *Induction: Some Current Issues*, pp. 114–149. Middletown, Conn.: Wesleyan University Press.

[72] A version of [63] appeared in M. Wartofsky (ed.), *Boston Studies in the Philosophy of Science*, vol. 1, pp. 178–195. Dordrecht: Reidel.

1964

[73] A British edition of [67] was published by Routledge and Kegan Paul, London.

[74] A German translation of [56] appeared as 'Geometrie, Zeitmessung und Empirismus,' *Archiv für Philosophie* **12**, 179–303.

[75] A version of [68] appeared in J. J. C. Smart (ed.), *Problems of Space and Time*, pp. 397–425. New York: Macmillan.

[76] A version of [38] appeared in *ibid.*, pp. 313–317.

[77] 'Popper on Irreversibility.' In M. Bunge (ed.), *The Critical Approach, Essays in Honor of Karl Popper*, pp. 317–331. London: Macmillan.

[78] 'Questions for Brand Blanshard.' In S. C. Rome (ed.), *Philosophical Interrogations*, p. 205. New York: Holt, Rinehart and Winston.

[79] 'The Anisotropy of Time,' *Monist* **48**, 219–247.

[80] 'The Bearing of Philosophy on the History of Science,' *Science* **143**, 1406–1412.

[81] 'Is a Universal Nocturnal Expansion Falsifiable or Physically Vacuous?', *Philosophical Studies* **15**, 71–79. *Errata* in this article are listed in *Philosophical Studies* **16**, 47–48.

1966

[82] A version of [8] appeared in R. Ulrich, T. Stachnik and J. Mabry (eds.), *Control of Human Behavior*, pp. 3–10. Glenview: Scott, Foresman and Co.

[83] A version of [63] appeared in P. K. Feyerabend and G. Maxwell (eds.), *Mind, Matter and Method: Essays in Philosophy and Science in Honor of Herbert Feigl*, pp. 273–305. Minneapolis: University of Minnesota Press.

[84] A version of [62] appeared in L. Z. Hammer (ed.), *Value and Man*, pp. 55–66. New York: McGraw-Hill.

1967

[85] *Modern Science and Zeno's Paradoxes*. Middletown, Conn.: Wesleyan University Press. (See also [95].)

[86] Versions of [8] and [62] appeared in M. Mandelbaum, F. W. Gramlich, A. R. Anderson, and J. B. Schneewind (eds.), *Philosophic Problems*, pp. 448–462. 2nd ed. New York: Macmillan.

[87] 'Theory of Relativity.' In P. Edwards (ed.), *The Encyclopedia of Philosophy*, vol. 7, pp. 133–140. New York: Macmillan.

[88] 'The Anisotropy of Time.' In T. Gold and D. L. Schumacher (eds.), *The Nature of Time*, pp. 149–177, 245–247, and Discussion on pp. 177–186. Ithaca: Cornell University Press. (See also [111].

[89] 'The Status of Temporal Becoming.' In R. Gale (ed.), *The Philosophy of Time*, pp. 322–353. New York: Anchor Doubleday Books. A *revised* British edition of this anthology was published in London by Macmillan in 1968. See pp. 322–354.

[90] 'Modern Science and Zeno's Paradoxes of Motion,' *ibid*., pp. 422–494. (See also [112].)

[91] A version of [66] appeared in S. Morgenbesser (ed.), *Philosophy of Science Today*, pp. 125–135. New York: Basic Books.

[92] 'The Denial of Absolute Space and the Hypothesis of a Universal Nocturnal Expansion: A Rejoinder to George Schlesinger,' *Australasian Journal of Philosophy* **45**, 61–91.

[93] 'The Status of Temporal Becoming,' *Annals of the New York Academy of Sciences* **138**, 374–395.

1968

[94] *Geometry and Chronometry in Philosophical Perspective*. Minneapolis: University of Minnesota Press.

[95] A revised version of [85] was published by Allen and Unwin, London.

[96] A version of [51] appeared in L. P. Williams (ed.), *Relativity Theory*, pp. 107–114. New York: John Wiley and Sons.

[97] 'Spatial and Temporal Congruence in Physics,' 'Has the General Theory of Relativity Repudiated Absolute Space?' and 'Is There a Flow of Time or Temporal Becoming?' pp. 142–144, 151–156 and 189–194. In P. R. Durbin (ed.), *Philosophy of Science*. New York: McGraw-Hill.

[98] 'Are Infinity Machines Paradoxical?' *Science* **159**, 396–406.

1969

[99] A Russian translation of [67], with revisions and additions, published by Progress Publishers, Moscow as *Filosofski problemy prostranstva i vremeni*.

[100] A version of [59] appeared in L. I. Krimerman (ed.), *The Nature and Scope of Social Science*, pp. 126–132. New York: Appleton-Century-Crofts.

[101] A version of [65] appeared in L. I. Krimerman (ed.), *The Nature and Scope of Social Science*, pp. 163–165. New York: Appleton-Century-Crofts.

[102] 'Reply to Hilary Putnam's "An Examination of Grünbaum's Philosophy of Geometry".' In R. S. Cohen and M. Wartofsky (eds.), *Boston Studies in the Philosophy of Science*, vol. 5, pp. 1–150. Dordrecht: Reidel.

[103] (With W. C. Salmon.) Introduction to 'A Panel Discussion of Simultaneity by Slow Clock Transport in the Special and General Theories of Relativity,' *Philosophy of Science* **36**, 1–4.

[104] 'Simultaneity by Slow Clock Transport in the Special Theory of Relativity,' *Philosophy of Science* **36**, 5–43. (See also [114].)

[105] 'Are Physical Events Themselves Transiently Past, Present and Future? A Reply to H. A. C. Dobbs,' *British Journal for the Philosophy of Science* **20**, 145–153.

[106] 'Can an Infinitude of Operations be Performed in a Finite Time?' *British Journal for the Philosophy of Science* **20**, 203–218.

[107] 'Free Will and Laws of Human Behavior,' *L'Age de la Science* **2**, 105–127. (A list of *errata* in this article is published in **4**, 1970, of *L'Age de la Science*. (See also [109, 116, 120, 124, 135, 136, 150].)

[108] 'Can We Ascertain the Falsity of a Scientific Hypothesis?' *Studium Generale* **22**, 1061–1093. (See also [119, 142].)

[109] An Italian translation of [107] appeared in R. Campa (ed.), *Implicazioni politiche della scienza*, pp. 87–123. Rome: Edizione della Nuova Antologia.

1970

[110] A version of [5] published by University Microfilms, Ann Arbor, 1970.

[111] A German translation of [88] appeared as 'Die Anisotropie der Zeit.' In L. Krueger (ed.), *Erkenntnisprobleme der Naturwissenschaften*, pp. 476–508. Cologne: Verlag Kiepenheuer und Witsch.

[112] A version of [90] appeared in W. C. Salmon (ed.), *Zeno's Paradoxes*, pp. 200–250. New York: Bobbs-Merrill.

[113] 'The Meaning of Time.' In N. Rescher (ed.), *Essays in Honor of Carl G. Hempel*, pp. 147–177. Dordrecht: Reidel. (See also [118, 138, 169].)

[114] A version of [104] appeared in P. Weingartner and G. Zecha (eds.), *Induction, Physics and Ethics*, pp. 140–166 and Discussion remarks, *passim*. Dordrecht: Reidel.

[115] 'Zeno's Metrical Paradox of Extension.' In W. C. Salmon (ed.), *Zeno's Paradoxes*, pp. 176–199. New York: Bobbs-Merrill.

[116] A Russian translation of [107] appeared in *Voprosy Filosofii* **6**, 62–74.

[117] 'Space, Time and Falsifiability,' *Philosophy of Science* **37**, 469–588.

1971

[118] A version of [113] appeared in E. Freeman and W. Sellars (eds.), *Basic Issues in the Philosophy of Time*, pp. 195–228. LaSalle, Ill.: Open Court; and in J. Zeman (ed.), *Time in Science and Philosophy*, pp. 67–87. Amsterdam: Elsevier.

[119] A version of [108] appeared in M. Mandelbaum (ed.), *Observation and Theory in Science*, pp. 69–129. Baltimore: Johns Hopkins Press.

[120] A version of [107] appeared in *American Philosophical Quarterly* **8**, 299–317.

[121] 'Why I Am Afraid of Absolute Space,' *Australasian Journal of Philosophy* **49**, 96.

[122] 'Are Spatial and Temporal Congruence Conventional?' *General Relativity and Gravitation* **2**, 281–284.

1972

[123] 'Abelson and Feigl's Mind-Body Identity Thesis,' *Philosophical Studies* **23**, 119–121.

[124] A version of [107] appeared in H. Feigl, K. Lehrer and W. Sellars (eds.), *New Readings in Philosophical Analysis*, pp. 605–627. New York: Appleton-Century-Crofts.

1973

[125] A version of [8] appeared in A. C. Kamil and N. Simonson (eds.), *Patterns of Psychology: Issues and Prospects*, pp. 18–25. Boston: Little, Brown and Co.

[126] A Spanish translation of [51] appeared in L. P. Williams (ed.), *La theoria de la relatividad*, pp. 119–125. Madrid: Alianza Editorial.

[127] 'The Ontology of the Curvature of Empty Space in the Geometrodynamics of Clifford and Wheeler.' In P. Suppes (ed.), *Space, Time and Geometry*, pp. 268–295. Dordrecht: Reidel.

[128] 'Reply to J. Q. Adams' "Grünbaum's solution to Zeno's Paradoxes",' *Philosophia* (Israel) 3, no. 1.

[129] 'Geometrodynamics and Ontology,' *Journal of Philosophy* 70, 775–800.

[130] A much enlarged version of [67] published by D. Reidel, Boston and Dordrecht. *Boston Studies in the Philosophy of Science*, vol. 12.

1974

[131] An Italian translation of [68] in *La filosofia di Rudolf Carnap*, pp. 583–658. Milan: Il Saggiatore.

[132] 'Karl Popper's Views on the Arrow of Time.' In P. A. Schilpp (ed.), *The Philosophy of Karl Popper*, vol. II, pp. 775–795. LaSalle, Ill.: Open Court.

[133] 'Is the Coarse-Grained Entropy of Classical Statistical Mechanics an Anthropomorphism?' In B. Gal-Or (ed.), *Modern Developments in Thermodynamics*, pp. 413–428. New York: John Wiley; Jerusalem: Israel Universities Press and Keter Publishing House. (See also [137, 139].)

[134] 'Introductory Remarks' for the symposium on 'Space, Time, Matter.' In K. Schaffner and R. S. Cohen (eds.), *PSA 1972* (*Boston Studies in the Philosophy of Science*, vol. 20, pp. 335–337. Dordrecht: Reidel).

[135] A version of [107] appeared in L. Goldberger and W. H. Rosen (eds.), *Psychoanalysis and Contemporary Science*, vol. III, pp. 3–39. New York: International Universities Press.

1975

[136] A version of [107] appeared in P. R. Struhl and K. J. Struhl (eds.), *Philosophy Now*, pp. 211–228. New York: Random House.

[137] A version of [133] appeared in J. Zeman and L. Kubat (eds.), *Entropy and Information in Science and Philosophy*, pp. 173–186. Amsterdam: Elsevier.

1976

[138] A version of [113] appeared in M. Capek (ed.), *Concepts of Space and Time*, pp. 471–500. *Boston Studies in the Philosophy of Science*, vol. 22. Dordrecht: Reidel.

[139] A version of [133] appeared in E. Laszlo and E. B. Sellon (eds.), *Vistas in Physical Reality*, pp. 11–29. New York: Plenum.

[140] 'Is Falsifiability the Touchstone of Scientific Rationality? Karl Popper vs. Inductivism.' In R. S. Cohen, P. K. Feyerabend and M. W. Wartofsky (eds.), *Essays in Memory of Imre Lakatos*, pp. 213–252. *Boston Studies in the Philosophy of Science*, vol. 39. Dordrecht: Reidel.

[141] A version of [50] appeared in S. Harding (ed.), *Can Theories Be Refuted?* pp. 116–131. Dordrecht: Reidel.

[142] A version of [108] appeared in *ibid.*, pp. 206–288.

[143] 'Can a Theory Answer More Questions than One of Its Rivals?' *British Journal for the Philosophy of Science* **27**, 1–23.

[144] 'Is the Method of Bold Conjectures and Attempted Refutations *Justifiably* the Method of Science?' *British Journal for the Philosophy of Science* **27**, 105–136.

[145] '*Ad Hoc* Auxiliary Hypotheses and Falsificationism,' *British Journal for the Philosophy of Science* **27**, 329–362.

[146] 'Is Pre-acceleration of Particles in Dirac's Electrodynamics a Case of Backward Causation? The Myth of Retrocausation in Classical Electrodynamics,' *Philosophy of Science* **43**, 165–201. (See also [159, 164].)

1977

[147] 'Remarks on Arthur I. Miller's Review of *Philosophical Problems of Space and Time*,' *Isis* **68**, 447–448.

[148] 'Absolute and Relational Theories of Space and Space-Time.' In J. Earman, C. Glymour and J. Stachel (eds.), *Foundations of Space-Time Theories*, pp. 303–373. Minneapolis: University of Minnesota Press.

[149] 'How Scientific Is Psychoanalysis?' In R. Stern, L. Horowitz and J. Lynes (eds.), *Science and Psychotherapy*, pp. 219–254. New York: Haven Press.

[150] A version of [107] appeared in M. Lipman (ed.), *Discovering Philosophy*, pp. 421–431. Englewood Cliffs: Prentice Hall.

[151] 'Is Psychoanalysis a Pseudo-Science? Karl Popper versus Sigmund Freud,' *Zeitschrift für philosophische Forschung* **31**, 333–353. (See also [154].)

[152] (With Allen I. Janis.) 'The Geometry of the Rotating Disk in the Special Theory of Relativity,' *Synthese* **34**, 281–299. (See also [162].)

[153] (With Allen I. Janis.) 'Is There Backward Causation in Classical Electrodynamics?' *Journal of Philosophy* **74**, 475–482.

1978

[154] Second installment of [151] in *Zeitschrift für philosophische Forschung* **32**, 49–69.

[155] (With Allen I. Janis.) 'Can the Effect Temporally Precede Its Cause in Classical Electrodynamics?' *American Journal of Physics* **46**, 337–341.

[156] 'Poincaré's Thesis that Any and All Stellar Parallax Findings are Compatible with the Euclideanism of the Pertinent Astronomical 3-Space,' *Studies in History and Philosophy of Science* **9**, 313–318.

[157] 'The Role of Psychological Explanations of the Rejection or Acceptance of Scientific Theories,' *Humanities in Society* **1**, 293–304. (See also [163, 171, 172].)

[158] 'Popper versus Inductivism.' In G. Radnitzky and G. Andersson (eds.), *Progress and Rationality in Science*, pp. 117–142. *Boston Studies in the Philosophy of Science*, vol. 58. Dordrecht: Reidel. (See also [170].)
[159] A version of [146] appeared in *Epistemologia* **1**, 353–396.
[160] 'Hans Reichenbach's Definitive Influence on Me.' In M. Reichenbach and R. S. Cohen (eds.), *Hans Reichenbach Selected Writings: 1909–1953*, vol. 1, pp. 65–67. *Vienna Circle Collection*, vol. 4. Dordrecht: Reidel.
[161] 'How Is Science Distinguished from Pseudo-Science? Two Views,' *Proceedings of the Pennsylvania Academy of Science* **52**, 96–98.

1979

[162] A version of [152] appeared in W. C. Salmon (ed.), *Hans Reichenbach, Logical Empiricist*, pp. 321–339. Dordrecht: Reidel.
[163] A version of [157] appeared in J. Barmack (ed.), *Perspectives in Metascience: A Festschrift for Hakan Tornebohm*, pp. 95–115. Gothenburg: University of Göteborg Press.
[164] A German translation of [146] appeared as 'Der Mythos kausaler Rückkopplung in der klassischen Elektrodynamik,' *Allgemeine Zeitschrift für Philosophie* **4**, 1–39.
[165] 'Is Freudian Psychoanalytic Theory Pseudo-Scientific by Karl Popper's Criterion of Demarcation?' *American Philosophical Quarterly* **16**, 131–141.
[166] (With Allen I. Janis.) 'Retrocausation and the Formal Assimilation of Classical Electrodynamics to Newtonian Mechanics: A Reply to Nissim-Sabat's "On Grünbaum and Retrocausation",' *Philosophy of Science* **46**, 136–160.
[167] 'Epistemological Liabilities of the Clinical Appraisal of Psychoanalytic Theory,' *Psychoanalysis and Contemporary Thought* **2**, 451–526. (See also [174].)

1980

[168] 'Psychoanalysis,' *Commentary* **70**, 21–23.
[169] A Greek translation of [113] appeared in *Deucalion* **32**, 343–382.
[170] A version of [158] appeared in G. Radnitzky and G. Andersson (eds.), *Fortschritt und Rationalität der Wissenschaft*, pp. 129–156. Tübingen: Mohr.
[171] A version of [157] appeared in M. P. Hanen *et al.* (eds.), *Science, Pseudo-Science and Society*, pp. 29–53. Waterloo, Ontario: Wilfrid Laurier University Press.
[172] A version of [157] appeared in *The Role of Psychological Explanations of the Rejection or Acceptance of Scientific Theories – A Festschrift for Robert Merton. Transactions of the New York Academy of Sciences, Series II* **39**, 75–90.
[173] Preface in Henry Mehlberg, *Time, Causality, and the Quantum Theory*, vol. I, pp. xiii–xiv. *Boston Studies in the Philosophy of Science*, vol. 19. Dordrecht: Reidel.
[174] A version of [167] appeared in *Noûs* **14**, 307–385.
[175] (With Allen I. Janis.) 'The Rotating Disk: A Reply to Gron,' *Foundations of Physics* **10**, 495–498.

1981

[176] 'The Placebo Concept,' *Behaviour Research and Therapy* **19**, 157–167.

[177] 'How Valid is Psychoanalysis? An Exchange,' *The New York Review of Books* **28**, (March 5) 40.

1982

[178] 'Can Psychoanalytic Theory Be Cogently Tested "On the Couch"?' *Psychoanalysis and Contemporary Thought* **5**, 155–255, 311–436. (See also [180].)

[179] 'Retrospective versus Prospective Testing of Aetiological Hypotheses.' In J. Earman (ed.), *Minnesota Studies in Philosophy of Science*. Minneapolis: University of Minnesota Press.

1983

[180] A revised version of [178], entitled 'The Foundations of Psychoanalysis', appears in L. Laudan (ed.), *Mind and Medicine: Explanation and Evaluation in Psychiatry and the Biomedical Sciences*, pp. 143–309. *Pittsburgh Series in the Philosophy and History of Science*, vol. 8. Berkeley: University of California Press. (See also [181].)

[181] A much expanded version of [180] will appear as *The Foundations of Psychoanalysis: A Philosophical Critique*. Berkeley: University of California Press.

[182] 'Freud's Theory: The Perspective of a Philosopher of Science,' Presidential Address to the American Philosophical Association (Eastern Division) to appear in Proceedings and Addresses of the American Philosophical Association.

[183] 'Is Object Relations Theory Better Founded than Orthodox Psychoanalysis?' *Journal of Philosophy* **80**, 46–51.

[184] 'Logical Foundations of Psychoanalytic Theory.' In W. K. Essler and H. Putnam (eds.), *Festschrift for Wolfgang Stegmüller*. Dordrecht: Reidel. It will appear simultaneously in *Erkenntnis* **19**, no. 1. It will be reprinted in J. Reppen (ed.), *Future Directions of Psychoanalysis* (New York: Lawrence Erlbaum) as well as in T. Millon (ed.), *Theories of Psychopathology and Personality* (3rd ed., New York: Holt, Rinehart and Winston).

[185] 'Retrospective versus Prospective Testing of Aetiologic Hypotheses in Psychoanalytic Theory.' In J. Earman (ed.), *Testing Scientific Theories. Minnesota Studies in the Philosophy of Science*, vol. 10. Minneapolis: University of Minnesota Press.

[186] 'What are the Clinical Credentials of the Psychoanalytic Compromise-Model of Neurotic Symptoms?' In G. Klerman and T. Millon (eds.), *Contemporary Issues in Psychopathology*. New York: The Guilford Press.

INDEX OF NAMES